STRENGTH OF MATERIALS

STRENGTH OF MATERIALS

GUSTAVO MENDES
AND
BRUNO LAGO
EDITORS

Nova Science Publishers, Inc.
New York

For permission to use material from this book please contact us:
Telephone 631-231-7269; Fax 631-231-8175
Web Site: http://www.novapublishers.com

LIBRARY OF CONGRESS CATALOGING-IN-PUBLICATION DATA

Strength of materials / edited by Gustavo Mendes and Bruno Lago.
p. cm.
Includes bibliographical references.
ISBN 978-1-60741-500-8 (hardcover)
1. Strength of materials. 2. Composite materials. I. Mendes, Gustavo. II. Lago, Bruno.
TA405.S765 2009
620.1'12--dc22

2009013535

Published by Nova Science Publishers, Inc. ✦ New York

CONTENTS

PREFACE

The strength of a material refers to the material's ability to withstand an applied stress without failure. The applied stress may be tensile, compressive, or shear. A material's strength is dependent on its microstructure. The engineering processes to which a material is subjected can alter this microstructure. This book provides a variety of material strength research including an extensive overview on the state of the art ceramic composite material BIOLOX delta which, since 2001, has successfully implanted more than 500,000 artificial hip joints. Due to the unique strength and toughness of this material, the risk of fracture has been substantially reduced when compared to conventional ceramic materials. Several different aspects of ionomer research from a physical property standpoint is discussed as well, including the history and current trends in ionomer research and a discussion on the immediate needs in this field. Furthermore, particle modeling (PM) as an innovative particulate dynamics based modeling approach is examined as a robust tool for simulating fracture problems of solids under extreme loading conditions, including situations of collapse, impact, blasting or high strain rate tension/compression. This book includes research on the ability of particle modeling to correctly predict dynamic fragmentation of materials with good accuracy.

Ceramic-matrix composites (CMCs) have been considering as potential structural materials for advanced energy-generation systems and propulsion systems. SiC fibers with low oxygen content and high crystallinity, which derived from polycarbonsilane, are the backbone as reinforcements in load-bearing CMCs. For high temperature application, the most desired critical properties of SiC fibers are high strength and stiffness as well as the reliable retention of these properties throughout the service life of application. Low fiber strength and thermal stability could result in low fracture toughness and accelerate sub-critical crack propagation in CMCs. Thus, the mechanical durability and microstructure stability of SiC fibers are major concerns under severe environments. Furthermore, in practical service environments, rarely is one degradation mechanism operative, but several mechanisms operate simultaneously, leading to the environment-pertinent degradation mechanism is complex for the SiC materials. In order to enhance the understanding of reliability and durability of CMCs applied to high temperature and oxidative environments, the investigations on the high temperature mechanical properties and microstructure for SiC-based fibers subjected to severe environments were integrated into Chapter 1.

The field of ionomers is an often overlooked and under-utilized branch of polymer research. Although ionomers can be broadly described as a class of polymers that contain any number of ionic groups, from a structural property standpoint only a low percent of ionic groups are necessary to impart significantly improved properties over the nonionic version of the same polymer. Current trends in the field of ionomers are highly focused on the field of fuel cell technology. There appears to be a significant hole remaining in the study of imparting strength to materials using ionic groups. This hole is very significant from an industrial point of view, and has a large commercial potential. There are very few commercially available ionomers, which shows how little this field has been explored to date.

Chapter 2 will focus on several aspects of ionomer research from a physical property standpoint: 1) A history of ionomer research, 2) Current trends in ionomer research - a) stand-alone polymers, b) nanocomposites, c) blends; 3) A commentary on the immediate needs in the field of ionomer research.

Layered composites subject to impacts can fail by delamination, i.e. by debonding between laminae, if the stress waves cause damaging phenomena to take place mainly within the resin-enriched interlaminar phases. To simulate delamination at the structural level, processes dissipating energy are lumped onto fictitious zero-thickness interlaminar surfaces, and softening interface constitutive laws are adopted to describe the progressive failure of the interlaminar phases.

Since delamination occurs inside very narrow regions, results of experimental testing on whole composites need to be accurately and reliably filtered to calibrate the interface constitutive laws. To this aim, Chapter 3 proposes a sigma-point Kalman filter approach. The performances of the proposed methodology, in terms of constitutive parameter estimations and dynamic delamination tracking, are assessed through pseudoexperimental testings on a two-layer composite, and real testings on multi-layer glass fiber reinforced plastic composites.

An extensive overview about the state of the art of the ceramic composite material BIOLOX®delta is given. The unique properties rely on a well defined alumina based fine composite microstructure which is mainly achieved by high temperature solid body reaction of the different ceramic phases during sintering. Zirconia comprises 17 % of the total volume. The tetragonal phase of zirconia is stabilized chemically and mechanically.

The high strength and toughness of the material depend on transformation toughening of the zirconia which is clearly shown by various experimental results. The excellent mechanical properties are reproduced batch by batch with a very low scatter.

As presented in Chapter 4, the outstanding properties of the material BIOLOX®delta support advantageous properties of the final product, e.g. ceramic hard-hard bearings for hip arthroplasty. The burst load of the components is significantly increased. It is shown that the design of the components is also very important for the reliability and the ultimate properties of the system. Wear properties at severe conditions are significantly improved by using the new composite material BIOLOX®delta in comparison to pure alumina.

Phase transformation of zirconia from the tetragonal to the monoclinic phase due to hydrothermal aging is extensively discussed. Due to the particular distribution and stabilization of the zirconia particles instable aging effects are not possible in this material. After very long time of accelerated aging conditions an increase of monoclinic phase is found – however, it is shown that dynamic and static properties of BIOLOX®delta are not influenced by this effect.

Chapter 5 studies particle modeling (PM), which is an innovative particulate dynamics based modeling approach. It has been demonstrated as a robust tool for simulating fracture problems of solids with dynamic fragmentation under extreme loading conditions. These loading conditions can include situations of collapse, impact, blasting or high strain rate tension/compression, as well as thermally-induced breakage problems.

Initially, PM was developed for the purpose of mimicking the microscopic material process at macroscopic level. This method can be conceptually illustrated by fully dynamic particles (or "quasi-particles") placed at the nodes of a lattice network without explicitly considering their geometric size. The potential can be specified for particle-particle interactions via axial springs. Theoretically, PM is an upscale of the molecular dynamics (MD) model applicable to various length scale problems. This is possible if a proper equivalent macroscopic potential is found, and, in case of lattice spacing decreasing to a few Angstroms, a MD model at zero Kelvin with, say, Leonard-Jones potential is recovered. In its current form, PM has been developed as a tool applicable to real engineering problems.

The advantages of PM over the existing discrete element based methods can be summarized as follows: (1) Sample in theory. Four conservative/equivalent rules (mass, potential energy, Young's modulus and tensile/compression strength) are applied to preserve the equivalent material properties. (2) Easy for implementation. Since the physical size of each particle is ignored other than its equivalent mass, the algorithm of coding a PM computation is fairly easy.

Current research work has exhibited that PM is able to correctly predict dynamic fragmentation of materials with a good accuracy. In modeling an epoxy plate with randomly distributed holes in tension, the PM result of the final crack pattern compared favorably with the associate experiment; for the simulations of impact study of two polymeric materials (nylon, 6-6 and vinyl ester) subject to a rigid falling indenter, the modeling results of resistant force, energy, deflection and drop speed of indenter vs. time quantitatively agree fairly well with the according empirical observations.

Electrical steels are the core materials for electrical motors or transformers. Those materials for motors are played an energy conversion roll from electricity to motion. However, energy losses are accompanied with this conversion. To minimize these losses is a key technology to conserve our environment.

Numerous researches on the grain-oriented electrical steels reported. Those researches especially for transformers are focused on the reducing the losses at supplying the electricity from power plants. On the other hand, home or industrial appliances are the power consuming devices, and the most effective point on the energy loss reduction. These home appliances are used small motors using non-oriented electrical steels.

In Chapter 6, several researches on the non-oriented electrical steels are discussed and focused on the metallurgical control of the steels to reduce the core loss for generating waste heats and motor building innovation technologies for decreasing the building factor of the core losses.

In the metallurgical part, some additive elements as phosphorus, aluminum and manganese for improving magnetic properties reviewed. Moreover some contaminating elements as vanadium, titanium and zirconium are discussed especially for precipitation studies in the steels have been done. These precipitations are inhibited the grain growth at final annealing or stress relief annealing. These inhibited small grains increase the core losses.

For studying motor building technologies, compression stress effect, shearing stress effect are discussed. Even though the best core materials are used for manufacturing motors, those building deteriorations make worse for the motor efficiency. Therefore, those technologies are also important for reducing the carbon dioxide emission.

The aim of Chapter 7 was to investigate regional root canal push-out bond strengths for a fiber-reinforced post system varying the application method of the luting agent.

Recently extracted maxillary incisors (n=30) were sectioned transversally at the labial cemento-enamel junction, and the roots treated endodontically. Following post space preparations, fiber-reinforced posts (Endo Light-Post; RTD) were placed using adhesive system and resin cement provided by the manufacturer. Three equal groups (n=10) were assessed according to the technique used to place the luting agent into post space: using a lentulo spiral, applying the cement onto the post surface, injecting the material with a specific syringe. Each root was sliced into three discs (2 mm thick) representing the coronal, middle and apical part of the bonded fiber post. Push-out tests were performed for each specimen to measure regional bond strengths. Results were statistically analyzed using two-way ANOVA and Tukey tests ($\alpha = 0.05$). All fractured specimens were observed using a scanning electron microscope to identify the types of failure.

The results indicated that bond strength values were significantly affected by the application method of the resin cement ($p < 0.05$). The "syringe technique" and the "lentulo technique" showed higher bond strength values compared with the "post technique". No significant differences were recorded among the post space thirds. Microscopic analysis revealed a prevalence of post/cement and mixed failures.

The best performance in terms of push-out bond strengths for the post system tested was obtained when the luting agent was applied into the post space either with a specific syringe or using a lentulo spiral. There were not differences in bond strength among root thirds.

In Chapter 8, the microstructures of a ceromer (Artglass®) reinforced by either glass fibers (GlasSpan®) or polyethylene fibers (Connect®) were characterized and compared and the influence of the fiber reinforcement on the flexural strength of the resulting products evaluated. With this objective, seven bars of each material were produced. One bar of each material was separated for microstructural analysis. The microstructural samples were subjected to metallographic polishing and finishing, and then analyzed using optical microscopy at different magnifications. The images obtained were treated using an image processing computer program (Image Pro Plus) in order to quantify the microstructure by calculating the mean diameter and mean volume fraction of fibers. The flexure tests were made by three-point bending, using six samples of each material. After statistical analysis, the results showed that the mean diameter of the glass fibers (4μm) was smaller than the polyethylene ones (23.6 μm). The mean volume fraction of glass fibers (0.42) was larger than that of the polyethylene fibers (0.28) and the mean center-to-center distance between fibers was smaller in the glass fibers material (33 μm) than in the polyethylene fibers material (61 μm). The flexural strength of both glass and polyethylene fiber-reinforced materials was statistically equal, despite the fiber volume fraction being statistically larger in the fiber glass material.

The purpose of Chapter 9 is to describe the influence of anisotropy on the geomechanical strength properties of two Spanish slates with different chemical and physical characteristics. From laboratory testing results of slates under point load and uniaxial compression and the use of indirect methods, as it is the measurement of P velocities, principal parameters were calculated for this rock material. As it is well known under uniaxial compressive strength slates are strong and also very strong rock when loading is parallel (90°) o perpendicular (0°) to the main anisotropic planes. In contrast it is a weak rock with minimum strength values for angles between 45 to 60° of inclination of anisotropy planes. The correlation equations were calculated between different parameters. Despite weak correlation between different geotechnical properties were found and when all lithologies are considered together correlation of geomechanical properties is weak. However when each lithology is considered separately the geomechanical properties can be coherently defined. Linear and polynomial equations were found for the point load and uniaxial strength correlations with the inclination of the anisotropy. Different strength fields were calculated when uniaxial strength and point load test plot and its comparison include the inclination of the anisotropy planes on slates. Uniaxial compressive strength and P wave velocity appears to be strongly influenced by uniaxial strength and good polynomial correlations resulted. Plots of slates with other sedimentary type of rocks from Cantabrian Zone, CZ, revealed the hardness and highest strength of slates when loading is perpendicular to the main anisotropy planes.

In: Strength of Materials
Editors: G. Mendes and B. Lago, pp. 1-60

ISBN: 978-1-60741-500-8

Chapter 1

HIGH TEMPERATURE MECHANICAL PROPERTIES AND MICROSTRUCTURE OF SIC-BASED FIBERS UNDER SEVERE ENVIRONMENTS

Jianjun Sha[1,2*]

[1] Shool of Aeronautics and Astronautics, Dalian University of Technology
116024, Dalian, China

[2] Ceramic Materials Engineering, University of Bayreuth,
D-95440 Bayreuth, Germany

Abstract

Ceramic-matrix composites (CMCs) have been considering as potential structural materials for advanced energy-generation systems and propulsion systems. SiC fibers with low oxygen content and high crystallinity, which derived from polycarbonsilane, are the backbone as reinforcements in load-bearing CMCs. For high temperature application, the most desired critical properties of SiC fibers are high strength and stiffness as well as the reliable retention of these properties throughout the service life of application. Low fiber strength and thermal stability could result in low fracture toughness and accelerate sub-critical crack propagation in CMCs. Thus, the mechanical durability and microstructure stability of SiC fibers are major concerns under severe environments. Furthermore, in practical service environments, rarely is one degradation mechanism operative, but several mechanisms operate simultaneously, leading to the environment-pertinent degradation mechanism is complex for the SiC materials. In order to enhance the understanding of reliability and durability of CMCs applied to high temperature and oxidative environments, the investigations on the high temperature mechanical properties and microstructure for SiC-based fibers subjected to severe environments were integrated into this review.

* Corresponding address: School of Aeronautics and Astronautics, Dalian University of Technology, 116024, Dalian, China

1. Introduction

Non-oxide Ceramic-matrix Composites (CMCs), have been extensively studied during the last two decades. Currently, CMCs have been proposed to use as the structural materials in application of high temperature technologies, such as advanced nuclear energy systems [1-3], various stationary gas turbine engines and aerospace propulsion systems [4-6].

As structural engineering materials, one of the important advantages of CMCs is the significant improvement of toughness by using continuous ceramic fibers compared to their monolithic ceramics. In continuous ceramic fiber reinforced CMCs, the improved toughness is attributed to several energy dissipating mechanisms when the matrix crack is occurring under applied stress, such as crack deflection, fiber bridging and fiber sliding [7]. The energy dissipation procedure can be illustrated schematically in Figure 1. These energy dissipations enhanced the fracture toughness and resulted in a non-catastrophic failure mode. If the conditions for de-bonding are satisfied, fibers bridge the crack faces in the wake of the crack tip, subsequent to matrix cracking. The stress bore by bridge fibers applies traction forces to the crack faces that reduce the stress intensity at the crack tip. Under specific condition crack propagation does not occur without additional applied stress. The performance of CMCs could be improved through the optimization of fiber/interface/matrix, and now the fabrication process is still in developing and progressing.

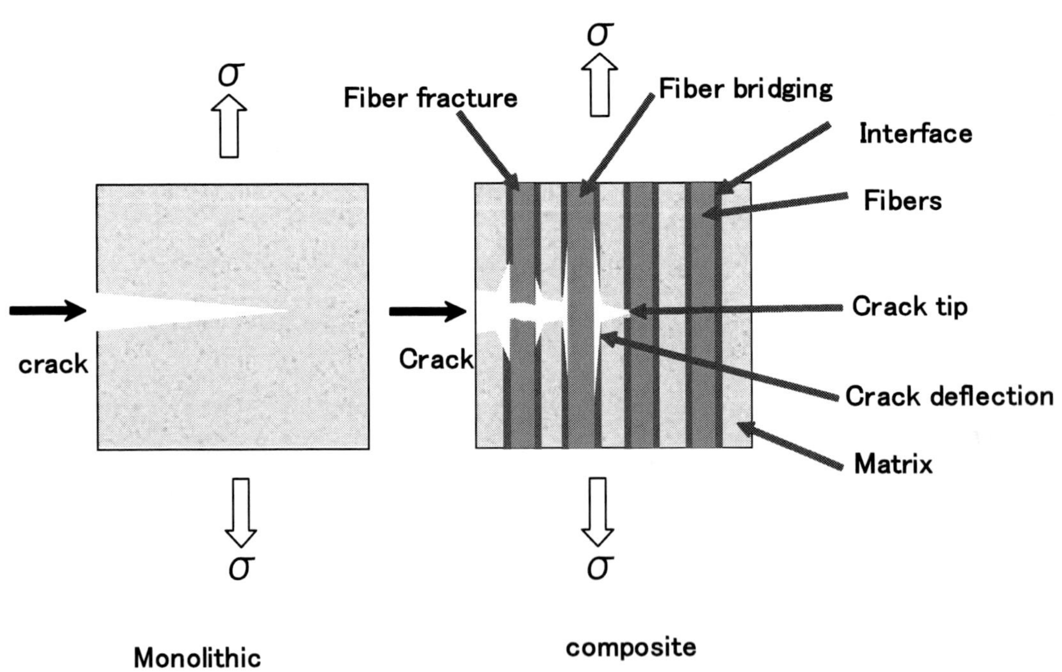

Figure 1. Schematic of crack propagation and principle for improved toughness in CMCs.

Based on the simple theory of mixture for the ultimate tensile strength (UTS) calculation of continuous fiber reinforced ceramic matrix composites [8], in the case of CMCs with I-D reinforcement alignment, when the reinforced fiber has a smaller modulus and a similar or even large strength, significant matrix cracks must be occurred before reaching the UTS of fiber. In other words, when the applied stress is beyond that of the crack initiation of matrix, the residual stress is mainly carried by the fiber alone. Pull out of the fiber can significantly improve the fracture toughness. The fibers are backbone in CMCs and play a very important role on the mechanical properties of CMCs.

Among of CMCs, SiC fiber reinforced SiC matrix composite (SiC/SiC) has been considered as one of the most potential candidate materials, because it possessed many attractive properties for structural engineering applications under severe environments, such as excellent mechanical and chemical stability. It is well know, the fracture behavior of monolithic silicon carbide is brittle and fails catastrophically. However, if the SiC fibers with an appropriate coating as reinforcement are incorporated with the silicon carbide matrix by specific fabrication process to form the ceramic matrix composite, the fracture characteristics of silicon carbide materials can be significantly improved. Figure 2 shows the typical polished morphologies and the fracture surface of near-net shape SiC/SiC composite. It is apparent that reinforcing fibers incorporated with dense matrix through an appropriate interphase. The significant fibers pull out could improve the fracture toughness and result in a pseudo-ductile fracture behavior. The tough ceramics as structural materials have the potential for being used up to about 1500 °C which is much higher than the operation temperature of superalloy (maximum 1100 °C close to the melting point). In different fields, such as advanced nuclear energy system, gas turbines for power/steam co-generation, heat exchangers and so on, they give different requirements for materials performance, but common features for high temperature technologies are excellent mechanical performance and environmental durability.

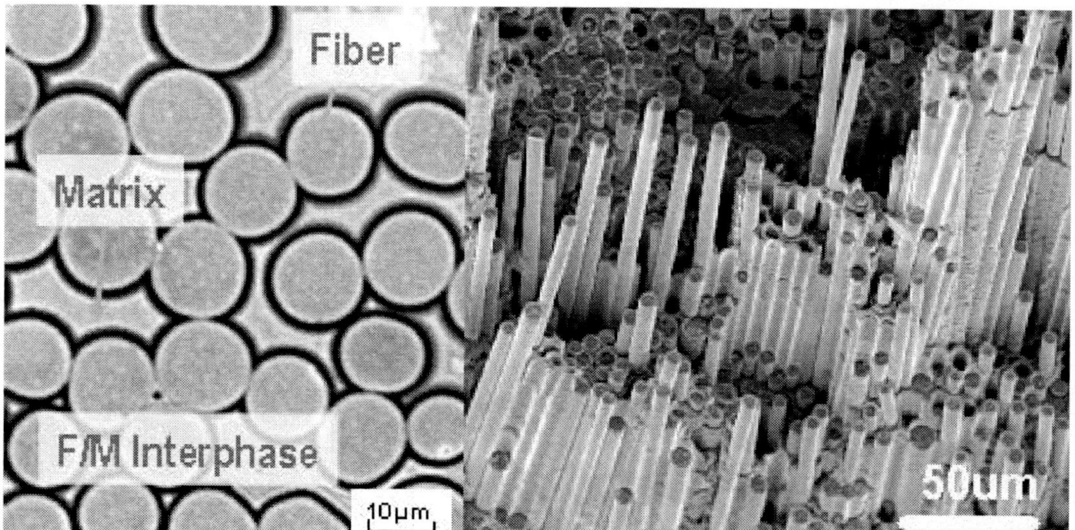

Figure 2. Typical polished morphology of SiC/SiC composite and its fracture surface with long fiber pull out.

The major advantages for SiC/SiC composite applied to engineering are: (i) high specific strength, (ii) superior high temperature strength and creep resistance, (iii) low thermal expansion coefficient and high thermal conductivity, (iv) low neutron irradiation-induced radioactivity in nuclear environments.

SiC fibers with low oxygen content and high crystallinity, which derived from polycarbonsilane, are the backbone as reinforcements in load-bearing CMCs. For high temperature application, the most desired critical properties of SiC fibers are high strength and stiffness as well as the reliable retention of these properties throughout the service life of application. Low fiber strength and thermal stability could result in low fracture toughness and accelerate sub-critical crack propagation in CMCs. The key to the successful application of high temperature ceramic matrix composites (CMC) is the judicious selection and incorporation of ceramic fiber reinforcement with the proper chemical, physical and mechanical properties. Thus, the mechanical durability and the microstructure stability of SiC fibers are major concerns under severe environments. In practical service environments, rarely is one degradation mechanism operative, but several mechanisms operate simultaneously, which leading to the environment-pertinent degradation mechanism is complex for the SiC materials.

For understanding the environmental durability and describing the response of reinforced fibers to service environments and further evaluation of reliability of CMCs, the investigation of thermal mechanical properties on SiC-based fibers in complex situation is essential. In this chapter, the investigations on mechanical properties and microstructure of SiC-based fibers are reviewed in terms of varied environments; some issues concerning the environment-pertinent properties are discussed.

2. Materials System and Characterization Technique

2.1. Materials System

The first SiC-based Nicalon fiber was produced by Nippon Carbon which allowed non-oxide ceramic matrix composites to be developed. It made possible to use the SiC fiber as the reinforcement for high temperature structural materials in very severe environments. In order to improve the flexibility and strength so that the preform can be woven in complex shape, the fine diameter SiC fibers were developed. However the first SiC-based fibers are inherently limited by oxidation at very high temperatures. As a result of this limitation a renewal of interest has occurred in oxide resistance of SiC fibers by approaching to the near stoichiometric composition or addition of small amount additives to improve thermal stability at elevated temperatures. Efforts have been made to improve the high temperature properties of fine diameter SiC fibers by making them with compositions increasingly approaching stoichiometry. Based on the chemical composition and the fabrication process, the development of SiC-based fibers could be categorized into three generations as illustrated in Figure 3.

The first fine diameter SiC fiber (Nicalon NL200) was synthesized by Yajima in Japan in 1970 [9]. The Nicalon fiber could be viewed as the representative of the first generation of SiC fibers (Figure 3). This fiber is thermodynamically unstable at high temperature, because it consists of SiC-nanocrystals (average size: 1-2 nm) and free carbon embedded in an amorphous SiC_xO_y matrix. The amorphous SiC_xO_y phase decomposes at temperature beyond 1300 °C [10-11], with a significant gaseous species evolution and SiC crystal growth [11-12]. In order to improve the high temperature resistance of the Si-C-O fiber, a new fabrication process of the fiber, radiation curing method [13-15], has been developed. Irradiation curing with an electron beam was applied to make the fiber infusible and cross linking.

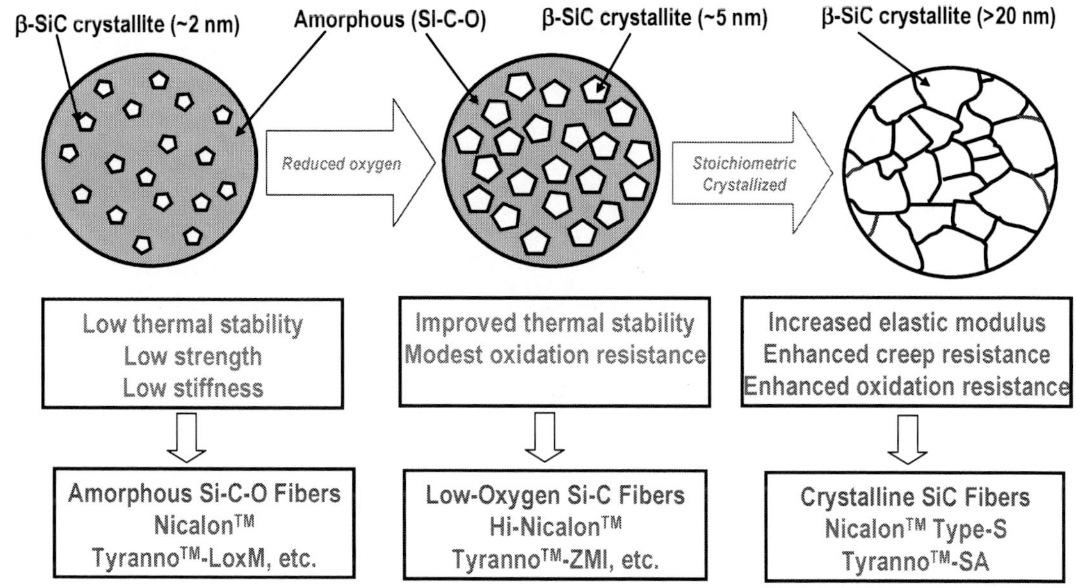

Figure 3. Illustration of R&D of SiC-based Fibers.

To avoid the thermal instability caused by the decomposition of oxycarbide phase (SiC_xO_y), in 1990, a nearly oxygen-free SiC fiber, Hi-Nicalon (Nippon-Carbon) was developed by melt spinning, electron beam curing and pyrolysis of a polycarbosilane precursor (PCS) under anaerobic conditions [15-16]. This fiber had a much higher thermal stability than the standard Nicalon fiber and was viewed as the representative of the second generation. However, the Hi-Nicalon fiber consists of not only SiC nanocrystals (average crystal size: 5 nm) but also excess of free carbon which affects its oxidation and creep resistance.

To reduce the free carbon content and eventually improve the high temperature properties of the fibers, extensive efforts have been devoted to develop near stoichiometric and high crystallized SiC fibers. The precursor fiber can be sintered at high temperatures that excess carbon and oxygen are lost as volatile species to yield polycrystalline and near-stoichiometric SiC fiber. These fibers are advanced SiC fibers and generally called the third generation of SiC fibers (Figure 3), including Hi-Nicalon type S fiber [17], Tyranno SA fibers [18] and Sylramic SiC fiber [19]. The third generation of SiC fibers is oxygen-free and near-stoichiometric (atomic ratio: C/Si=1.00–1.08). Furthermore, their grain size is relatively large (20–200 nm) and their thermal stability is excellent.

For enhancing the environmental durability of CMCs, SiC-based fibers with high crystallinity and near stoichiometry would be preferential. Based on this standing point, the following SiC-based fibers were used for the work presented in this chapter (Table 1).

Table 1. SiC-based fibers used for the work in this chapter and their properties provided by manufacture

SiC fiber	C/Si	Oxygen (wt%)	Strength (Gpa)	Modulus (Gpa)	Density (g/cm3)	Diameter (μm)
HNL	1.39	0.5	2.8	270	2.74	14
HNLS	1.05	0.2	2.6	420	3.1	12
TySA	1.07	<0.5	2.6	400	3.0	7

Note*: HNL=Hi-Nicalon™ fiber (500 fiber/yarn), HNLS=Hi-Nicalon™ Type S (500 fiber/yarn) and TySA=Tyranno™ SA fiber (1600 fiber/yarn).

Both HNL [20] and HNLS [21] fibers were fabricated by Nippon Carbon Co., Japan. TySA [18] fibers were fabricated by Ube Industry Co. Ltd., Japan. It is clear that: the HNL fiber ($SiC_{1.39}O_{0.014}$) consists of a mixture of SiC nano-crystals and free carbon; the HNLS ($SiC_{1.05}O_{0.007}$) and TySA ($SiAl_{0.02}C_{1.07}O_{0.03}$) fibers have near stoichiometry and high crystallinity. Noting the TySA fiber contains small amount of alumina (less than 1 wt%) in order to improve its thermal stability.

2.2. Methodology

Due to varied manufacturing approaches, SiC fibers are being produced with different surface morphologies and internal microstructures, particularly regarding size and population of defects, grains and grain boundary phases. The strength and creep resistance of SiC fibers, which are dependent on intrinsic and extrinsic factors, such as material itself and service environment, are first properties examined. The conventional evaluation methods developed for engineering materials, could not be applied to the SiC fibers with fine diameter and brittle nature. The methodology for the evaluation of these mechanical properties in precise way is crucial, and it is described as follows in detail.

2.2.1. Single Fiber Tensile Test Technique

Single fiber tensile test technique is used to evaluate the tensile properties of SiC fibers. Generally, in each condition, about 30 single fibers were selected at random from the fiber yarn and cut into 50 mm lengths. The 50 mm length fiber is mounted on a paper cardboard frame. Tensile tests were carried out at ambient conditions using an Instron test machine equipped with a 2.5 Newton load cell (Figure 4). The tensile test generally followed ASTM-recommended procedures [22]. The individual fiber was carefully separated and selected randomly from the yarn of each fiber type. The fiber diameters were determined from one end of projecting fragments. To do this, each 50 mm length fiber was mounted by centering and fastening its ends with Aradi glue onto a paper cardboard frame with a 25.4 mm distance between bonding points that defined the fiber gauge length (Figure 4). To prevent the fiber pull-out from the bonding point of Aradi glue before reaching the failure load, the glue coated at least a 5 mm length of the fiber ends and set for several days for complete drying. Before tensile test, the paper cardboard was cut very carefully along the center line across the hole so that the load was completely applied to the fiber. Load was applied at a constant displacement rate of 0.3 mm/min (equivalent to a strain rate: 2.2×10^{-4} s^{-1}). However, it is very difficult to collect the fiber fragments after tensile test in normal way, because the sudden release of very high stored energy at fracture will make the fragment breaking into many pieces. Therefore, a procedure to decay the release of stored energy for capturing the broken fragments was developed. The developed procedure was very effective to capture the fiber fragments. In the case of fiber fractured at the edge of bonding point, the fracture of fiber might be caused by bending moment due to poor alignment, which doesn't reflect the true strength of a fiber. Thus, this test was viewed as invalid.

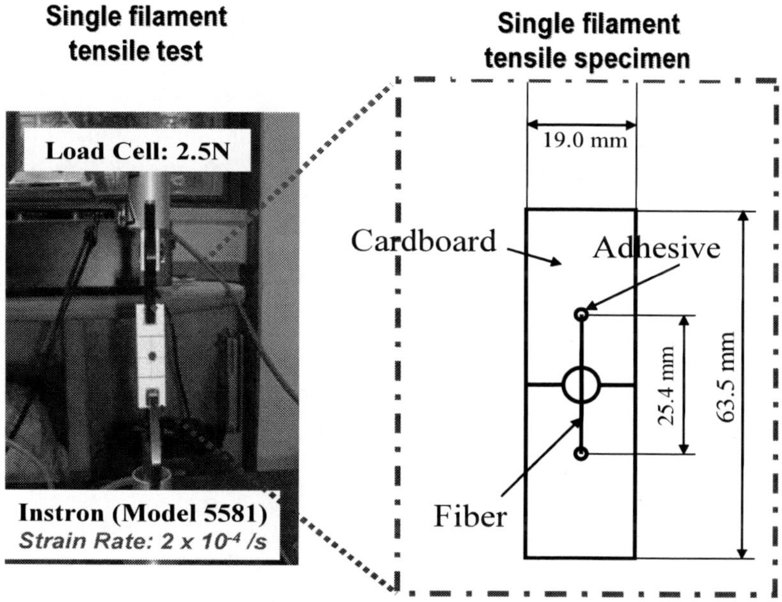

Figure 4. Test method for measuring the tensile strength of single fiber and specimen geometry.

In the capturing of fracture fragments, a small rectangular plastic film (6.0×8.0 mm^2) was used and coated on one side with glycerin. This plastic film with glycerin on one side was carefully bridged across the center hole of cardboard (Figure 4), and the fiber to be tested was completely wetted by glycerin. The glycerin effectively damped the shock wave in the fracture of fiber and it usually fractured only at one location. Each fiber segment remained to its half of the mounting frame, which is important for later SEM examination.

Because the fracture of ceramic materials generally originates from the critical flaws, assuming those flaws in the fiber are distributed randomly in location, then the strength of the fiber is determined by the strength at its weakest point (weakest link rule). Test on randomly selected fibers will show a considerable dispersion in failure strengths because of the presence of flaws. The strength of fibers can be shown generally to follow the classical two-parameter Weibull distribution.

The two-parameter Weibull theory of statistical fracture was applied to characterize the fracture behavior of brittle SiC fibers [23].

According to weibull's statistical theory, the probability of failure F_i, of fiber subjected to nominal tensile strength, σ, is given as

$$F_i = 1 - \exp[-L(\frac{\sigma - \sigma_u}{\sigma_0})^m] \quad (1)$$

where m is the Weibull modulus of the fiber, L is the gage length of the fiber, σ_u the stress below which fiber is assumed to have zero failure probability and σ_0 the Weibull scale parameters. Both σ_0 and m are constant for a given material and assuming σ_u=o. The Weibull modulus, m, of the fiber can be determined by rearranging equation (1) into the form

$$\ln\ln(\frac{1}{1-F_i}) = m\ln(\frac{\sigma}{\sigma_0}) + c \quad (2)$$

where c is constant. Actually, m is the slope in a two parameter weibull plot, which can be obtained by least squares fitting to the linear relationship of equation (2). In equation (2), the probability of fiber failure F_i at the nth ranked sample from a total of N specimens is obtained from the mean rank method as F_i=n/(N+1).

The Weibull average strength (σ_{avg}) was calculated from the relation $\sigma_{avg}=\sigma_0\Gamma(1+1/m)$, where $\Gamma(1+1/m)$ is a Gamma function [24].

$$\Gamma(\zeta) = \int_0^\infty e^{-x} X^{\zeta-1} dx. \quad (3)$$

2.2.2. Bending Stress Relaxation Test

For evaluating the creep and rupture strength of individual fibers, the conventional tensile creep test procedure is to subject an individual fiber specimen of length L and diameter D to a constant tensile load P at a constant test temperature T and to measure fiber elongation ΔL versus time t until the fiber finally fractures at rupture time t_R. Creep strain is then determined from

$$\varepsilon_c = \Delta L(t, T, \sigma, G)/L \quad (4)$$

where $\sigma=4P/\pi D^2$ is the applied stress and G symbolizes effects from the environments. Rupture time typically is also a function of temperature, stress, length and environment,

$$t_R = t_R(T, \sigma, L, G). \quad (5)$$

Unfortunately, measuring the creep of SiC fibers under tensile loading is difficult. This is especially true with fine diameter fibers which are often degraded by an air test environment and can be easily fractured during grip and strain sensor attachment. Another problem is the accuracy of the creep strain, because in many cases it is hard to define the gauge length (including the cold grip and hot grip) during the tensile test. To avoid these problems, in this study, a modified bend stress relaxation (BSR) method was utilized to evaluate the creep resistance of SiC fibers, and attempts were made to relate the BSR with tensile creep for fine-diameter fibers. An schematic illustration of the BSR test jig was shown in Figure 5. For evaluating the environmental effect on the creep resistance of SiC-based fiber, a modification was made on the conventional method [25] as shown in Figure 5. This improvement makes the tested specimen to be sufficiently exposed to the test environment.

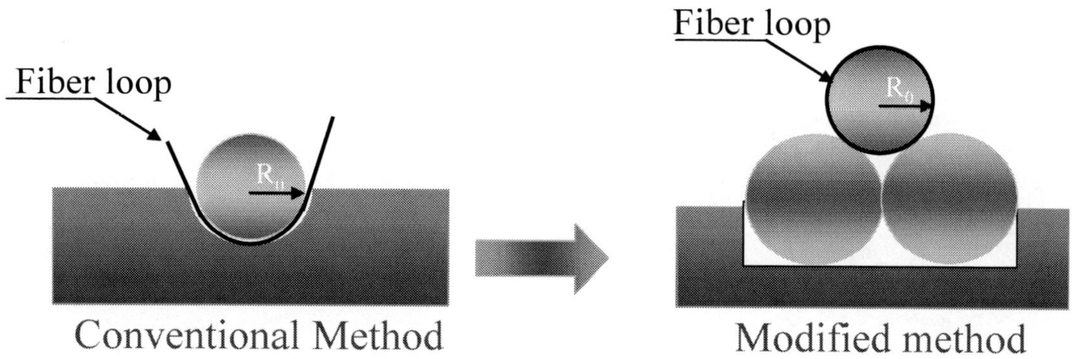

Figure 5. Comparison between conventional and modified bend stress relaxation test method.

In this method, the fiber with a length of 2-5 cm are wound around the rod at a constant surface strain and held at desired temperature for given times in controlled environment. For small diameter fibers in ambient conditions, the bending modes of different applied strain can be achieved by tying the fiber into small loops with different radius, R_0. The fiber loop is then subjected to a specific time (t), temperature (T), and environmental treatment. After treatment, the applied stain is then removed by release the fiber loop from the test jig or broken the fiber loop at one point at room temperature. The stress relaxation-induced effects are measured in terms of the residual radius of fiber loops, R_a. If the fiber remains completely elastic during treatment, the broken loop will be straight with no curvature, i.e., $R_a=\infty$. If the creep-induced stress relaxation occurs, the R_a will be finite and typically will decrease with increasing the treatment time and exposure temperature.

To quantify the stress relaxation occurred during thermal exposure, a parameter m, stress relaxation parameter was defined, which is the ratio of final to initial stress at any local position in the fiber as illustrated in Figure 6. That is

$$m = \sigma(t,T,\varepsilon_0)/\sigma_0(0,T,\varepsilon_0) = E \bullet \varepsilon_e(t,T,\varepsilon_0)/(E \bullet \varepsilon_0) = [\varepsilon_0 - \varepsilon_c(t,T,\varepsilon_0)]/\varepsilon_0 \quad (6)$$

where ε_0, ε_e, and ε_c are the local initial strain, final elastic strain and total creep-induced strain, and all of these strains vary within the fiber. For convenience, one can assume that (1) ε_c is linearly proportional to the ε_0 regardless of the stress direction and (2) it can be measured at room temperature by relation $\varepsilon_c=z/R_a$. z is the distance from the neutral axis in the fiber loop plane. The first assumption of linear strain dependence is generally valid for polycrystalline materials which stress relax due to grain boundary sliding mechanisms that are either elastically or diffusionally accommodated. That is, a stress power dependence of $n \approx 1$ ($\varepsilon_c \propto \sigma^n$) is typically observed throughout both the primary and secondary creep stages. The second assumption implies that at each local position within the fiber, stress relaxation not only is proportional to ε_0 but follows the same time-temperature dependence. This typically requires a fiber with a uniform isotropic microstructure that creeps with an $n \cong 1$ power dependence. If these assumptions apply, the BSR m ration is independent of position and initial applied strain. It is then only a function of treatment time and temperature and can be determined by the simple relation:

$$m\ (t, T)=1- R_0/R_a \tag{7}$$

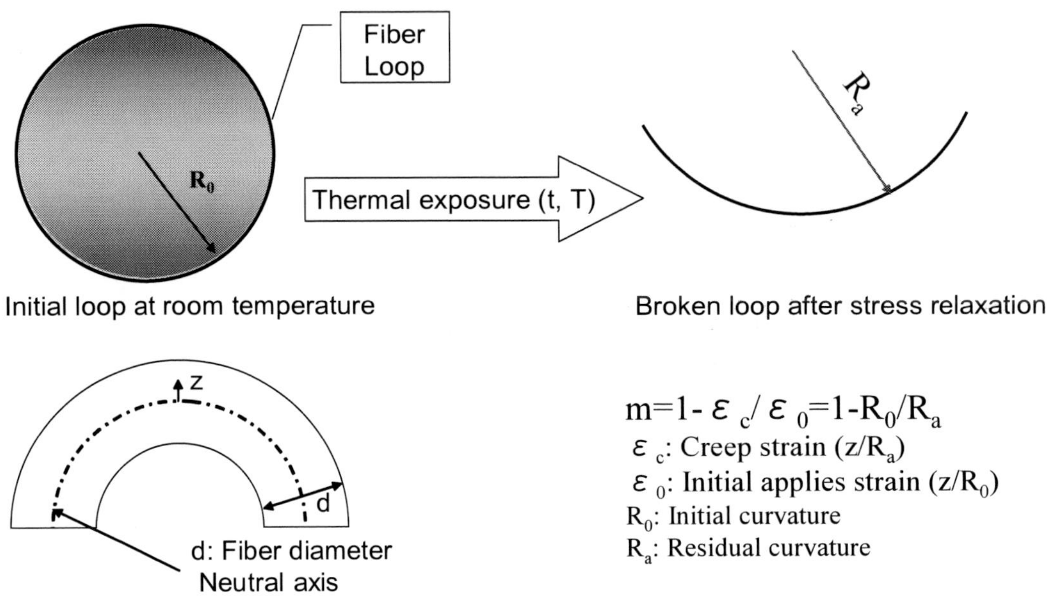

Figure 6. Schematic representation of the test principle of the bend stress relaxation originally developed in Ref. [25].

In comparison to tensile creep test which conducted under a dead load with accessories for strain measurement and a defined gauge length, the BSR offers many advantages including the ability to simultaneously study many fibers of small diameter and short length under same time, temperature, and controlled environmental conditions.

Here, it is obvious that stress relaxation parameter, m, can be determined based on the extent of permanent deformation occurred during stress relaxation. An m value which approaches 1 indicates that no permanent deformation occurred during the high temperature exposure, while a m value of 0 indicates that the stress completely relaxed. Hence, fibers are considered more thermally stable against creep as m values increase from 0 to 1 [25].

Practically, the BSR test also offers insight into the behavior of bent fibers in woven performs as well as conditions for "creep-forming" fibers into complex shapes. And also, it eliminates the need for furnace with long uniform hot zones, for mechanical grips, for remote sensors and for multiple experimental runs that are often required to establish time, temperature and stress dependencies and also to determine statistical variations. Second, for polycrystalline fibers, which generally creep with stress power dependencies near unity, if the BSR m-ratios are independent of applied strains, and thus equal to those stress relaxation ratios that would be measured in a pure tensile test. Furthermore, by BSR test, it will be beneficial to understand the basic mechanisms which controlled the creep behavior of SiC fibers with fine diameters [25].

2.2.3. Microstructural Characterization

This section described some techniques that will help to clarify why the mechanical properties were changed and how the microstructure influenced the mechanical properties. Among these techniques, the facilities frequently used are optical microscopy (OM), scanning electron microscopy (SEM) and X-ray diffractometer (XRD).

Optical microscopy with a video was used to examine the macrostructure of materials. It is also useful in the determination of fiber loop diameter in the BSR test. In BSR test, a photograph was taken of the loops before and after thermal treatment. The initial applied curvature R_0 or residual curvature R_a was measured by fitting a circle on the fiber in the photograph, and then the curvature could be obtained by a graphic technique.

XRD is very useful in the identification of the crystal phase and the estimation of the crystallite size. The X-ray diffraction (XRD) patterns were recorded by means of X-ray diffractometer (Rigaku) with a rotating anode (Cu-Kα radiation).

X-ray scattering of the atom planes in the crystals gives a diffraction pattern characteristic of the crystal structure. Diffraction peaks correspond to scattering of specific planes which are defined by the structure factor and Bragg's law,

$$n \bullet \lambda = 2d \bullet \sin\theta \tag{8}$$

where n=1, 2, 3,..., λ is the X-ray wavelength, d is the planar spacing and θ is the diffraction angle. The relative intensity of the diffraction pattern varied with the diffraction plane to aid in structure identification. Comparing the experimental diffraction pattern to a known pattern allows the crystal structure to be identified.

The specimen was prepared by attaching the powder sample on the glass slide with double-side adhesive tape. The powder was obtained by pulverizing the fiber tow of about 0.1 g in a mortar. During pulverizing, in order to prevent the spray of fiber fragments from pulverizing, the alcohol was mixed with powder to make viscous slurry. After careful milling and drying, powder was put on the glass slide with double-side adhesive tape, and it was pushed to be attached tightly.

During scanning, the XRD operated at 40 kV and 20 mA was used to identify the crystal phase in the fibers. All of the scans were run at 2°/min with a time interval of 0.05 s for sampling. The range of 2θ was 10°-90°. The apparent crystallite size (D_{111}) of the β-SiC crystalline phase present in the samples was calculated from the half-value width of (111) diffraction peak using Scherrer's formula:

$$D = K \bullet \lambda / (H_w \bullet \cos\theta) \tag{9}$$

where K is a constant (taken as 0.9), λ the CuKα wavelength (i.e., λ=0.154056), H_w the half-value width of β-SiC (111) peak and θ the Bragg angle (θ=17.5° for β-SiC (111)).

FE-SEM (Field-Emission Scanning Electron Microscope, model; JEOL JEM-2010), which provides narrow probing beams as well as high electron energy resulting in both improved spatial resolution and minimized sample charging and damage, is a powerful weapon in the characterization of dimension and microstructure such as examination of surface morphologies and fractograph.

The FE-SEM was employed to determine the typical diameter variation across a cross section of fiber yarn and along the fiber length. The selected fiber was attached on the specimen holder with the double-sided carbon tape. The fiber diameter was determined from SEM image with high magnification (x5000). Special care was taken in the register of fiber fragment so that the diameters represented the fibers that we want to investigate.

To examine the fracture surface, firstly a technique described in section 2.2.1 was adopted to obtain the fracture fragments, and the fracture location was noted. In order to take a high quality picture, the clean fracture surface is needed and it can be gotten by washing the fragment in ultrasonic bath contained alcohol for about 30 s. Each fiber segment for the successful tests was gripped with a narrow tip tweezers and broken off at the bonding point.

The clean segments were mounted on double-sided carbon tape applied to the circular side surface of cylinder specimen holder (10 mm copper cylinder in diameter). Usually about 10 segments were mounted with each pair of matching fiber fracture surfaces, and keep the fracture surface with a protruding length about 2 mm above the specimen holder surface. And also, the fragments should be perpendicular to the horizontal surface of holder. Then, the fracture surfaces of the aligned fragments could easily be located, identified and imaged by SEM.

3. Basic Characteristics

3.1. Fiber Diameter Variation Analysis

Accurate determination of fiber diameter is necessary for the estimation of fiber's strength, because the use of a nominal/mean fiber diameter to determine individual fiber strength is not precise. This is the likely situation for most polymer-derived SiC fibers that are processed using a spinning method [26]. For instance, a 10% error in diameter would result in about a 21% error in the strength calculation. Such errors cause additional scatter in the Weibull strength distribution, which results in a low value of Weibull modulus. In such cases, to properly determine the fiber diameter in the assessment of fiber strength is important.

The fiber diameter variation from fiber to fiber across a tow and along the single fiber length was assessed by image analysis from SEM.

3.1.1. Fiber Diameter Variation within a Tow

To investigate the fiber diameter variation across a tow, a yarn of each fiber type was scattered and mounted on the plane surface of copper specimen holder, and then carbon tape was used to fix this fiber bundle. The picture was taken on these fibers one by one and the number of selected fibers is as large as possible. Figure 7 showed the fiber diameter variation across a fiber tow. Mean diameter and standard deviation were also calculated in Table. 2. From this result, the HNLS fiber type showed smallest diameter variation across the tow, which indicated this fiber type has more uniform diameter within a cross section of its tow. The HNL fibers displayed a relatively wide fiber diameter variations within a tow (10.78-16.60 um). Noting the average diameter for each fiber type is possibly different from batch to batch. In these fiber types, the measured mean fiber diameter values given in Table 2 agree fairly well with the manufacture's value in Table 1. The HNL and HNLS fiber in diameter size distribution have a standard deviation, 9.6% and 4.8%, respectively, and the TySA fiber has a relatively high standard deviation, 13.2%.

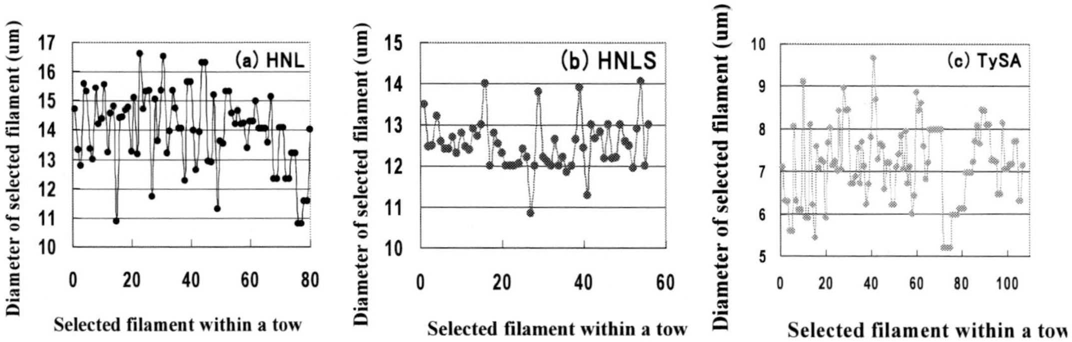

Figure 7. Fiber diameter variation within a tow: (a) HNL fiber, (b) HNLS fiber, (c) TySA fiber.

Table 2. Diameter variation within a tow determined by image analysis from SEM

Fiber Types	Hi-Nicalon™	Hi-Nicalon™ Type S	Tyranno™ SA
Number of fibers counted	80	56	112
Mean Diameter (um)	13.99	12.5	7.13
Standard Deviation (um)	±1.34	±0.6	±0.94
Minumum/Maximum	16.60/10.78	14.04/10.85	9.67/5.20

3.1.2. Fiber Diameter Variation along the Fiber Length

To investigate typical fiber diameter variation along a fiber length, three individual fibers with a length of 30 cm were pulled out randomly from a tow and cut sequentially into 1 cm segments. Before pulling out the individual fiber, soaking the fiber tow in acetone for 2 days and followed by washing in boiling water for 1 minute. This step is quite necessary in aiding the fiber separation, pulling, and reduction of the handling damage to the fiber. For viewing by SEM, the 1 cm length segments were fastened sequentially on the flat specimen holder by carbon tape. For reducing the charging effects, the segments should be connected well with the specimen holder. The diameter was determined directly from image taken by SEM. It should be noted that the ends of each segment were carefully retained in register so that the diameters represented the variation of the diameter along the fiber length at 1 cm interval. The fiber diameter variations at 1 cm interval along 30 cm length fiber for the randomly selected fibers were shown in Figure 8. The typical diameter variation exhibited by the 30 cm fibers was about ±1~3 um for HNL fiber. The HNL fibers exhibited a cyclic diameter variation with a repeat distance of about 15 cm. The HNLS fiber did not exhibit strong cyclic diameter variation, but a rather abrupt rate of change in diameters of about ±0.25 um/cm was observed. The two of three TySA fibers exhibited a very similar variation tendency in diameter, and the least variation ration of about <0.25 um/cm was also observed along the fiber length. But for the third set of TySA fiber, it exhibited high diameter variation rate of >0.25 um/cm.

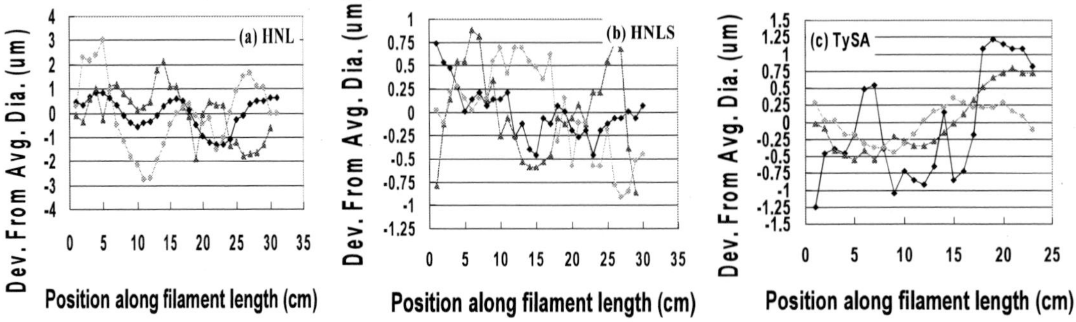

Figure 8. Fiber diameter variation along the fiber length of 30 cm at a 1 cm interval: (a) HNL fiber, (b) HNLS fiber, (c) TySA fiber.

3.2. XRD Patterns

Figure 9 shows XRD patterns for HNL, HNLS and TySA fiber. Obvious β-SiC peaks were observed in these patterns, but HNLS and TySA fiber showed a relatively sharp peak. This indicated the HNLS and TySA fiber have large crystallite size and high crystallization. This is in agreement with the manufacture's information that these fibers are near stoichiometric and high crystallization. In the case of HNL fiber, it has been reported by manufacturer that excess free carbon and amorphous phase existed at the grain boundaries.

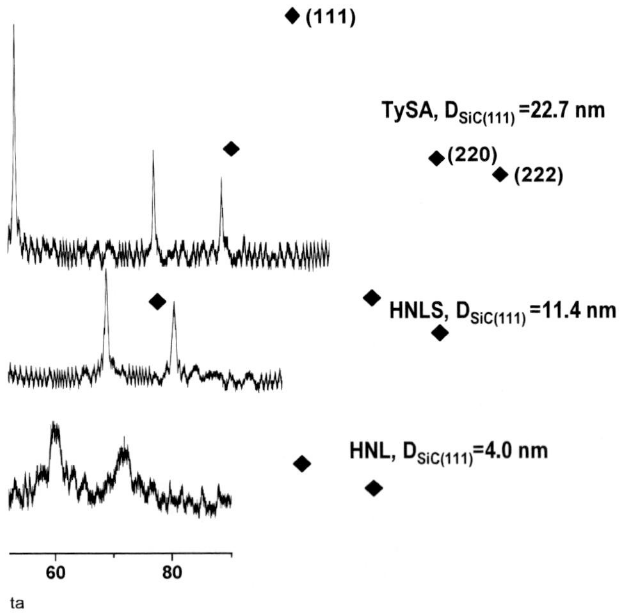

Figure 9. XRD patterns of as-received SiC-based fibers, ◆ (β-SiC).

Using Scherrer's equation described in section 2.2.3, the apparent crystallite size of β-SiC, was calculated from the half-value width of the (111) peak as shown on top right in Figure 9.

3.3. Tensile Properties and Fracture Surface

In Figure 10, the tensile strength distribution and related tensile properties of three fiber types were shown in a two-parameter Weibull plot.

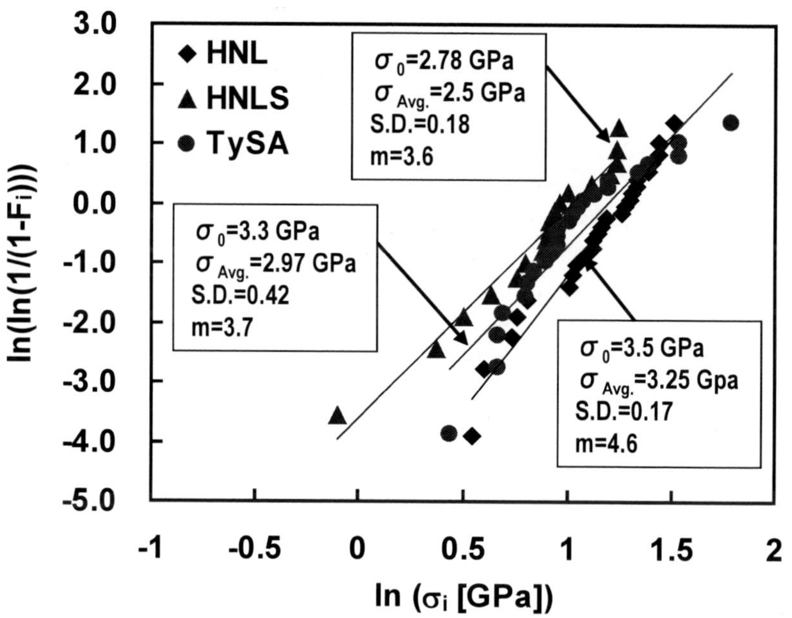

Figure 10. Two-parameter Weibull plot for three fiber types indicating the related tensile properties.

The m values listed in Figure 10 are slightly lower than that of in Refs. [27-28], but they are very similar to the value obtained in Ref. [29]. As we know, the strength of ceramic fibers is associated with the gauge length (weakest link rule) and fiber diameter. Long gauge length and poor uniformity of fibers might be responsible for low Weibull modulus. Furthermore, the performance of fibers also varied from batch to batch.

To examine the fracture surface of individual fiber segments, each fiber segment for the successful test was gripped with narrow tip tweezers and broken off at the frame edge. Generally, for the brittle materials, such as ceramics and glass, their fracture originated from the critical flaw surrounded by the mirror zone, mist zone and hackle zone. Figure 11 is an schematic illustration of crack initiation and propagation route of the SiC-based fiber showing fracture mirror zone surrounding the critical flaw.

During microstructure observation, special care was taken on the features of critical flaw size (r_c), flaw type and mirror size (r_m). The obvious fracture mirror zone was observed on the fracture surface of most of HNL and HNLS fiber fragments except for TySA fiber.

For the set of HNL fiber successfully tested, the critical flaws with different dimension were observed on the surface or inner area of fibers, but the locations of most of critical flaws are near the surface of fibers. In the case of HNLS fiber, the critical flaws were mainly identified as the inner flaws (inner pore or inclusion).

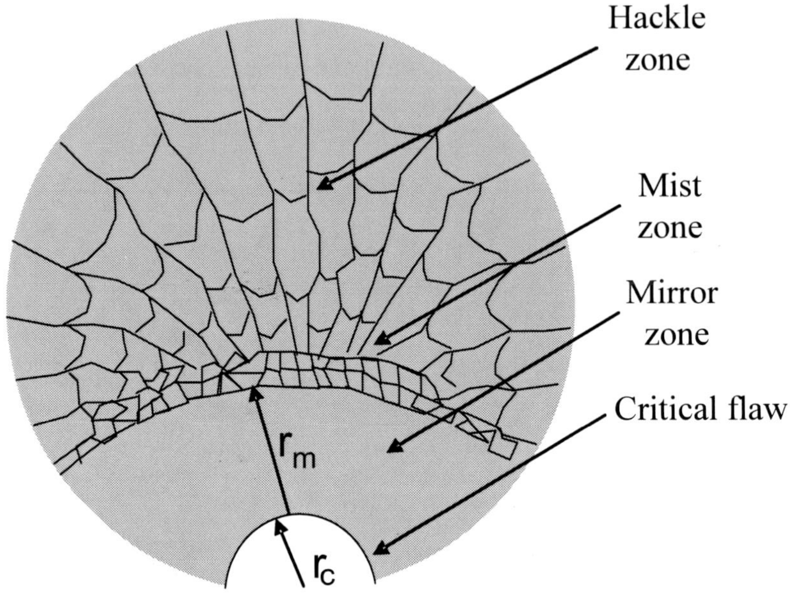

Figure 11. Schematic illustration of fracture surface of fiber, showing fracture originated from critical flaw surrounding by mirror and hackle regions.

Typical features of fracture surfaces for the SiC-based fibers are shown in the following SEM micrographs.

In Figures 12 (a)-(b), a mating pair of fracture surfaces shows a surface critical flaw and the surrounding mirror, mist and hackle regions. Both sides of the mating fracture surfaces exhibited a well defined void. For this particular fiber, the diameter (d), the mirror radius (r_m) and the critical flaw radius (r_c) were measured to be 15, 2.05 and 0.74 um, respectively. The tensile strength (σ_f) is 2.59 GPa.

Figures 12 (c)-(d) are a typical pair of mating fracture surfaces showing a critical flaw of inner pore type. It was very clear that the inner pore was observed in each surface. For this sample, d=13.0 um, σ_f =4.22 GPa, r_m=1.05 um and r_c=0.34 um.

More attention was paid to HNLS fiber. As observed on the fracture surface of HNLS fiber, most of critical flaws are inner flaws and identified as the second inclusions. Figure 12 (e)-(f) shows a pair of mating fracture surface of HNLS fiber exhibited a pore (Figure 12 (e)) and inclusion (Figure 12 (f)) on the opposite fracture surface. For this type of flaw, it is possible that a second inclusion was pulled out freely from a pore. For this HNLS fiber with a diameter of 14 um, it gave an strength (σ_f) of 3.34 GPa. The mirror radius and the critical flaw radius are r_m=2 um and r_c=0.47 um, respectively.

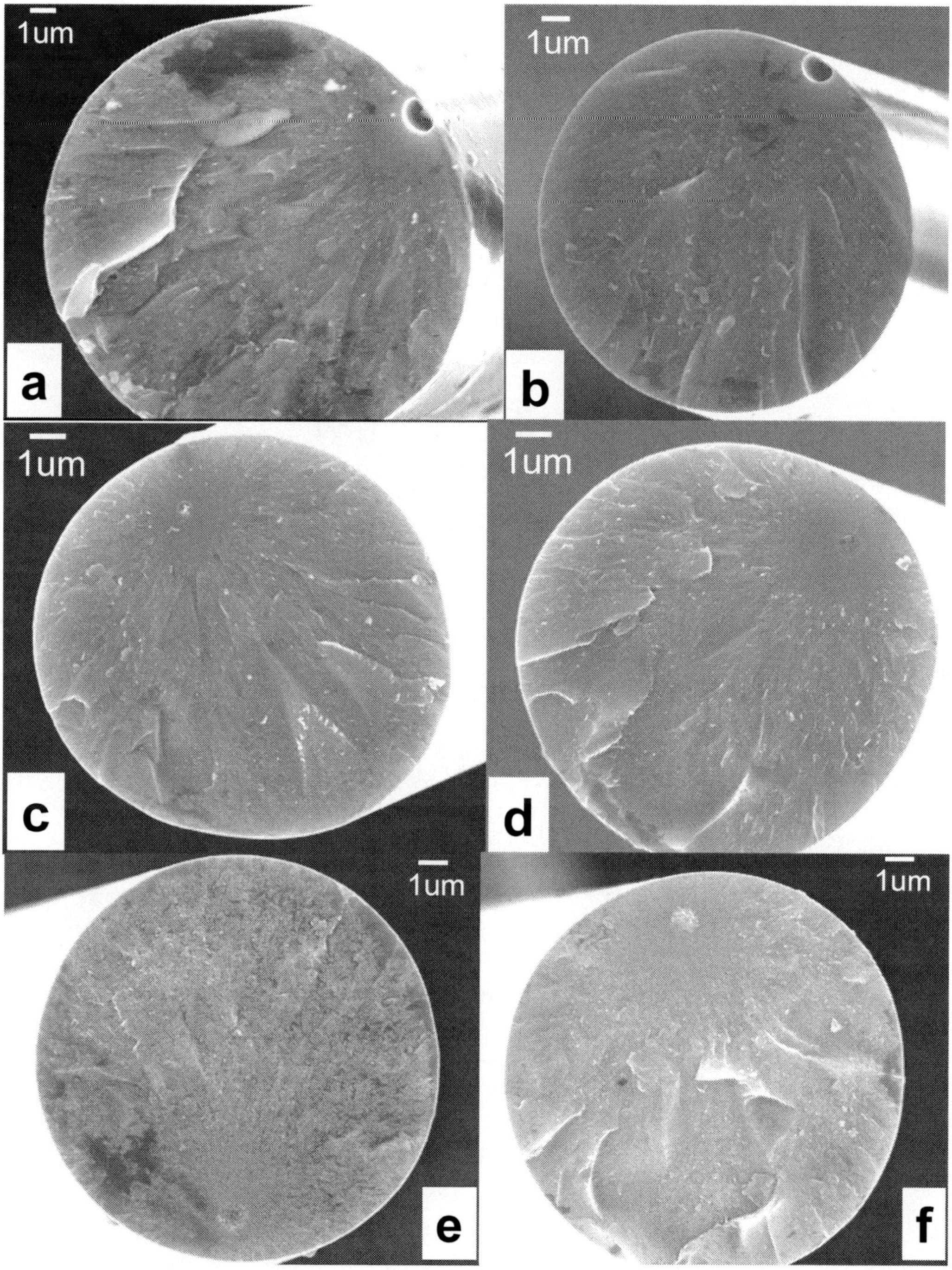

Figure 12. A typical pair of mating fracture surfaces: (a) and (b) for HNL fiber, showing surface critical flaw is a void, for this sample, d=15 um, σ_f=2.59 GPa, r_m=2.05 um and r_c=0.74 um; (c) and (d) for HNL fiber, showing an internal pore (critical flaw) in each surface, for this sample, d=13.0 um, σ_f=4.22 GPa, r_m=1.05 um and r_c=0.34 um; the opposite fracture surfaces of HNLS fiber appeared a remaining pore (e) and second phase inclusion (f) surrounded by the mirror and hackle regions, for this HNLS sample, d= 14 um, σ_f=3.34 GPa, r_m=2 um and r_c=0.47 um.

In order to know the composition of this type of flaw, a higher magnification of a typical inclusion is shown in Figure 13, and energy-dispersive spectroscopy (EDS) scan was performed across this inclusion. The three line scans qualitatively show the variation of the C, O and Si concentrations in the region of the critical flaw. The relative concentrations of C, O and Si are almost unchanged until the inclusion is encountered by the scan, where the C concentration abruptly increases and the Si concentration abruptly decreases at the region of the inclusion. The O concentration within the region of inclusion appears to somewhat lower than its base-line level. The pertinent data for this particular fiber are given in the caption of Figure 13. Based on this information, the inclusion region (critical flaw) appears to be a carbon-rich region. Inclusion with similar characteristics was also observed in previous study of HNL fiber [27].

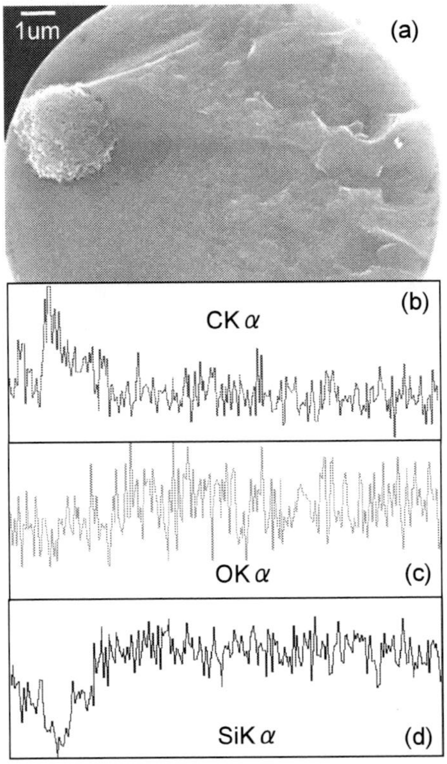

Figure 13. An enlarged SEM view of: (a) an inclusion-type critical flaw, (b), (c) and (d) corresponding to the C, O and Si EDS line scan across the inclusion.

Linking these micrographs to the production process, the defects such as voids or inclusions from impurities or un-melted precursors may exist in polycarbonsilane-derived fibers. These defects may generate local internal stress concentration during the process and lead to the crack formation under tension. And also, under the same processing parameters, it is likely that the stress concentration will vary with fiber diameter, since it is easier to relax the stress concentration in a fiber with a smaller diameter. Wanger [30] suggested that the spinneret hole has laminar flow properties which change with diameters during the fabrication. This may also result in the flow density variation with varying fiber diameters.

The variation in flaw size is certainly a factor affecting the ceramic fiber's strength and thus needs to be incorporated into the fracture statistics.

In the case of TySA fiber with the small diameter (about 7.5 um), no critical flaw was observed on the fracture surface (Figure 14). The fracture surfaces showed a trans-crystallite fracture behavior. This fracture behavior could be partially related to the residual stresses caused by the addition of alumina in this fiber. Existence of residual stresses in the grain boundary of TySA fiber is quite possible because of significant mismatch in the coefficient of thermal expansion between SiC and Alumina (SiC: 3.3×10^{-6}/K; Alumina: $9.1\text{-}9.9\times10^{-6}$/K) and high sintering temperature (higher than 1700 °C). In TySA fibers, the change in the extension stability of micro-crack in the residual stress filed might improve the grain boundary strength. The increase in grain boundary strength could explain the trans-crystalline fracture surface of TySA fiber.

Figure 14. Typical fracture surface observation of TySA fiber showing trans-crystallite fracture.

3.4. Correlation between Tensile Strength and Fiber Diameter

The dependencies of the fiber tensile strengths on diameter for the HNL, HNLS and TySA fibers are illustrated in Figure 15, respectively.

In Figure 15, the fiber tensile strengths exhibit significant scatter. Nevertheless, the general tendency that fibers with larger diameters have lower strengths is consistent. The fiber with large diameter will be easy to cause the stress concentration around the defects. The similar phenomenon for Nicalon fibers was also observed in Ref. [31], and quantitative explanation has been given from the fracture mechanics. The tensile strength σ vs. normalized diameter (d/d_0) data were fit with an empirical power law dependence of the form

$$\sigma=K_0(d/d_0)^{-n} \tag{10}$$

where d_0 is the average diameter of the data set, n is the power law exponent and K_0 is the average strength expected for a fiber with diameter d_0. The values of K_0 and n, determined from linear least squares fitting of the fiber tensile strength data to Equation (10), are included in the plot for each fiber type.

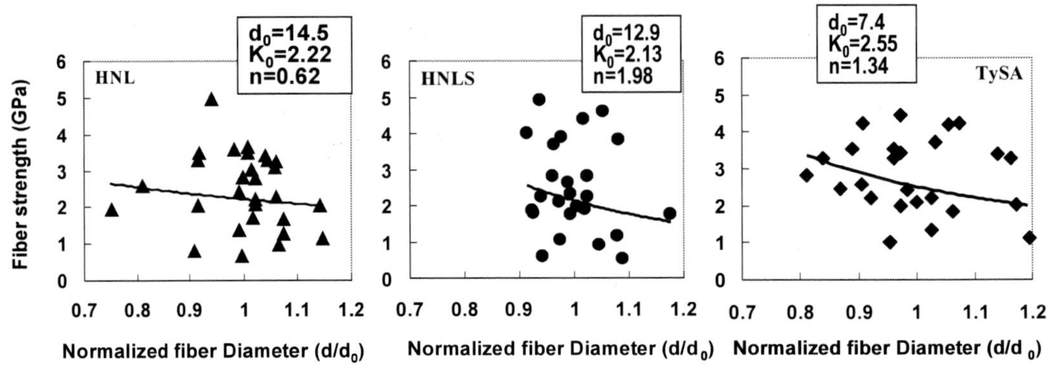

Figure 15. The dependence of tensile strength on the fiber diameter for: (a) HNL, (b) HNLS, (c) TySA fiber types. The solid curve was obtained by fitting the data point in each plot with Equation (10). Although the data is significant scatter, the general tendency that fibers with larger diameters have lower strength is consistent.

3.5. Correlation between Tensile Strength and Mirror Size

Prior to understanding the correlation between the tensile strength and the mirror size, the dependence of critical flaw size on its corresponding mirror size was examined as shown in Figure 16.

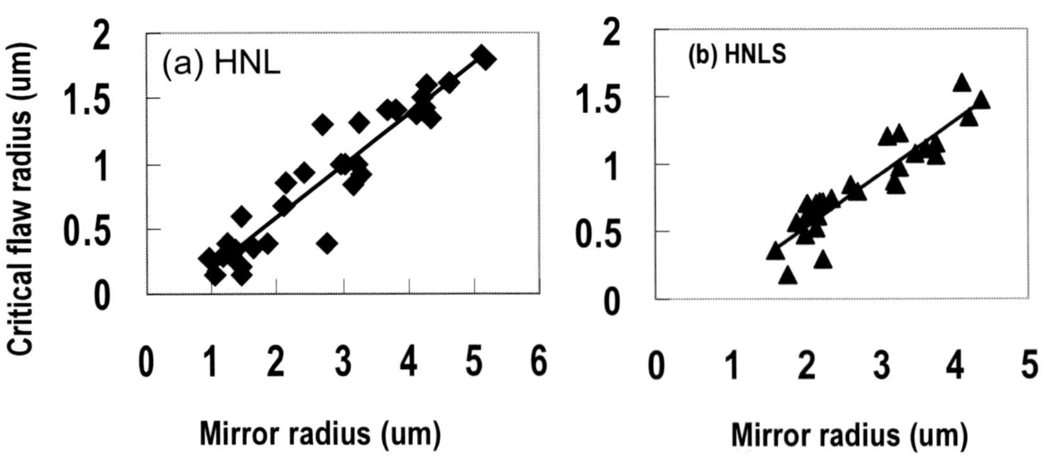

Figure 16. The critical flaw radius vs. the mirror radius: (a) HNL fiber, the slop of fitting straight line is 0.38; (b) HNLS fiber, the slop of fitting straight line is 0.39.

Although the data is a little scattering in Figure 16, surprisingly, the slops of fitting line in HNL and HNLS fibers have a very similar value. This gave us an information that critical flaw size (r_c) was linearly related to the mirror size (r_m). A linear relationship between the critical flaw size (r_c) and the mirror size (r_m) is commonly observed for brittle fracture of ceramics and glass [26-27,32-33]. The ratio r_c/r_m in this study is about 0.39 which is somewhat larger than the 0.33 for Hi-Nicalon fiber in previous study [27], and 0.19–0.22 range of values observed for ceramic grade (CG) Nicalon fiber [26]. Possibly, this difference could be attributed to the definition of mirror size and the crystallization of PCS-derived fibers. On the other hand, it could also be associated with the accuracy in estimating the actual flaw sizes from the SEM micrographs.

In Figure 17, the individual fiber strengths are compared to their corresponding mirror sizes for the HNL and HNLS fibers. The best fit straight line has an slope of -0.48 for HNL fiber, and -0.49 for HNLS fiber (Figure 17). These values are very close to the -0.5 for brittle ceramic materials according to the Griffith theory [34-35]. The data scatter appeared in the plots of tensile strength vs. mirror size for the HNL and HNLS (Figure 17) could be consist with the trend that fiber with larger diameters are weaker as reported previously for the Hi-Nicalon™ fiber [35] as well as for other polymer-derived SiC fibers [24]. In general, for the ceramic fibers, the critical flaw density varied with varying fiber diameters, and the flaw within the fiber with larger diameter is more easily to cause the stress concentration. Furthermore, the error in the diameter measurement of fibers will also cause the data scatter in the calculated fracture strength. We have observed that the diameter of fiber varied along it's length. And also, the strength of SiC fibers is sensitive to the surface critical flaw. For this batch of HNL and HNLS fibers, the fraction of critical flaws occurred at the fiber surface is relatively low, but mostly were distributed internally with only a slight preference for being located nearer to the fiber surface than to the fiber center.

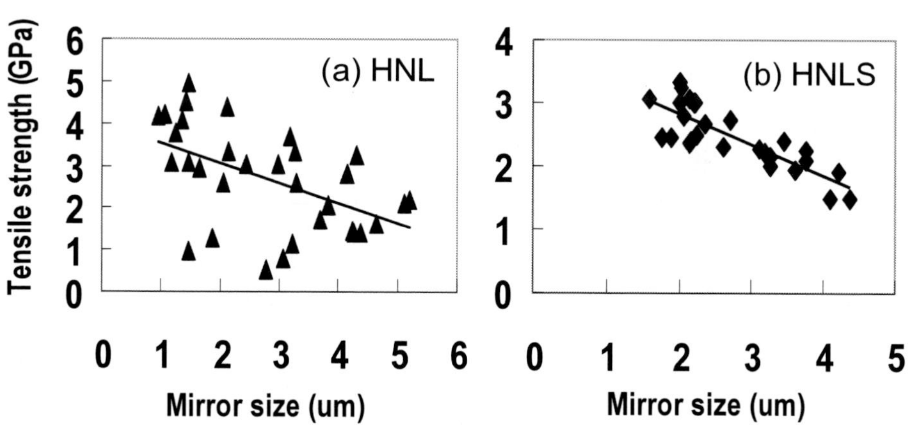

Figure 17. Tensile strength vs. mirror size: (a) HNL fiber, (b) HNLS fiber. Fitting straight line represented slops of approximate 0.5.

3.6. Fracture Toughness and Critical Fracture Energy

Because the relation between the fracture strength and the critical flaw size observed the Griffith theory according to the results in Figure 16 and Figure 17, thus, the fracture mechanics principle could be applied to calculate the fracture toughness and the critical fracture energy.

3.6.1. Fracture Toughness

Fracture mechanics predicts a relation between flaw radius, fracture strength (σ_f) and the fracture toughness (K_{1c}) for brittle materials [26,31-32], where K_{1c} is the mode 1 fracture toughness of the SiC fiber.

$$\sigma_f(r_c)^{1/2}=YK_{1c}=\text{constant} \quad (11)$$

In Equation (11), Y is a geometric factor which depends on the critical flaw shape and location and its relative size compared to the fiber dimension. Y is 1.56 for a small, centrally located penny-shaped flaw in a plane normal to the tensile axis given in the Ref. [31].

Additionally, it has been extensively demonstrated that the product of strength, σ_f, and the square root of mirror size obeyed following formula [26,31-33,36]

$$\sigma_f=A_m(r_m)^{-0.5} \quad (12)$$

where A_m is the mirror constant.

Substituting σ_f in Equation (11) with Equation (12), the fracture toughness, K_{1c}, could be expressed as:

$$K_{Ic}=A_m(r_c/r_m)^{0.5}/Y \quad (13)$$

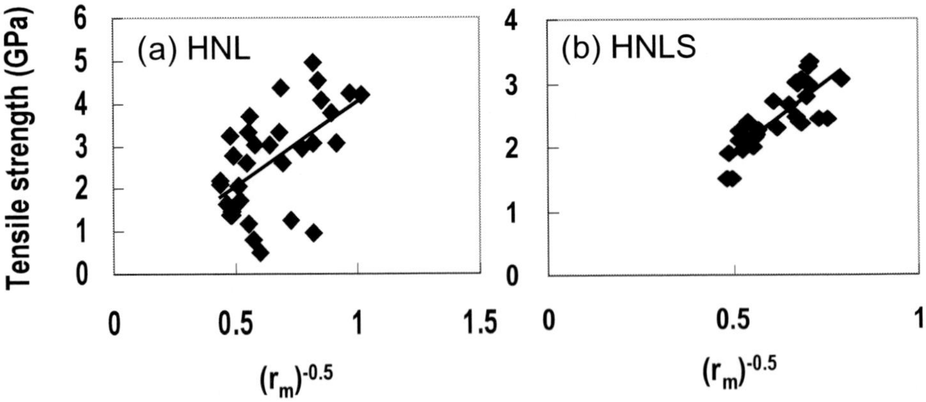

Figure 18. The fiber tensile strength vs. the square root of the fracture mirror radius: (a) HNL fiber, the slope yields the mirror constant A_m=3.93 MPam$^{1/2}$; (b) HNLS fiber, the slope yields the mirror constant A_m=4.33 MPam$^{1/2}$.

In Figure 18, the tensile strength σ_f vs. $(r_m)^{-0.5}$ are plotted for HNL and HNLS fiber, respectively. The data were fit to a linear relation by regression analysis. The mirror constant A_m, defined as the slope of the fit straight line, was determined to be 3.93 $MPam^{1/2}$ for HNL fiber, 4.33 $MPam^{1/2}$ for HNLS fiber, respectively. From $K_{1c}=A_m(r_c/r_m)^{0.5}/Y$, using $r_c \approx 0.39\ r_m$ and Y=1.56, the calculated K_{1c} is 1.56 $MPam^{1/2}$ for HNL fiber, 1.74 $MPam^{1/2}$ for HNLS fiber. Since A_m is an average value, the K_{1c}-value determined for these fibers also is an average value.

The K_{1c} for polycrystalline SiC is ≈ 2 $MPa \cdot m^{1/2}$, while that for most amorphous ceramics is ≈0.5 to 1 $MPa \cdot m^{1/2}$ [36].

3.6.2 Critical Fracture Energy

Attempts [32-33] have been made to relate the critical flaw radius to the critical fracture energy, γ_c, which can be obtained from the following equations,

$$r_c = Y^2 2E\gamma_c / \sigma_f^2 \tag{14}$$

$$\sigma_f \bullet r_c^{1/2} = Y\sqrt{2E\gamma_c} \tag{15}$$

where Y is a geometric factor, E the modulus of elasticity. By substituting the fracture strength in Equation (15) with Equation (11), the critical fracture energy could be simplified as:

$$\gamma_c = \frac{K_{1c}^2}{2E} \tag{16}$$

The critical fracture energy calculated with Equation (16) is 4.5 J/m^2 for HNL fiber, 3.6 J/m^2 for HNLS fiber. The low critical fracture energy for HNLS fiber could be attributed to the low strain to failure (HNL: 1%, HNLS: 0.65%).

The Griffith theory presents a criterion for propagation of preexisting flaws that generally determines the failure of brittle materials. After enough energy has been supplied to the crack, it will propagate at velocity which increases as its length increases. Since the driving force depends on crack length, crack velocity will increase until it approaches a terminal velocity. As the crack approaches the terminal velocity, the sum of the potential energy resulting from its increasing length and the kinetic energy resulting from its motion becomes greater than the energy that can be used to increase the velocity of the crack. Small cracks are nucleated around the tip of the main crack, forming mist, but there is insufficient energy to propagate these secondary cracks very far. Limited velocity increases allow propagation of such secondary cracks to form hackle. Finally, when enough energy is available, the crack can branch macroscopically.

4. Mechanical Properties and Microstructure Under Various Environments

4.1. Heat Treatment at Elevated Temperatures

Because the CMCs may be fabricated above the fiber's processing temperature [37-38], in which case, the performance of fibers could be changed by high temperature treatment. Thus, the identification of correlation between performance and heat treatment temperature is essential for exploring the optimum condition for high performance CMCs fabrication and application.

4.1.1. Correlation between Tensile Strength, Crystal Size and Heat Treatment Temperatures

Figure 19 shows average room temperature tensile strengths and apparent crystallite sizes for three types of fibers after heat treatment in Ar for 1 h at elevated temperatures. The apparent crystallite size of β-SiC, D_{SiC}, was calculated from the half-value width of the (111) peak by using the Scherrer's formula. From the dependence of crystallite size on the heat treatment temperature as shown in Figure 19, following features were observed: (i) the grain coarsening of HNL fiber started at 1400 °C; (ii) the crystallite size of β-SiC in HNLS and TySA fiber remained almost constant as heat treatment temperature <1600 °C, while higher temperature heat treatment caused an continuous coarsening in crystallite size of SiC in HNLS. The crystallite size of β-SiC in TySA fibers appears to be little dependent on the heat treatment temperature. The crystallite sizes for as-received HNL, HNLS and TySA fibers are 4.0 nm, 11.4 nm, 22.7 nm, while they are 15.5, 30.3, and 25.5 nm for fibers heat treated at 1900 °C for 1 h, respectively.

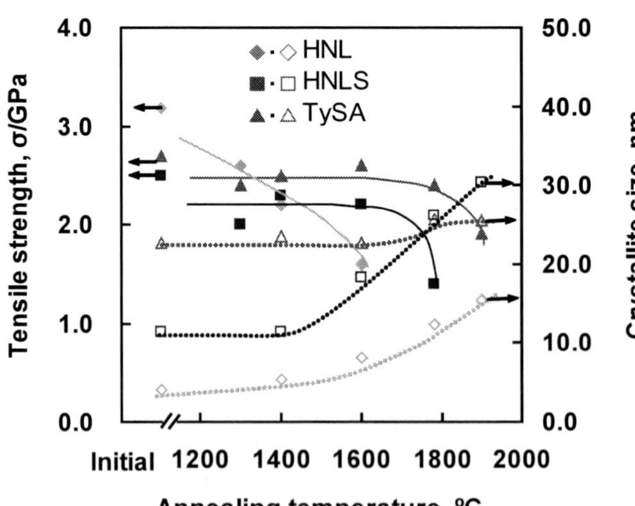

Reprinted by permission of Elsevier from [69].

Figure 19. Tensile strength and its relation to the crystallite size of SiC fibers heat treated at elevated temperatures in Ar for 1 h.

The grain coarsening could be attributed to the coalescence of β-SiC nanocrystals due to either decomposition of amorphous phase or diffusion of Si and C atoms at grain boundaries during exposure at high temperatures. For the bulk materials with clean grain boundaries, the grain growth proceeds through large grains incorporating the small one by grain boundary diffusion. Especially, for grains with small size, the grain boundary diffusion operates much more readily. As observed in two Nippon Carbon fibers, the grain coarsening is more obvious in heat treated state than those of as-received state (TEM observations have revealed that grain size is about 5 nm for HNL fiber [39], 20 nm for HNLS fiber [40], 200 nm for TySA fiber [41]). On the other hand, the residual trace oxygen may play a role in the Si and C grain boundary transport by accelerating diffusion [39], because the oxygen is not necessarily eliminated from the fiber as reported in the literature [42], even for the HNLS fiber which was fabricated at very high temperature.

Considering the starting temperature for grain coarsening in Figure 19, the grain size might be related primarily to the maximum temperature at which the fibers were fabricated. The fabrication temperatures have been presented for two Nippon Carbon fibers (HNL:1350 °C, HNLS: 1600 °C) and for TySA fiber (about 1800 °C) in the literature [43]. From the Figure 19, it can be seen that the crystallite size increased when heat treatment temperature is above the fabrication temperature as expected. For HNL fiber, the grain coarsening occurred at relatively low temperature is due to the decomposition of amorphous phase at about 1300 °C. On the other hand, the thermally activated diffusion play an important role on the grain coarsening of SiC materials at high temperatures.

If we make a further comparison in crystallite size between two Nippon Carbon fibers again, we can see that a large difference in crystallite size was observed for two fibers het treated at same temperature. As above mentioned, the HNL fiber has an small starting grain size, which was expected to have a high diffusivity at grain boundaries and result in a large grain size as heat treating at high temperatures. However, an unexpected phenomenon was observed between two Nippon Carbon fibers. This can be attributed to the excess carbon in HNL fiber. TEM observation revealed that heat treatment of the HNL fiber results in a gradual organization of the free carbon phase in terms of the size of the carbon layer and the number of stacked layers as increasing temperatures [39]. Takeda et al. [44] have investigated the properties of polycarbosilen-derived silicon carbide fibers with various C/Si compositions, and revealed that microstructure and mechanical properties are quite dependent on the C/Si composition. Grain growth is suppressed with increase in excess carbon. In other studies [45-46], the carbon suppressing growth and coalescence of the SiC microcrystals was also observed. Sasaki [46] found that carbon disappeared above 1500 °C heat treatment in SiC fiber using Raman study. And then an abrupt increase of crystal size at 1500 °C was observed.

The grain growth has a significant effect on the strength of SiC-based fibers. For the HNL fiber, crystallization degraded its strength at all HTT. In both near stoichiometric fibers, strength degradations occurred at the temperatures where crystallite size began to increase. Fibers with larger grain size generally have relatively lower strengths, but it should be noted that HNL fiber showed more rapid strength degradation than HNLS fibers above 1400 °C heat treatment as shown in Figure 19. HNL fiber has smaller crystal sizes comparing to that of HNLS fiber. The growth of SiC crystals reduces the bonding forces at the grain boundaries. Since the manufactures are always seeking the optimal fabrication temperature at which the superior thermal stability and excellent mechanical strength can be obtained simultaneously, thus, the upper fabrication temperatures are typically fixed by those temperature conditions above which performance degradation of the fibers occurred. The dependence of strength on temperature in Figure 19 is in agreement with those of previous studies [18,39-40,46]. Ichikawa et al. [40] reported that HNLS was quite stable chemically after one hour exposure in an argon gas at 1800 °C, since no structural decomposition occurred and it exhibited a good strength of 1.9 GPa. TEM observation shows that this heat treated HNLS fiber has a SiC grain size of approximately 200 nm, which is about 10 times larger than that of the as-received fiber.

As for TySA fiber, this is a sintering fiber, which is prepared by the reaction of a polycarbosilane (PCS) with aluminium acetylacetonate, and subsequently converted into the Tyranno SA fiber, by decomposition with an evolution of CO and SiO (1500 °C $<T<$1700 °C) and sintering (about 1800 °C). TySA fiber retained most of its initial strength, because no significant grain coarsening was observed even heat treated at 1900 °C. Excellent strength retention has been observed in a former work [18].

Fiber strength is controlled not only by grain size, but also by critical flaw size and the residual stresses etc, which were generated from fabrication process and gas evolution during heat treatment at elevated temperatures. The mismatch in thermal expansion coefficient between excess carbon and SiC grain, could cause strength loss, and the contribution of residual stresses from the gas evolution to strength loss could increase with increasing the β-SiC grain size.

On the other hand, the HNL fiber has smaller crystal size comparing to that of HNLS fiber, but it showed more rapid strength degradation than HNLS fiber above 1400 °C heat treatment as shown in Figure 19; both HNLS and TySA fiber have near-stoichiometric composition and high-crystallite structure, but they showed different strength retention. This observed phenomenon implied that other mechanisms must be responsible for strength degradation of SiC fibers besides the coarsening of crystallite size.

One of the most possible reasons for low strength retention of HNL fibers is residual stresses which were generated from phase transformation and the mismatch in the coefficient of thermal expansion between excess carbon and SiC grain. Sacks [47] produced a laboratory fiber (UF fiber) with the similar composition as HNL fiber. There was no loss in strength with heat treatments up to 1700 °C and then the strength decreased rapidly with further heat treatments up to 1900 °C. He believed that strength is controlled by the residual tensile stresses which developed as a result of the mismatch in the coefficient of thermal expansion between SiC and C. This situation should be true. The coefficient of thermal expansion of carbon/graphite ($2.0\text{-}3.0\times10^{-6}$/K) is less than SiC ($3.9\text{-}4.0\times10^{-6}$/K). When fibers were cooled from high heat treatment temperature to room temperature, the SiC grains want to contract, while carbon grain will resist their contraction. This action-reaction will put SiC in tension and carbon in compression. This residual tension stresses could have a contribution to the total stress loss. This case can be applied to each fiber type which contains excess carbon, but here it should be more significant in HNL fiber because of high excess carbon (C/Si=1.38). In both HNL fiber and TySA fiber, the size of carbon grain increased with increasing heat treatment temperature [39,48-49]. For the TySA fiber, this fiber originally has a very large crystallite size. In previous studies [42,48-49], a carbon-rich core was revealed in TySA fiber, which results from the production process. Colomban et. al. estimated carbon grain size and SiC grain size in TySA fiber from Raman spectroscopy [48-49]. The Carbon grains appear approximately 2-3 times smaller on the fiber's core (0.9-1.7 nm) than on its periphery (1.7-2.6 nm). The grain size of SiC in fiber core is much smaller than edge region. Likely, this is due to that carbon suppressed the grain growth of β-SiC. Furthermore, this fiber contains the small amount of alumina (less than 1 wt %) as sintering additive, which will also inhibit the grain growth of SiC. As a result, the TySA fiber showed an excellent thermal stability in Ar atmosphere. The growth of the carbon corresponds to a decrease of localized spin centers [50]. The growth of the carbon grain might result in an increase of residual stress, however, this evidence is insufficient because the magnitude of residual stresses is strongly dependent on the volume fraction of carbon phase in a bulk material.

4.1.2. Microstructure

Figure 20 shows SEM morphologies of the fibers after heat treatment at high temperatures in Ar for 1 h. The HNL fibers heat treated at temperatures below 1400 °C had a smooth surface, which is almost no difference from that of as received fibers. Heat treatment at 1400 °C caused slight coarsening of fiber surface. Obvious changes in appearance were observed for the HNL fibers heat treated at 1780 °C as shown in Figure 20 (a). The fibers showed a porous microstructure and large grains deposition on the fiber surface (Figure 20 (a)). Such large grains are not observed within bulk of the fiber, due to the presence of free carbon which inhibit the grain boundary or/and gaseous diffusion. For the HNLS fibers heat treated below 1600 °C, their microstructure did not vary compared to the as-received fibers. After heat treated at 1600 °C, although the individual SiC grain grown on the fiber surface, but fiber surface still remained smooth and it appeared no structure degradation. The fiber heat treated at 1780 °C exhibited a rough surface with deposition of bulk SiC grains, but it still remained a relatively dense structure as shown in Figure 20 (b). TySA fibers showed outstanding thermal stability in microstructure comparing with other SiC fibers and didn't exhibit obvious structure damage even heat treated at 1900 °C shown in Figure 20 (c).

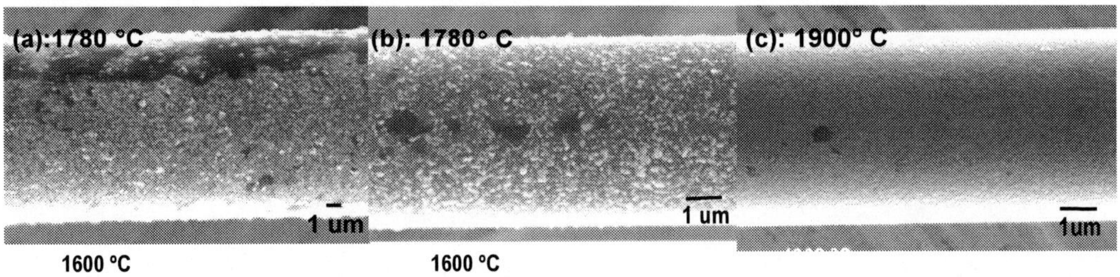

Figure 20. SEM photographs for SiC fibers heat treated in Ar for 1 h: (a) HNL fibers at 1780 °C, (b) HNLS fiber at 1780 °C, (c) TySA fibers at at 1900 °C; typical fracture surface observation for: (d) HNL fiber at 1600 °C, (e) HNLS fiber at 1600 °C, (f) TySA fiber at 1900 °C.

Obvious differences were found in subsequent observations of fracture surfaces. The fracture of as-received HNL and HNLS fibers mainly originated from the inner critical flaw (inclusion-type or inner pore-type critical flaw). After heat treatment at 1600 °C, most of the examined HNL and HNLS fibers fractured at surface flaw as shown in Figure 20 (d)-(e), and flaw size slightly increased with increasing heat treatment temperature. The critical flaw size and mirror size were measured. The critical flaw sizes (r_c) are: 0.90 μm for as received HNL fiber, 1.07 μm for 1600 °C heat treated HNL fiber. In the case of HNLS fiber, the critical flaw sizes (r_c) are: 0.84 μm for as-received fibers, 0.90 μm for 1600 °C heat treated fibers. Figure 20 (f) showed that fracture surface of the TySA fibers after annealing at 1900 °C did not reveal obvious difference in the fracture mode comparing with the as-received fibers. The fracture origin and mirror zone are invisible on the fracture surface and fracture surface showed a trans-crystallite fracture behavior. The trans-crystallite fracture behavior could be partially related to a high compression residual stresses in SiC caused by addition of alumina in this fiber. Existence of compression residual stresses in the grain boundary of TySA fiber is quite possible because of significant mismatch in the coefficient of thermal expansion between SiC and Alumina (SiC: 3.9-4.0$\times10^{-6}$/K; Alumina: 8.0-9.0$\times10^{-6}$/K) and high sintering temperature (higher than 1800 °C). In TySA fibers, the change in the extension stability of micro-crack in the compression residual stress filed might improve the grain boundary strength. The increase in grain boundary strength could explain the trans-crystalline fracture behavior of TySA fiber.

Linking the tensile strength data in Figure 19 with the microstructure examination in Figure 20 again, the decomposition of amorphous phase, grain coarsening and residual stress at high temperatures in HNL fiber could be responsible for strength and microstructure degradation. Above 1600 °C, the outward growth of huge grains was observed as shown in Figure 20 (a) and (b), and these huge grains might act as the critical flaw during the fracture of fiber. Observation of surface morphologies (Figure 20 (a)-(c)) and fracture surface (Figure 20 (d)-(f)) provided some information for the strength degradation of SiC fibers. Due to near-stoichiometric composition in HNLS fiber, it's damage was limited on the surface of fiber.

The formation of porous structure in HNL fiber could be attributed to the rapid evolution of gases at the earlier stage of high-temperature exposure according reaction (17).

$$SiC_xO_y \rightarrow SiC(s) + C(s) + SiO(g) + CO(g) \quad (17)$$

It should be noted that large grains grown outward from the surface of fibers at 1780 °C appear to be β-SiC crystals, which were produced by following gas-phase reactions [51-52].

$$SiO(g) + 3CO(g) \rightarrow SiC(s) + 2CO_2(g) \quad (18)$$

$$SiO(g) + 2C(s) \rightarrow SiC(s) + CO(g) \quad (19)$$

Reaction (19) could occur because of presence of free carbon in surface and body of HNL fibers [20]. According to above result, the reaction (17)-(19) are quite dependent on the quantity of amorphous phase and content of carbon in SiC fibers. The use of graphite crucible in this work might also cause the reaction:

$$CO_2(g) + 2C(s) \rightarrow 2CO(g) \quad (20)$$

Combing the reaction (18) and (20) indicating the gas-phase reaction proceeded mainly by reaction (19). Additionally, the CO-CO_2 gas mixture might modify the microstructure of SiC fibers at high temperatures [53].

Considering the surface degradation of HNLS fibers heat treated above 1600 °C (Figure 20 (b)), as we know, the quantity of amorphous phase and oxygen content should be very small in this fiber. Thus, thermal decomposition of the amorphous phase is almost negligible in this fiber. As for the large grains deposited on the surface of HNLS fiber, it can be explained by reactions (18)-(20), because the carbon layer on the surface of HNLS fiber (80 nm) is thicker than that of HNL fiber (20 nm) [42]. Concerning the origin of gas species, other mechanism could be responsible for it. Active oxidation is quite possible. The transition from passive oxidation to active oxidation occurs at oxygen partial pressure of 10-25 Pa and 1-2.5 Pa at 1500 °C for HNL and HNLS fiber [52], respectively. At same oxygen partial pressure level, the increase of heat treatment temperature will accelerate the transition from passive-to-active oxidation [51].

For the TySA fiber, the excellent microstructure stability could be attributed to the high processing temperature (over than 1800 °C) and addition of alumina [18]. The small amount of alumina addition could inhibit the grain growth and enhanced the oxidation resistance. This higher stability can also be linked to the silica protective layer formed on the surface of fiber [48,54].

Combining the fracture properties with microstructure characterization of SiC-based fibers, we can not deny the existence of other degradation mechanisms such as contaminants during heat treatment and metallic impurities introduced during process [55-57]. The existence of metallic impurities within the fibers is possible, because all these fibers are polymer derived. The metallic impurities can easily enter the fibers during the various steps of polymer handling and can cause rapid or abnormal grain growth in local areas. There are at least two indirect observations supporting above mentioned mechanism: (i) Observation of fracture surface in Figure 20 for the HNL and HNLS fiber showed that the strength-limiting flaws after heat treatment are larger than the average grain size, indicating rapid defect growth in selected areas of the fiber and thus suggesting the possible existence of metallic impurites; (ii) the UF fiber showed high strength retention than HNL fiber [47]. This suggests that the UF fiber during processing did not introduce the metallic impurities to the degree that employed for the HNL fiber.

4.1.3. BSR Creep Resistance

Figure 21 shows dependence of 1-h BSR creep resistance m on HTT, which was tested at 1400 °C. Heat treatments of the fibers above the processing temperature resulted in improved creep resistance as shown in Figure 21. The creep resistance of heat treated HNL fiber above 1400 °C was significantly improved. Likely, this could be attributed to the increased grain sizes, high crystallization of β-SiC. Such microstructural changes are expected to inhibit diffusion-controlled creep processes. For the 1600 °C heat treated HNL fiber, the creep resistance was better than those of as-received near stoichiometric fibers although the fact that the grain sizes were much smaller than those of the latter fibers. This result indicated that the improved creep resistance depended on not only the crystallization and grain growth, but also the composition at grain boundaries. The excess carbon distributed at the grain boundary of the HNL fiber inhibits the coalescence of β-SiC, which results in a stable grain boundary structure. This implies that stability of Grain boundaries (GB) plays an important role on the creep resistance of SiC fiber. This assumption was also demonstrated by TySA fiber. The enhanced creep resistance of the TySA fiber was obtained prior to increase its crystallite size. As a result of Al addition to TySA fiber, the complex oxide would be formed at GB by heat treatment and they can stabilize the grain boundary to improve the creep resistance. The stability of GB could be affected by GB composition.

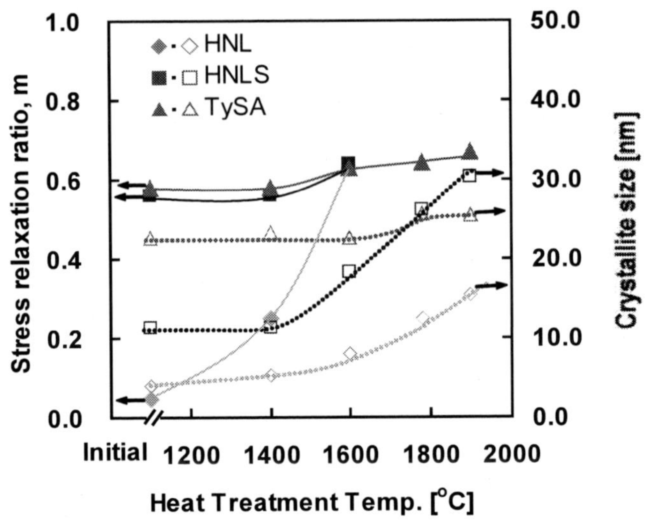

Reprinted by permission of Elsevier from [69].

Figure 21. 1-h stress relaxation ratio and crystallite size for heat treated HNL, HNLS and TySA fibers, creep tests were performed at 1400 °C in Ar.

4.1.4. Fracture Toughness and Critical Fracture Energy

Based on the principles described in section 3.6, the fracture toughness and critical fracture energy was calculated for the heat treated fibers, the resultant value of fracture toughness was shown in Figure 22. The fracture toughness decreased with increasing the heat

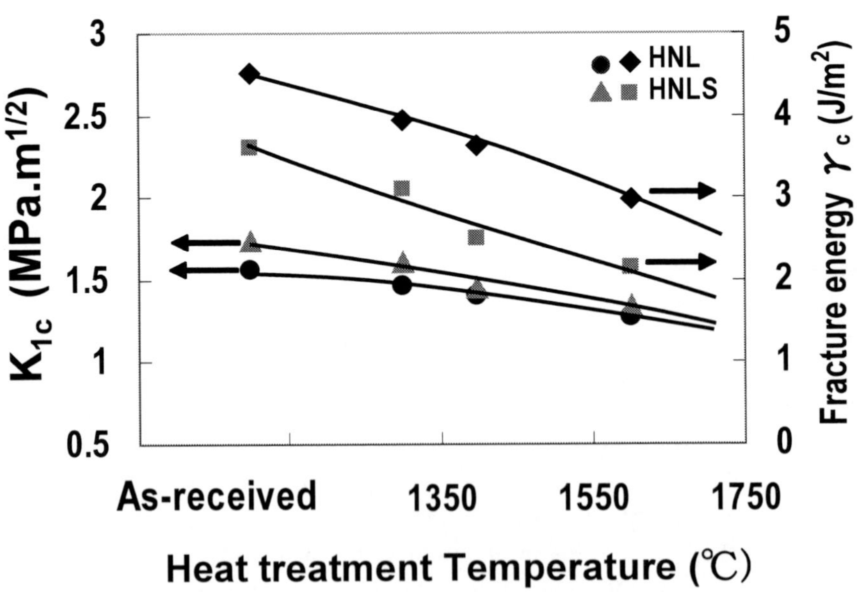

Figure 22. Dependence of fracture toughness and critical fracture energy on heat treatment temperatures.

treatment temperature, but it did not show strong dependence on the heat treatment temperature. For the as-received fibers, the carbon layer covered on the surface of fibers, can blunt the critical flaw and reduce the stress concentration on the surface flaw. However, this carbon layer can be removed by reaction with residual oxygen from fiber itself and atmosphere. In this case, the propagation of preexisting surface flaws will become easy. In addition, flaws produced by decomposition, active oxidation and large grain deposition can exist on the fiber's surface at high temperature resulting in the low fracture toughness and critical fracture energy. At fairly high strain rate (0.3 mm/min) at which the strengths were measured, these flaws would propagate gradually until they become critical because of stress concentration around the flaws.

4.2. Annealing and Creep in Various Oxygen Partial Pressures

SiC fibers with high thermal stability are considering as the promising reinforcement in CMCs. However, the mechanical and thermal stabilities of SiC fibers as reinforcements in CMCs are very sensitive to their purity, crystallinity and service environments [58-60] including thermal and loading history.

For high temperature applications, the CMCs are often subjected to oxidative environments with different oxygen partial pressures. In such case, the SiC materials would be oxidized in passive/active oxidation regimes [51]. As we know, the performance degradation of SiC materials in oxidizing environments strongly depends on the oxidation mechanism. Jacobson [61] has discussed the oxidation degradation mechanism of SiC materials in varied environments, but it is still insufficient because of the complexity of service environments. The key question concerns the oxidation kinetics: passive and active oxidation. This topic has given rise to much controversy for SiC materials, because the temperature boundaries for the oxidation kinetics are quite dependent not only on the materials themselves (purity and crystallinity), but also on the specific service environment (exposure temperature, oxygen partial pressure and mechanical state). Furthermore, rarely is one mechanism operative in performance degradation of SiC materials. In practice, several mechanisms operate simultaneously.

Therefore, for understanding the mechanical and thermal stabilities and failure mechanism of SiC fibers over a wide range of temperatures and varied environments, the part of this chapter reviewed the microstructure features and high temperature properties of SiC fibers under annealing and creep in various oxygen partial pressures at elevated temperatures, and attempted to clarify the correlation between the environment with mechanical and thermal stabilities. Thus, SiC fibers were annealed and crept in air (O_2: 20%, dew point: 3 °C), high-purity Ar (HP-Ar, O_2: 2 ppm, dew point: -5.5 °C) and ultra high-purity Ar (UHP-Ar, O_2: 0.1 ppb: dew point: -5.5 °C) under flowing atmosphere with a pressure of 10^5 Pa and held for 1 hour at desired temperatures ranging from 1000 to 1500 °C. Furthermore, the surface morphologies of fibers under annealing and creep were compared by the observation of field-emission scanning electron microscope (FE-SEM).

4.2.1. Morphologies of Fibers

(a) Under Annealing and Creep in Air

Figure 23 shows the morphologies of SiC fibers under annealing and creep at 1500 °C for 1 h. The SiO_2 film formed uniformly on the surface of fibers during annealing at elevated temperatures (Figures 23). From the cross-section of fibers under annealing at 1500 °C (Figure 23 (c-1)-(c-3), we could see that the silica layer consisted of a concentric sheath and the fiber's surface was blanketed well by silica layer. And also, no further oxidation between SiO_2 film and SiC fiber surface was observed.

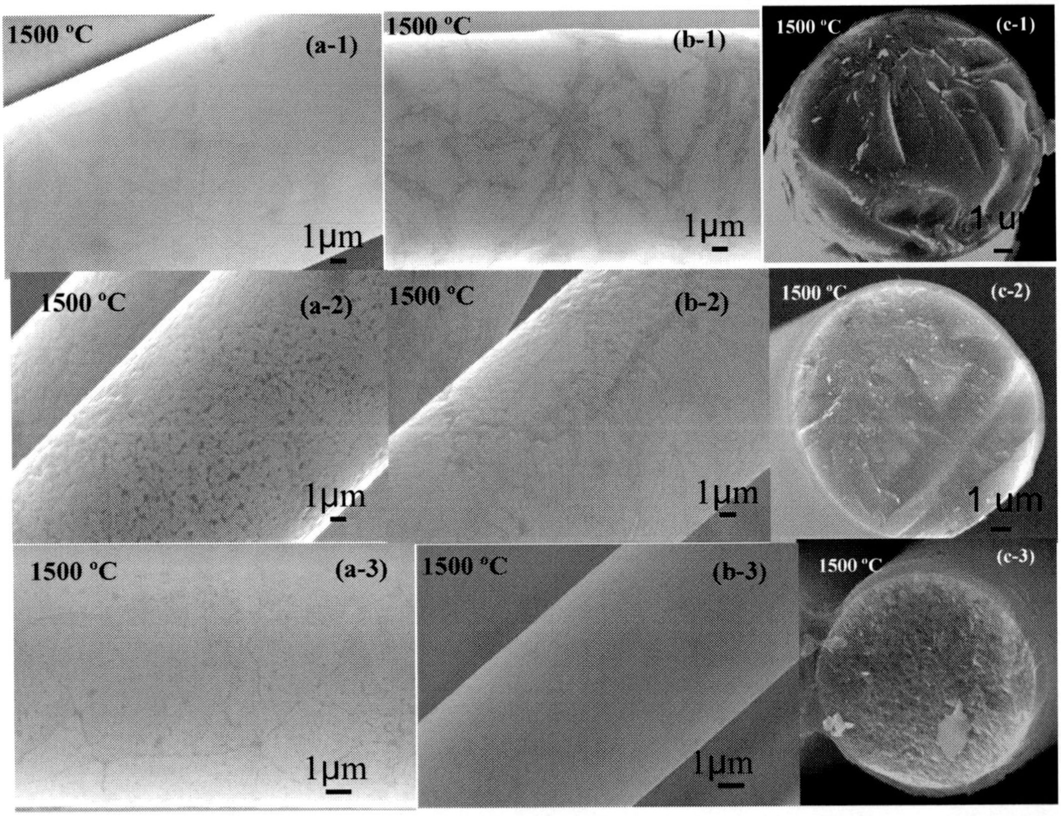

Figure 23. Morphologies of SiC fibers annealed: ((a-1) and (c-1) for HNL; (a-2) and (c-2) for HNLS; (a-3) and (c-3) for TySA fibers)) and crept ((b-1) for HNL; (b-2) for HNLS; (b-3) for TySA fibers) in air at 1500 °C.

For the HNL fibers under annealing (Figure 23 (a-1)) and creep (Figures 23 (b-1)) at 1500 °C, the cracks were found within silica film. Some patterns were also observed within the silica film formed on the surface of HNL fibers (Figures 23 (b-1) under creep condition.

Because of the near stoichiometric composition, the HNLS (Figure 23 (a-2)-(c-2)) and the TySA fiber (Figure 23 (a-3)-(c-3)) showed a relatively smooth surface compared with that of HNL fiber. No significant cracks were found within silica layer even under crept condition. A relatively rough surface of silica on the HNLS fiber crept at 1500 °C was observed (Figure 23(b-2)). In contrast, a smooth silica film coated on the surface of the TySA fiber (Figure

23(a-3)-(b-3)). Maeda et al. [62] have studied a sintered Al_2O_3-containing SiC for periods up to 3000 h. The actual kinetics involved at least four different parabolic stages for this material. They attributed these to the various microstructural changes in the scale: crystallization of amorphous silica, transformation of those crystalline phase, and viscosity changes in the oxide scale due to migration of the additives. This indicated the oxidation of TySA fiber is much complex because of addition of alumina in this fiber.

(b) Under Annealing and Creep in HP-Ar

Figure 24 shows the morphologies of SiC fibers under annealing and creep at 1500 °C for 1 h.

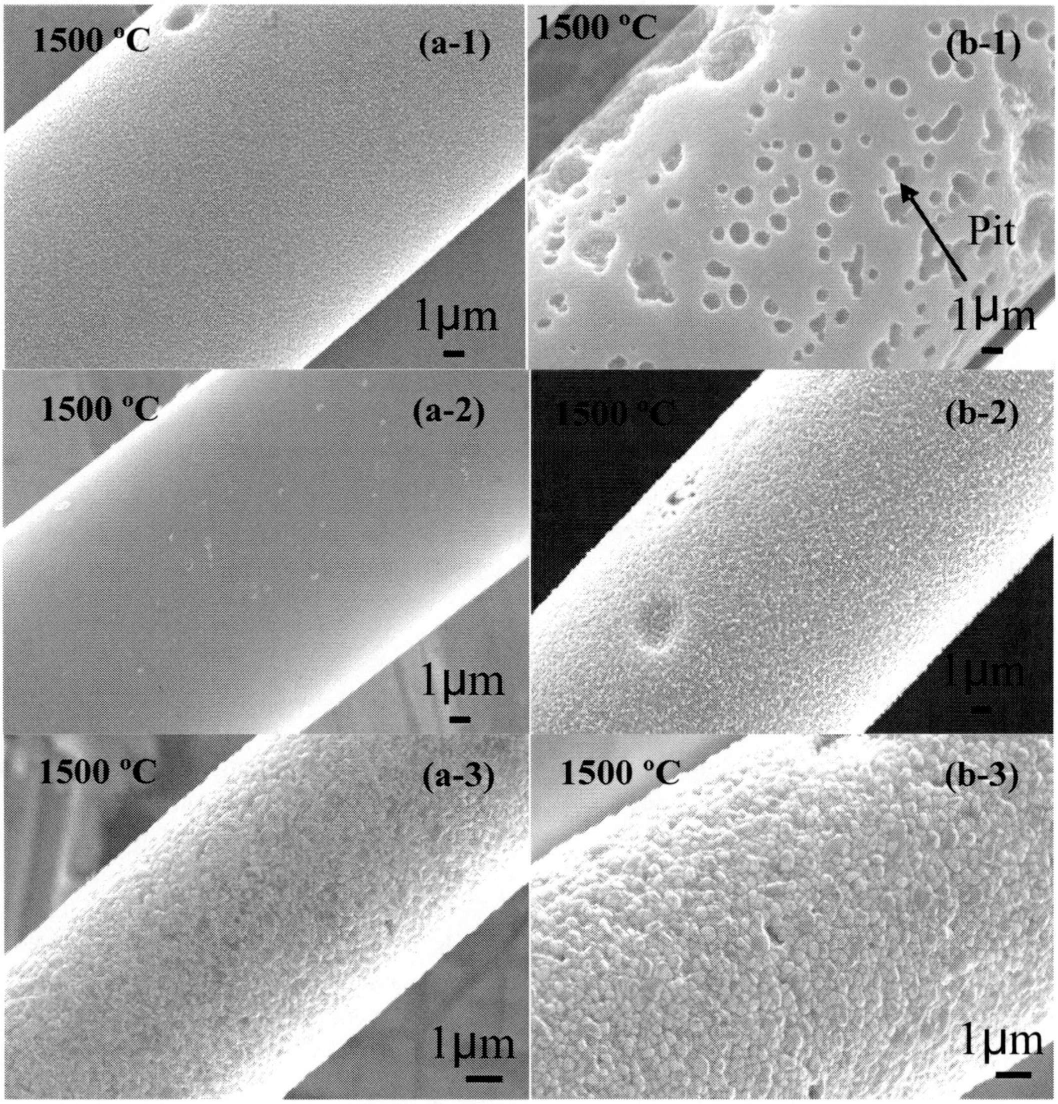

Figure 24. Morphologies of SiC fibers annealed ((a-1) for HNL; (a-2) for HNLS; (a-3) for TySA fibers) and crept ((b-1) for HNL; (b-2) for HNLS; (b-3) for TySA fibers) in HP-Ar at 1500 °C.

For the HNL fibers under annealing and creep at 1300 °C, fibers surface didn't show obvious oxygen attack. Under creep at 1400 °C, the bubbles were formed on the surface of fiber, but it didn't appear under annealing condition except for the slight grain growth. The HNL fibers annealing at 1500 °C have coarse surface accompanying with the formation of pits (Figure 24 (a-1)). In particular, for the fiber under creep at 1500 °C (Figure 24 (b-1)), it gave a porous microstructure with many pits formation.

Below 1500 °C, annealed HNLS fiber displayed a fine-grained and pore-free structure, namely, no obvious oxygen attack on surface was observed. At 1500 °C, few coarsening grains were sitting on the annealed fiber surface (Figure 24 (a-2)), but the fiber surface is still dense and smooth. Under creep condition, pits appeared on the surface of fibers (Figure 24 (b-2)). This result indicated that stress applied by BSR test could enhance the oxygen attack.

When TySA fibers were annealed in HP-Ar, a quite stable morphology was observed as shown in Figure 24 (a-3). From the observation of cross section of TySA fibers annealed in Ar at 1400 °C, a very thin silica film was formed on the surface. Under creep condition, the grain of TySA fibers showed a little coarsening at 1500 °C. In a previous study, the TySA fiber was oxidized in passive oxidation regime at 1500 °C even the oxygen partial pressure is much low (1 Pa). The passive oxidation of TySA fiber at extremely low oxygen partial pressure appears to be attributed to the addition of a minute of alumina. Alumina, as a oxidation product of aluminum, reacts with the SiO_2 film to form an alumino-silicates. At 1500 °C, the softening of the alumino-silicat film was observed because of low melting point. The alumino-silicates have a high oxygen permeability, presumably enhancing the passive oxidation of fibers [63-64]. On the other hand, SiO_2 can also be formed when SiC was annealed in low oxygen pressures by reaction of SiC with alumina [65]:

$$SiC+2/3Al_2O_3 = SiO_2 + 4/3\ Al + C \tag{21}$$

This reaction is slightly, but such development of SiO_2 film could suppress the active oxidation of TySA fiber at extremely low oxygen partial pressure.

(c) Under Annealing and Creep in UHP-Ar

Figure 25 shows the morphologies of SiC fibers under annealing and creep at 1500 °C for 1 h.

No obvious oxidation was observed for HNL fibers under annealing at 1300 °C, but for fibers under creep condition, bubbles and large grains were formed on the surface. At 1400 °C under annealing condition, many large pits was produced on the surface of fibers, meanwhile, the tensile side of crept fiber showed the needle-like grains with a length of about 5 um [66]. In particular, for the Hi-Nicalon fibers under annealing and creep at 1500 °C, the fibers were oxidized more severely and much coarse-grained surface was produced (Figure 25 (a-1) and Figure 25 (b-1)). It is clear these huge crystals grew outward from the fiber surface.

The HNLS fibers show a stable microstructure in annealed condition at temperature below 1400 °C. The grain coarsening was observed under crept at 1400 °C. At 1500 °C, the annealed fibers appeared the formation of pits (Figure 25 (a-2)), while in the crept fibers, the fibers showed a porous structure (Figure 25 (b-2)).

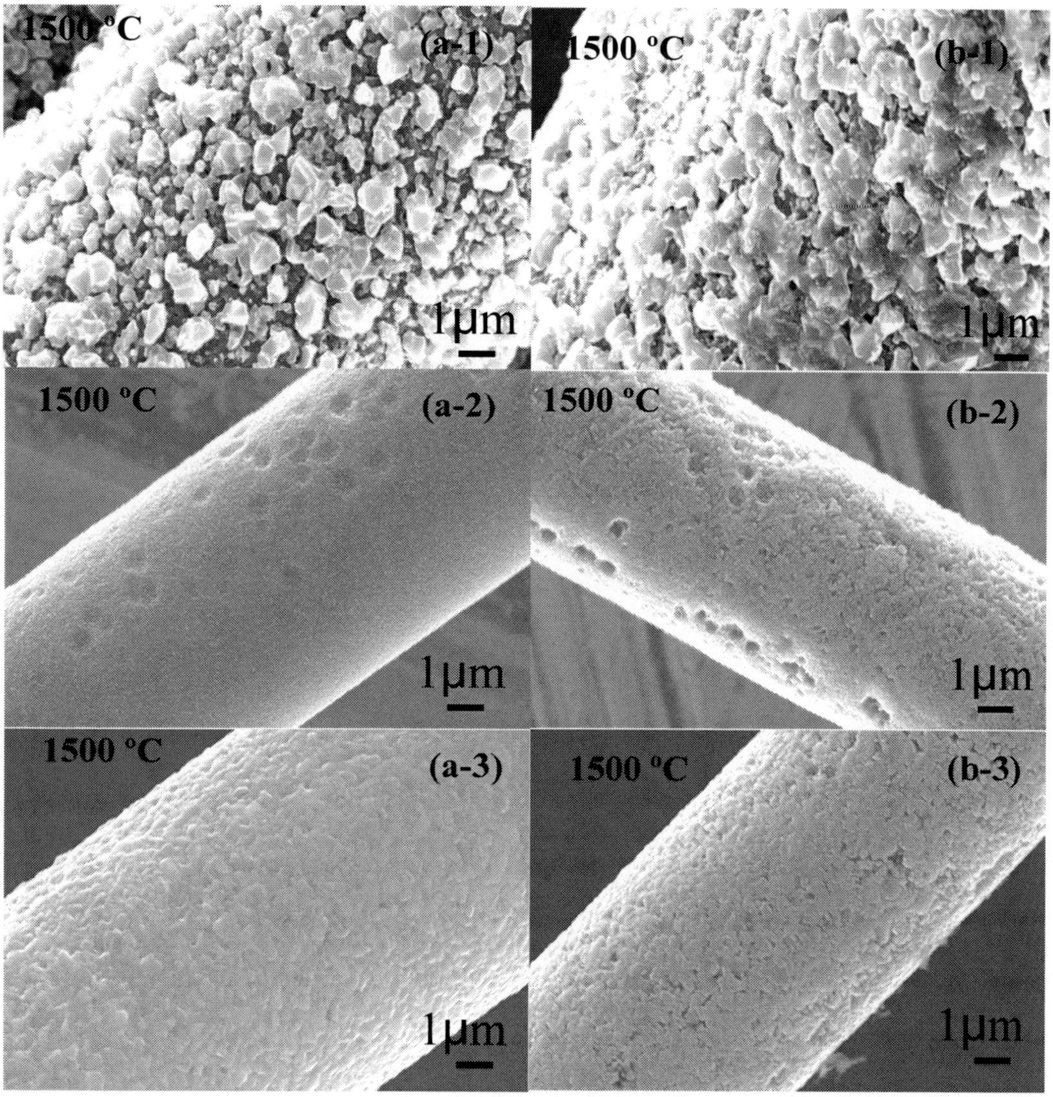

Figure 25. Morphologies of SiC fibers annealed ((a-1) for HNL; (a-2) for HNLS; (a-3) for TySA fibers) and crept ((b-1) for HNL; (b-2) for HNLS; (b-3) for TySA fibers) in UHP-Ar at 1500 °C.

For TySA fiber under creep at 1500 °C (Figure 25 (b-3)), the surface was also damaged slightly, but in annealed fibers no obvious change was observed (Figure 25 (a-3)).

As we know, SiC materials may be oxidized in passive and/or active oxidation regime, depending on their microstructure, the oxygen partial pressure and exposure temperature [67-70].

The thermochemical correlation and oxidation dynamics for active and passive oxidation of silicon carbide have been investigated experimentally and theoretically in the literatures [51,67-70]. These data were combined together and plotted into a new plot as shown in Figure 26. The passive-to-active oxidation transition is strongly affected by factors such as the type of silicon carbide, temperature and oxygen partial pressure. The oxygen partial pressures are, 2.1×10^4 Pa in air, 0.2 Pa in HP-Ar and 1×10^{-5} Pa in UHP-Ar, respectively. The oxygen partial pressures and test temperatures for this work were shown in Figure 26 by three lines. It is clear that proposed test conditions distributed in different regions.

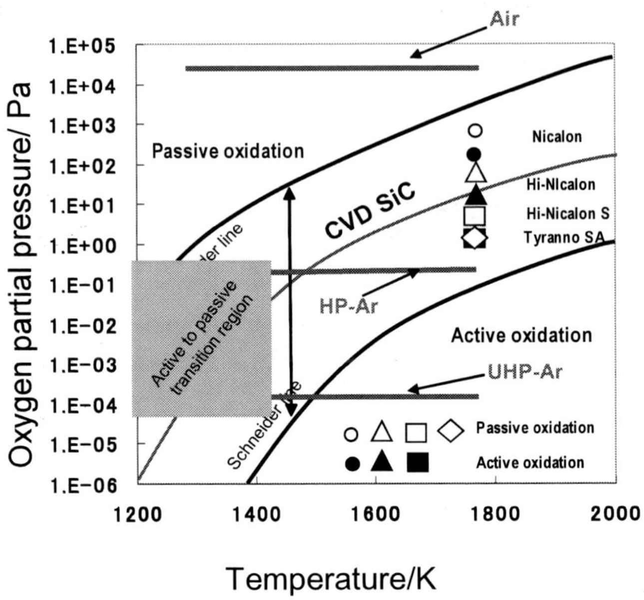

Data from [51,67].

Figure 26. Oxygen partial pressure for the transition from passive to active oxidation at elevated temperatures (at 1500 °C, p_{O2} for the transition is: Niclaon: 100-250 Pa, HNL:10-25 Pa, HNLS:1-2.5 Pa).

Based on the surface and cross section morphologies in Figure 23, it is thought that fibers were mainly oxidized in passive oxidation regime characterized by the formation of silica film when they exposed in air at high temperatures, but the passive oxidation was enhanced under crept condition due to increased oxygen permeation. Cracks in the silica layer observed in Figure 23 are mainly due to the difference in coefficient of thermal expansion (CTE) between the fiber core and the silica layer. Because the CTE of SiC fiber is less than that of Silica, on cooling, a tensile residual stress will be applied to silica layer. On the other hand, beta SiO_2 transforms into alpha SiO_2 at 300-370 °C with an accompanying volume change [71], can also generate the stress in silica layer resulting in the formation of cracks. The passive oxidation formed SiO_2 layer can refrain the further oxidation of SiC.

In low oxygen partial pressure atmosphere, the oxygen partial pressure for the transition from passive to active oxidation is a key point in the microstructure change of SiC materials. A previous study has found that the oxygen partial pressure for the transition from passive oxidation to active oxidation was $p_{O2} = 10^{-1} \sim 2$ Pa at 1300 °C [51]. The p_{O2} value was about 0.2 Pa (HP-Ar) and 10^{-5} Pa (UHP-Ar). Hence, these oxygen partial pressures are well below or fall into the range of $10^{-1} \sim 2$ Pa, indicating that the occurrence of active oxidation in the proposed conditions is possible.

From the observation of surface morphologies, we can see the active oxidation initialized at different temperatures in different atmospheres [66]. The temperatures for active oxidation in crept fibers, however, were shift to low values, indicating the transition from passive oxidation to active oxidation was enhanced under creep condition. The enhanced active oxidation in creep condition might be caused by the rupture of silica scale on the surface of fiber due to the stress applied by BSR test. Subsequently, a stress concentration would occur around the flaws generated by active oxidation or gas evolution, and then oxygen attack on the SiC fibers will be accelerated, leading to the formation of bubbles and pits. The formation of bubbles in the silica scale may provide some indications of pressure buildup [72], especially, when the passive and active oxidation proceeded concurrently.

Due to the near stoichiometric composition and high crystallinity of HNLS and TySA fibers, their active oxidation was gradual in comparison with other SiC fibers [58-60,67,72]. It is obvious that HNLS fibers crept in HP-Ar and annealed and crept in UHP-Ar at 1500 °C were slightly oxidized in active oxidation regime. Noteworthy is that the active oxidation is infinite slow if the oxygen partial pressure is very low [73].

4.2.2. Tensile Properties

Figure 27 shows the dependence of mean strength on the testing environments. The fiber's strength decreased with decreasing the oxygen partial pressure. It should be noted during the specimen preparation that fibers with low strength became very difficult to set without breaking them. The mean strength we gave will consequently not take the weakest fibers into account (no strength could be obtained). Due to this shortcoming, overestimation of tensile strength is likely. As observed for HNL fibers after annealing in UHP-Ar at 1500 °C for 1 h, the fibers are too fragile to measure the strength. However, still an enough strength for HNLS and TySA fibers was retained even after annealing in UHP-Ar as shown in Figure 27.

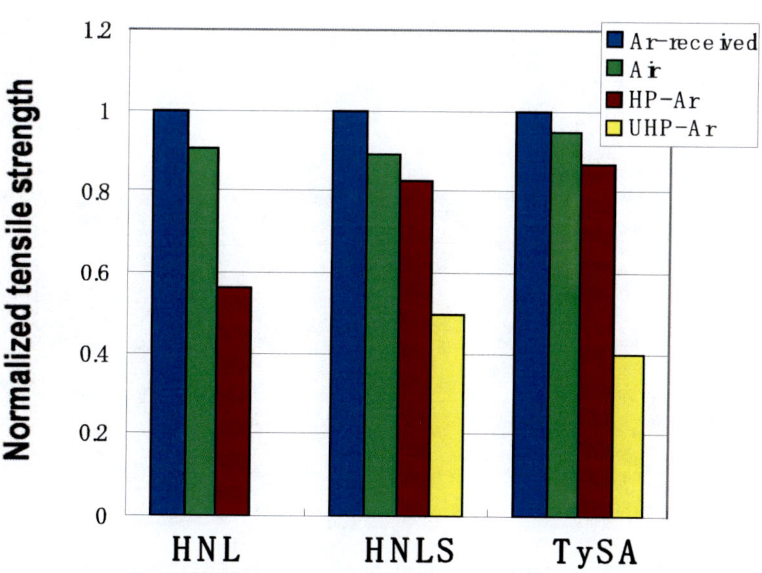

Figure 27. The tensile strength retention of SiC-based fibers annealed at 1500 °C under varied atmosphere.

The strength degradation of fibers in different oxygen partial pressures could be attributed to different microstructure and oxidative mechanism. The strength for the fiber annealed in air decreased only slightly (Figure 27). This is due to the SiO_2 layer, which acts as environmental barrier coating, stopping the further oxygen attack.

The fracture origins on the fracture surface of fibers were shown in Figure 28. The fracture mainly originated from surface flaws after annealing in different atmosphere. This means the surface of fibers was damaged by oxidation. The relative low strength retention for fiber annealed in UHP-Ar suggested changes in flaw population and flaw size. The low oxygen partial pressure accelerates the transition from passive to active oxidation resulting in a coarsening and pitting surface. At high temperature under low oxygen partial atmosphere, SiC is relatively easy to be oxidized actively.

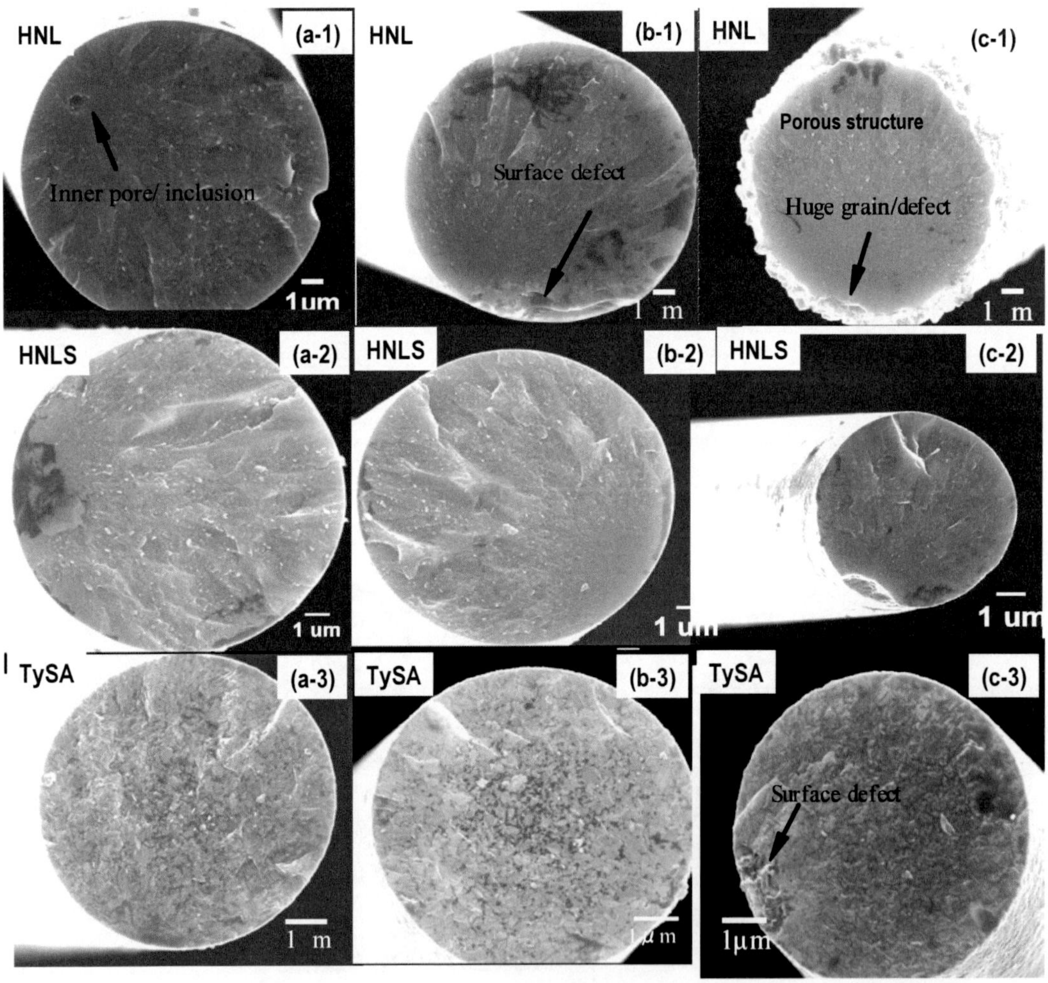

Figure 28. Fracture surface of fibers after annealing at 1500 °C: (a-1)-(a-3) in air, (b-1)-(b-3) in HP-Ar, (c-1)-(c-3) in UHP-Ar.

4.2.3. Creep Resistance

Figure 29 shows the dependence of 1-h BSR creep resistance m on temperatures under various oxygen partial pressures for HNLS fibers. HNLS fiber exhibited excellent creep resistance even exposed at 1400 °C. At temperatures above the 1300 °C, the BSR examinations indicate that creep resistance of HNLS fibers under high oxygen partial pressure is somewhat lower than that in low oxygen partial pressure. Namely, a weak dependence of BSR creep resistance on oxygen partial pressure is observed.

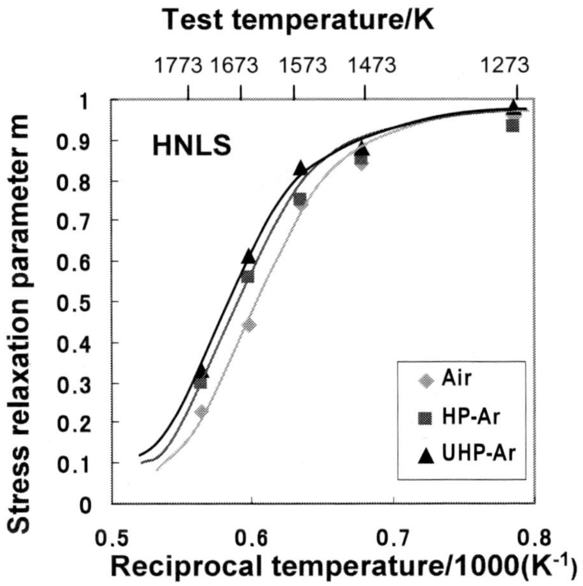

Figure 29. Effect of oxygen partial pressures on the BSR creep resistance of HNLS fiber.

As for the creep behavior of SiC fibers, generally, it can be explained by their oxygen content, grain size and second phase in the grain boundaries. The creep resistance of SiC fibers slightly decreased with increasing the oxygen partial pressure, which is likely controlled by the oxidation and grain coarsening. Under creep condition, grain coarsening of SiC has been observed for longer time creep experiments, which may contribute to a decelerating creep rate [74-75]. Especially, grain coarsening is relative easy in low oxygen partial pressure and at high temperature due to coalescence of β SiC grain [73]. The grain size increased with decreasing oxygen partial pressure has been observed in other studies [59,67]. Furthermore, under the creep test, the grain coarsening could be accelerated by applied stress [59]. The somewhat high creep resistance in low oxygen partial pressures might be attributed partially to the concurrent grain coarsening during BSR test. On the other hand, the low creep resistance in air could be partially explained by the resistance effect of silica layer. The outmost SiO_2 sheath formed in air after BSR test will counteract part of the initial applied stress leading to a low stress relaxation parameter m, especially, for the fibers with fine diameters.

4.3. Thermal Exposure Under Loading

Most previous studies have been concerned with the degradation in mechanical properties of SiC fibers after high temperature exposures [58,76-78]. There exist few studies on the performance change of SiC fibers in low oxygen partial pressure atmosphere under loading [79].

Therefore, for understanding the degradation mechanism of CMCs under loading in low oxygen partial pressure environments, the loading tests were performed on SiC fiber yarns by applying different dead loads at elevated temperatures in Ar atmosphere. After each loading test, the room temperature tensile properties and microstructure were characterized to clarify the performance degradation mechanism.

4.3.1. Tensile Properties

Figure 30 showed the room temperature tensile strength distributions of SiC fibers with different conditions. The details of tensile properties for each fiber type was indicated as follows: (i) The effect of applied load on the strength degradation of HNL fiber was observed at 1250 °C; both the retained strength and Weibull modulus decreased with increasing applied load in HNL fiber, but the effect of the applied load at 1250 °C was much smaller in HNLS fiber (Figure 30 (a) and Figure 30 (e)). (ii) In the strength distributions of fibers after yarn-loading test at 1250 °C and 1300 °C for 3 hours under an applied load of 201 g, the effect of exposure temperature on the strength degradation is obvious at same applied load of 201 g as shown in Figure 30 (b) and Figure 30 (e). (iii) A weak time effect on the tensile properties was observed within the limited time difference employed here (Figure 30 (c) and Figure 30 (e)). Likely, the time difference between 1h and 3h is too short. (iv) The strength retention and Weibull modulus decreased in both HNL and HNLS fibers under combined effect of applied load and exposure temperature as shown in Figure 30 (d) and Figure 30 (e). The combined effect of applied load and exposure temperature is more obvious in properties degradation of SiC fibers. Also, from the comparison of strengths between HNL and HNLS fibers after yarn-loading tests, the observed strength decreases are greater in HNL fibers. The near-stoichiometric and high-crystallite SiC fiber, HNLS, degraded gradually in tensile properties with increasing load and temperature, as shown in Figure 30 (e). This phenomenon seems to be related to the oxidation resistance and thermal stability of SiC fibers under loading.

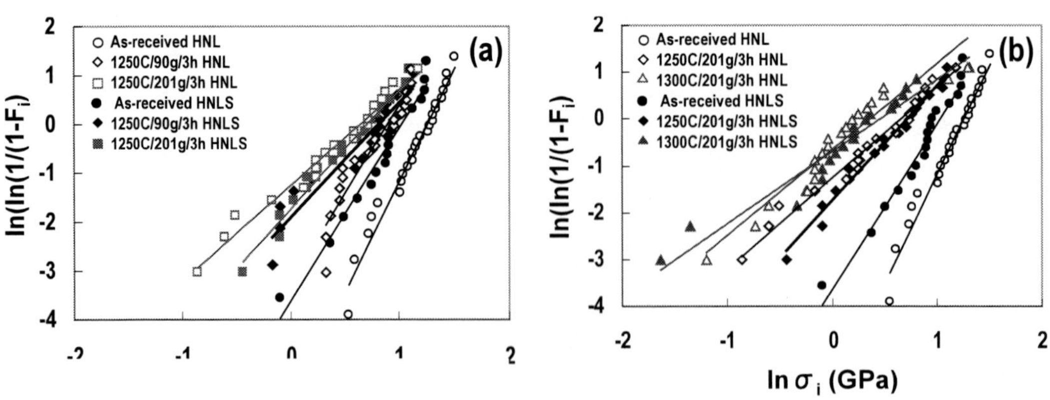

Figure 30. Continued on next page.

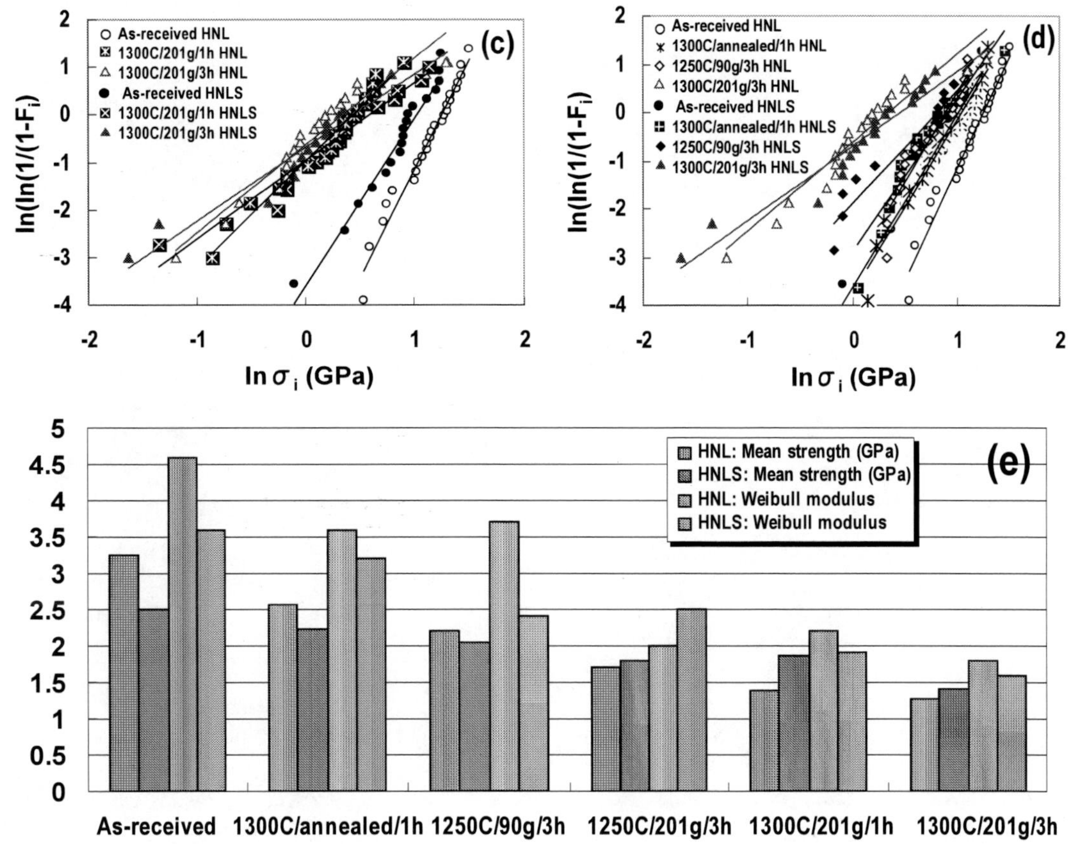

Reprinted by permission of Elsevier from [79].

Figure 30. Room temperature tensile strength distributions for HNL and HNLS fibers after loading test with different conditions: (a) effect of loading, (b) effect of exposure temperature, (c) effect of exposure time, (d) combined effect of exposure temperature and loading, (e) tensile properties of SiC fibers with different conditions.

4.3.2. Morphology

Figure 31 showed the microstructure features of as-received and annealed HNL fibers at 1300 °C for 1 h. No obvious differences in the surface morphologies were observed between the as-received and annealed fibers (Figures 31 (a) and (c)). For the as-received HNL fiber (Figure 31 (b)), the inner pore acted as the critical flaw during the tensile test (In the examined HNL fibers, about 74% of critical flaws were identified as inner flaws with different dimensions). But for the annealed HNL fibers (Figure 31 (d)), their fractures mainly originated from a surface flaw (about 78% of total fractures originated from the surface).

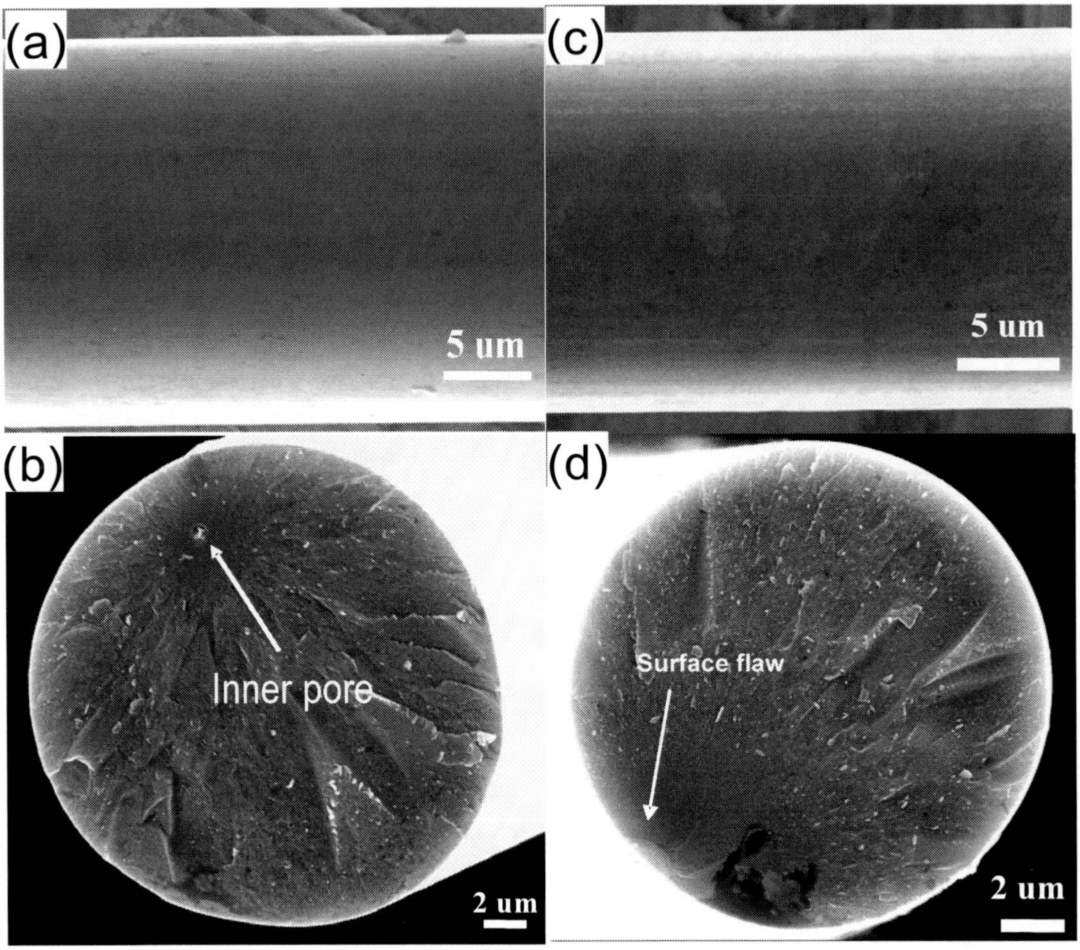

Reprinted by permission of Elsevier from [79].

Figure 31. SEM images of HNL fibers: (a) Surface morphology and (b) fracture surface of as-received fibers; (c) surface morphology and (d) fracture surface of annealed fibers at 1300 °C for 1 h.

Figure 32 showed the surface morphologies and fractographs of fibers tested in 1300 °C/201 g/3h. A quite different morphology is apparent for these fibers. A rough surface with extensive grain growth and micro pores was observed in HNL fibers as shown in Fig 32 (a). According to the observation of the fracture surface (Figure 32 (b)), the HNL fibers fractured at an irregular groove, which extended from surface to interior and was significantly different from that of annealed fibers in shape and size (Figure 31 (d)). There are only a few individual large grains grown on the surface of HNLS fibers (Figure 32 (c)), and the fracture of the HNLS fiber also originated from the irregular surface flaw (Figure 32 (d)), but the flaw size is smaller than that of HNL fiber.

In both HNL and HNLS fiber, the critical flaw was surrounded by a distinctive mirror zone and hackle zone.

In Figure 30, the tensile properties of SiC fibers decreased with increasing load and temperature, which suggested changes in flaw population and flaw size, as observed from surface morphologies and fractographs (Figure 31~Figure 32).

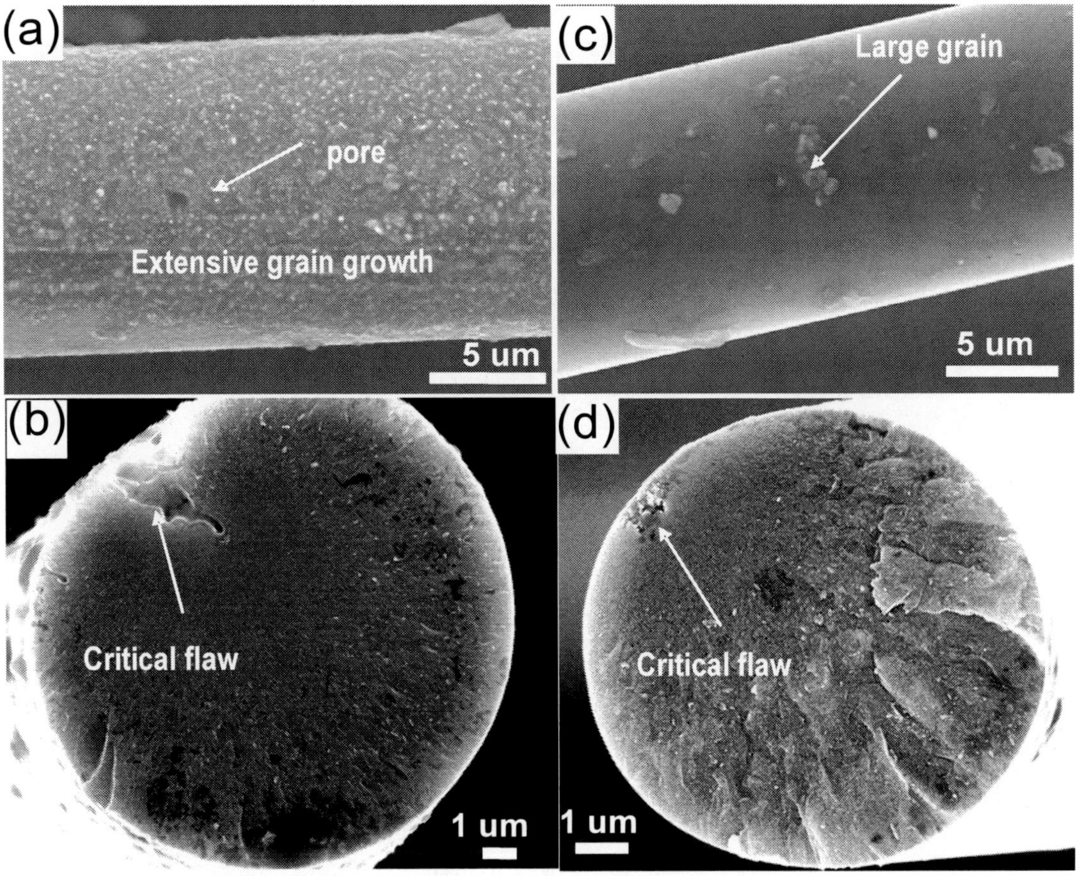

Reprinted by permission of Elsevier from [79].

Figure 32. SEM images of SiC fibers after loading test in 1300 °C/201 g/3 h: (a) Surface morphology and (b) fracture surface of HNL fiber; (c) Surface morphology and (d) fracture surface of HNLS fiber.

Fracture of HNL and HNLS fibers under loading mainly originated from an irregular surface flaw, and features are associated with brittle failure (Figure 32 (b) and Figure 32 (d)). By comparing the microstructure of the loaded fibers with that of the as-received and the annealed fibers (Figure 31), it is obvious that the fiber's surface was attacked by enhanced oxidation under loading (Figure 32). The irregular groove could be a stress corrosion crack (SCC) caused by combination of the oxidation and the applied load.

Generally, the oxidation behavior of silicon carbide at a high temperature depends on ambient oxygen partial pressure. Passive oxidation occurs at high oxygen partial pressures and results from the formation of SiO_2 that grows on the surface of exposed fibers. Passive oxidation could protect the materials from further oxidation attack. For active oxidation, pits and cracks occurred at low concentrations of O_2 where the SiO_2 formation rate is too low to seal the surface of materials. The passive-to-active oxidation transition is strongly affected by factors such as the type of silicon carbide, total gas pressure and gas flow rate, as well as temperature and oxygen partial pressure [51,70].

Thus, based on the surface morphologies and fractographs in Figure 31, it was thought that annealed fibers were mainly oxidized in passive oxidation mode characterized by formation of silica. The passive oxidation formed a thin silica film on the surface of annealed fibers and this thin silica film could protect the fiber from further oxidation attack resulting in no critical structural change. In this case, the thermal decomposition of amorphous phase which proceeded by reaction (17) should also be suppressed [67,80-81], because the gas species cannot be removed fast enough through the silica film by diffusion or migration [67,70].

Under loading conditions, the rupture of silica scale on the surface of fiber would be easier and a stress concentration occurs around the flaws. At that time, the oxidation was enhanced by loading, leading to the stress corrosion crack, as observed in Figure 32 (b) and Figure 32 (d). In this case, the active oxidation for SiC material could occur by following reactions [51]:

$$SiC(s)+O_2(g) \rightleftharpoons SiO(g)+CO(g) \tag{22}$$

$$SiC+3/2O_2(g) \rightleftharpoons SiO(g)+CO_2(g) \tag{23}$$

The oxygen partial pressure for the transition from passive oxidation to active oxidation was $p_{O2} = 10^{-1} \sim 2$ Pa at 1300 °C and total pressure of 100-800 Pa [51]. The total pressure in this study was 10^5 Pa; hence, the p_{O2} value in furnace chamber was about 0.2 Pa (oxygen concentration in Ar: 2 ppm). This oxygen partial pressure falls into the range of $10^{-1} \sim 2$ Pa, indicating that the occurrence of active oxidation is quite possible. A high total pressure will increase the limits for the transition from passive-to- active oxidation [51].

In addition to the HNL fibers, the oxidation of free carbon and the decomposition of amorphous intergranular phase by reaction (17) also yield a damaged structure [67,80], which accelerated the active oxidation because of the fiber's high permeability to oxygen. On the other hand, due to the acceleration of C/Si diffusivity at grain boundaries and reaction between SiO and free carbon, the grain growth of HNL fibers was enhanced by applied stress as shown in Figure 32 (a). The grain growth also has a contribution to the strength degradation [69]. Because of the near stoichiometric composition and high crystallinity of HNLS fibers, its active oxidation was gradual in comparing to that of HNL fiber.

In previous studies [26,36,58-59,76-78], many researchers have observed that the strength of ceramic fibers was associated with critical flaws. As observed in Figure 32, a clear mirror zone around the critical flaw on the fracture surface corresponds to the smooth propagation of a crack. In the present work, because most of the critical flaws are irregular in shape, it is very difficult to define the flaw size.

Thus, for clarifying the degradation mechanism, the fracture mirror size on the fracture surface was measured. There is a tendency that mirror size increased with increasing load and temperature. For fibers tested in 1250 °C/ 90 g/3 h and 1300 °C /201 g/3 h, the measured mean mirror sizes on their fracture surfaces were about 3.37 and 3.67 μm for HNL fiber, 2.97 and 3.34 μm for HNLS fiber, respectively.

The increased mirror size could be related to the increased critical flaw size. Critical flaw size could be associated with the oxidation and creep resistance of SiC fibers under loading. The creep resistance evaluated by bending stress relaxation method has indicated that applied stress of HNL fiber by bending at 1300 °C was nearly complete relaxation [69]. In contrast, the HNLS fibers showed excellent creep resistance at 1300 °C. Having a low oxidation and creep resistance could easily cause the new flaw nucleation and growth and slow crack propagation. Meanwhile, the excess carbon and amorphous phase in HNL fiber might also have a contribution to the flaw population and flaw size. Finally, the strength of fibers was limited by the degradation of fiber's microstructure, which could be attributed to the stress corrosion caused by oxidation and loading. For easily understanding the oxidation corrosion, following mechanism was proposed. For annealed fibers, a thin silica film formed on the surface of fibers and prevented the gas species passing through the fiber surface. This would be possible because of the use of the alumina furnace wall which resulting in the real oxygen partial pressure might be higher than the equilibrium pressure. In this case, the oxidation and the thermal decomposition of amorphous intergranular phase [67,82] were suppressed. Under loading condition, the grain growth of β-SiC and the decomposition of amorphous intergranular phase were enhanced by applied load. The SiO and CO gases were transported through the fiber surface. SiC crystals grew on the fiber surface due to the reaction between SiO and free carbon. Meanwhile, the rupture of silica scale on the surface of fiber would be easier during loading test [70,83]. Once the flaw was produced by oxidation and decomposition of amorphous phase, the stress concentration occurred around these flaws and preexisted crack tip. At that time, the active oxidation was enhanced by loading because of the rupture or removal of the protective silica film on the fiber surface, leading to the stress corrosion cracks, as schematically illustrated in Figure 33. The stress corrosion cracks acted as the critical flaws in the fracture of fibers. The active oxidation is very detrimental to properties of materials and it proceeded mainly by reactions proposed in literatures [51,58-59,67].

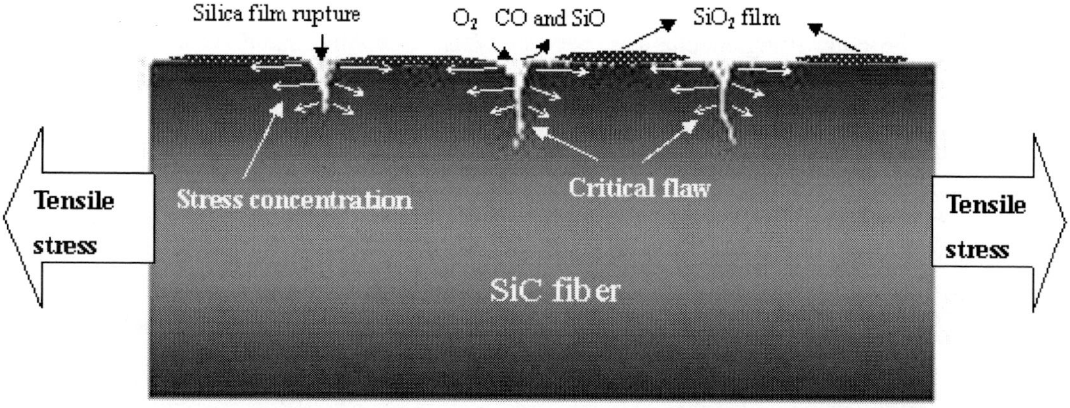

Figure 33. Schematic of strength degradation of SiC fibers under loading.

5. Tensile Creep Prediction by Long Time BSR Test

The understanding of creep behavior and creep mechanism of SiC materials is essential for real application. However, there are significant difficulties in the experimental measurement of tensile creep of advanced small diameter (about 7-14 μm) SiC fibers. In order to evaluate the creep resistance of these advanced SiC fibers, and also to clarify the creep mechanism to support continuing optimization of the properties, the apparent activation energy of creep was calculated by applying a cross-cut method to the results of the long-term BSR tests [84].

Sixta et al. [85] studied the flexural creep of sintered SiC between 1100 °C and 1500 °C for stress 200-100 MPa. The flexural creep rate exhibits linear stress dependence with an apparent activation energy of 230 ± 80 kJ/mol. The activation energy for grain boundary diffusion has been reported for high-purity SiC to be 563 kJ/mol [86] and 611 kJ/mol [87] for carbon and silicon in β-SiC, respectively. Given n=1 (linear stress dependence) and a relatively low activation energy, it is concluded that grain boundary sliding (GBS), accommodated by grain boundary transport of SiC, is the controlling creep mechanism in their test.

Besson et al. [88] investigated the compressive creep behavior of a Si_3N_4/SiC nanocomposite in the 1150 °C -1350 °C temperature range under stresses from 45 to 180 MPa. The stress exponent equals 1 and the apparent activation energy is 580 kJ/mol. Honda et al. [89] performed compression tests on Al-doped β-SiC at 2123-2223 K, the stress exponents were from 1.1 to 1.4 in the temperature range of 2123-2223 K. the apparent activation energy is 760 kJ/mol. In both above-mentioned studies, these authors considered GBS accommodated by diffusion as the critical creep mechanism.

Lane et al. [90] investigated a sintered polycrystalline α-SiC containing minor amounts of impurities. The creep was performed within a range of temperatures and stresses of 1547 °C-1747 °C and 138-148 MPa. The stress exponent increased from 1.44 to 1.71 with temperature. The activation energies were between 338-434 kJ/mol and 802-914 kJ/mol for temperatures below and above 1650 °C, respectively. These authors concluded: (1) the creep mechanism at low temperature is GBS accommodated by grain boundary diffusion, and at high temperatures the controlling mechanism becomes GBS accommodated by lattice diffusion; (2) the parallel mechanism of dislocation glide contributes increasingly to the total strain as the volume of precipitates declines as a result of progressive coalescence with increasing temperature.

As we know, creep of CMCs involves stress transfer between the matrix and fibers caused by their different creep rates, that may lead to fiber failures or matrix cracking [91-93]. When the matrix is elastic and creep resistant, fiber creep induces stress transfer from the fibers onto the matrix that may cause matrix cracking. Thus, the understanding of creep-related properties of fibers is essential in the identification of time-dependent failure mechanism of CMCs.

In literatures [25,94-97], the creep behavior of SiC-based fibers has been investigated. The apparent activation energies for creep obtained in these literatures are consistent with activation energy of self-diffusion for carbon [86,98] and silicon [99-100] in α-and β-SiC, which are 713-840 kJ/mol and 695-910 kJ/mol, respectively. Based on the results of these studies, creep is a thermally activated process and controlling creep mechanism is grain boundary sliding. Particularly, in the tensile creep behavior of SCS CVD SiC fibers, some authors observed that the fibers exhibited only primary creep, which was characterized by a continuously decreasing creep rate for progressively longer times, and that the creep rate was proportional to an exponential power of time [94-96].

The thermally activated creep can be described using traditional Bailey-type relationship [101].

$$\varepsilon_c = At^p\sigma^n \exp(-Q/RT) \quad (24)$$

where t is the creep time and A and p are creep parameters, σ elastic stress, n stress exponent, Q apparent activation energy, R universal gas constant, T absolute temperature. However, there are significant difficulties in the experimental measurement of tensile creep of advanced SiC-based fibers because of their fine diameters (about 7-14 μm). Fortunately, the bend stress relaxation (BSR) [25] test has been demonstrated to be an effective method for comparing the relative creep resistance of a wide variety of ceramic fibers. And also, some researchers have been attempted to relate BSR data with that of tensile creep [95-96]. Their results showed good correlation between BSR and tensile creep data.

5.1. Bend Stress Relaxation and its Relation to the Tensile Creep

In BSR test, the stress relaxation parameter, m, is defined as the ratio of final to initial stress at any local position in the fiber. That is:

$$m = \sigma(t, T, \varepsilon_0)/\sigma(0, T, \varepsilon_0) \quad (25)$$

For convenience, two assumptions were made [25]; one is assume that creep strain (ε_c) is linearly proportional to the initial strain (ε_0) regardless of stress direction, and the other one is that ε_c can be measured at room temperature by the relation $\varepsilon_c = z/R_a$ (z is the distance from fiber neutral axis, see Figure 6). The first assumption of linear strain dependence is generally valid for polycrystalline materials which relax stress by the grain boundary sliding (GBS) mechanism, because the stress power dependence of $n \cong 1$ is typically observed throughout creep stage [85,88,94-96,102]. The second assumption implies that at each local position within the fiber, stress relaxation is not only proportional to ε_0 but follows the same time-temperature dependence. This typically requires a fiber with a uniform microstructure that creeps with $n \cong 1$ power dependence. This assumption is suitable to the advanced SiC-based fibers with a near-stoichiometric composition and high-crystalline structure. The stress exponent of n=1 is indicative of a diffusional creep mechanism [103-105].

If these assumptions apply, the BSR ratio m, obtained by a method illustrated in Figure 1, is independent of position and initial applied strain. On the other hand, in a BSR test, the stress, σ(t), in a material in response to a constant bend strain, ε_0, as a function of time, t, can be described by the relaxation modulus in bending, $M_b(t)$, where

$$\sigma(t) = M_b(t)\varepsilon_0 \quad (26)$$

Substituting the σ (t) in Equation (25) with Equation (26), and then the Equation (25) could be expressed as:

$$m = M_b(t) \cdot \varepsilon_0 / E\varepsilon_0 = M_b(t)/E = 1 - \frac{R_0}{R_a} \quad (27)$$

where R_0 and R_a are the initial and residual curvature of fiber loops as indicated in Figure 6. It was also indeed that the negligible dependence of m ration on the initial applied strain was observed [66]. This result further supported that the above assumptions were appropriate. On the other hand, the negligible initial strain dependence of m also evidenced the grain boundary sliding (GBS) as the principal creep mechanism during the stress relaxation of SiC-based fibers. The rest of BSR tests in this study were performed at a constant initial strain for times ranging from 1-100 h at elevated temperatures in air. Furthermore, the apparent activation energy of thermally controlled creep was calculated by a cross-cut method from time-temperature dependence of stress relaxation parameter.

Since the individual fiber type with uniform microstructure displayed a strain-independent m ratio, the predictions of tensile creep of fibers from BSR data should be possible according to the previous results [25,94-96]. Because the ε_c is linearly stress-dependent, it is reasonable to relate tensile creep with the relationship between the tensile relaxation modulus, $M_t(t)$, and tensile creep compliance, $J_t(t)$ [96]:

$$J_t(t) = \frac{1}{M_t(t)} = \frac{\varepsilon_c(t) + \varepsilon_e}{\sigma} \tag{28}$$

If $M_b(t) = M_t(t)$ and substituting $M_t(t)$ in Equation (28) with Equation (27), so a normalized creep strain (NCS) could be defined as [96]:

$$NCS = \frac{\varepsilon_c(t)}{\varepsilon_e} = \frac{1}{m(t)} - 1 \tag{29}$$

If assuming the stress exponent, n, is equal to 1 in Equation (24) and combining the Equation (24) with Equation (29), the stress relaxation parameter could be correlated with the tensile creep described by the form:

$$NCS = \frac{\varepsilon_c(t)}{\varepsilon_e} = \frac{1}{m(t)} - 1 = A_0 \bullet E \bullet t^p \tag{30}$$

where $A_0 = A \bullet \exp(-Q/RT)$ is creep parameter (constant for the specific temperature). The time exponent, p, determined from a plot of log NCS vs log t, is constant.

5.2. BSR Tests at Elevated Temperatures

Figure 34 showed the stress relaxation parameter (m) as a function of temperature and time for three SiC fiber types. The duration of the stress relaxation tests was 1, 10, 25, 50 and 100 h. The stress relaxation follows an S-shaped curve. If we take m = 0.5 as an arbitrary value for which we can compare test results for these fibers, two trends are evident. First, the relaxation temperature for m = 0.5 in HNL fibers increased with an increase of the heat treatment temperature. Second, at the level of m = 0.5, the 1h BSR tests show a higher relaxation temperature than that of longer time BSR test. These results suggest that thermally activated process plays an important role in the creep behavior of SiC fibers.

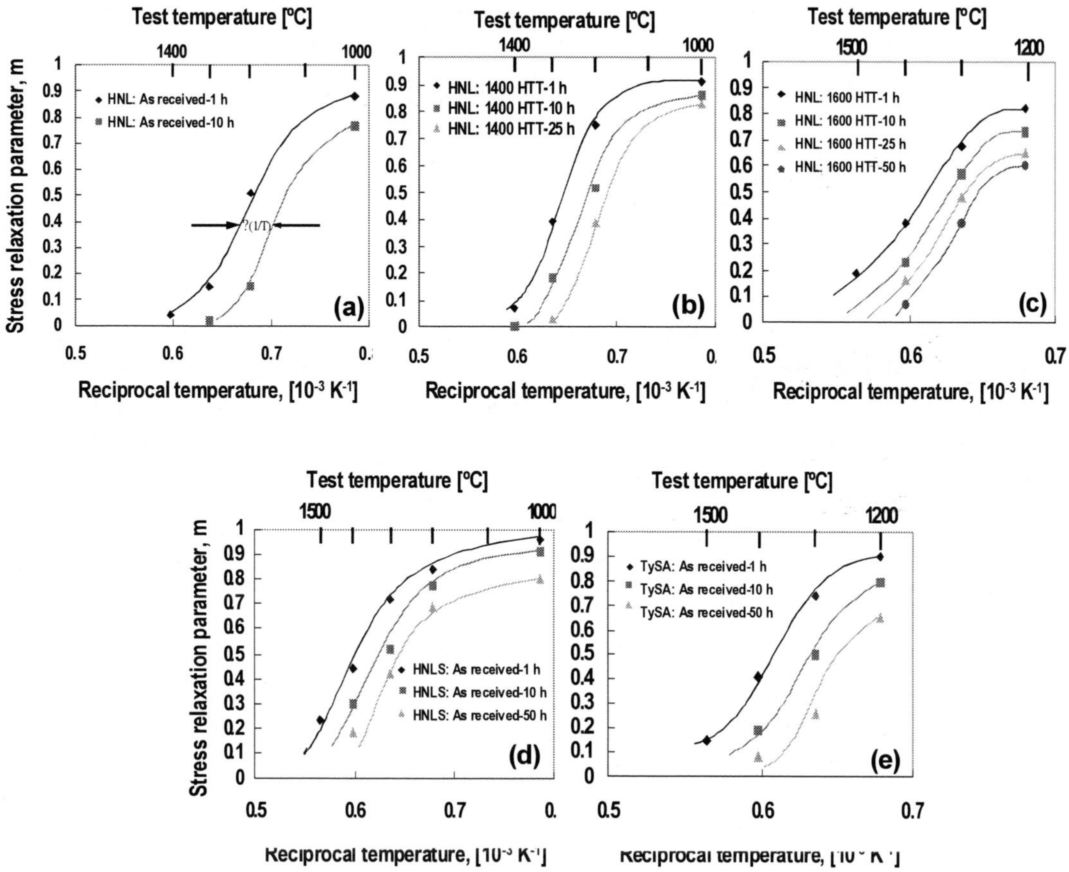

Reprinted by permission of Elsevier from [84].

Figure 34. Temperature dependence of m value of SiC fibers for (a) HNL fibers; (b) 1400 °C heat treated HNL fibers; (c) 1600 °C heat treated HNL fibers; (d) HNLS fibers; (e) TySA fibers.

Since the stress relaxation mechanism is thermally activated process for ceramic materials, the rate-controlling activation energy, Q, can then be determined from the $\Delta(1/T)$ spacing between the curves at a constant m value (cross-cut method) in Figure 34. The $\Delta(1/T)$ spacing corresponds to the apparent activation energy for stress relaxation in a cross-cut method.

The Q value for a given m value, can be calculated from the relationship:

$$Q = R.\frac{\ln(\frac{t_2}{t_1})}{\frac{1}{T_1} - \frac{1}{T_2}} \tag{31}$$

That is, for 1 order of magnitude change in time, $Q=2.3\ R/\Delta(1/T)$, where R is the gas constant (8.314 J/mol. K).

In present work, the activation energies were found (from m =0.3, 0.5, 0.7, which represent the high temperature, moderate temperature and low temperature regions.) to be 563 kJ/mol, 598 kJ/mol and 445 kJ/mol for the as-received HNL fibers (Figure 34 (a)); 707

kJ/mol, 692 kJ/mol, 500 kJ/mol for the as-received HNLS fibers (Figure 34 (d)); 774 kJ/mol, 707 kJ/mol and 524 kJ/mol for the as-received TySA fibers (Figure 34 (e)). From these results, we can see the Q value increases with increasing the test temperature. The same change in apparent activation energy was also found in earlier studies [90,96-97]. The large activation energy at high temperature regions could be related to the concurrent microstructure change during BSR test, such as grain growth, the crystallization and loss of oxygen due to the decomposition of amorphous phase (SiCxOy) at grain boundary. As we know, the creep behavior of ceramics can be explained by oxygen content and grain size. The crystallite sizes of as received fibers are 4.0 nm for HNL, 11.4 nm for HNLS, 22.7 nm for TySA, while the crystallite sizes are strongly dependent on the heat treatment temperature [69]. Furthermore, the grain growth could also be enhanced by applied stress. Thus, the effect of grain growth must be taken into account at high temperature test. All the grain boundary sliding mechanisms have a negative grain size exponent, which means that smaller grains will result in a faster creep rate. Further comparison in Q value for heat treated HNL fiber supports above result. The Q value for HNL fibers increased with increasing the heat treatment temperature (Figures 34 (b)-(c)), for instance, at the level of m=0.5, Q is 622 kJ/mol and 929 kJ/mol for HNL fibers heat treated at 1400 °C (crystallite size: 5.3 nm) and 1600 °C (crystallite size: 8.0 nm), respectively.

The activation energies obtained in our work are in acceptable agreement with the activation energies of carbon self-diffusion (713-840 kJ/mol) [86,98] and silicon self-diffusion (695-910 kJ/mol) [99-100] in SiC.

5.3. Prediction of Tensile Creep from BSR Data

Figure 35 shows the normalized creep strain (NCS) versus time in log-log form for a specific test condition. The NCS was calculated from BSR data with Equation (30). The creep parameters listed in Table 3 are the best fit in Figure 35. The parameter, A_0, is dependent on the temperature, while time exponent, p, is somewhat high in high temperature region. The fibers exhibit very similar creep behavior, which suggests that they all creep mainly via a similar mechanism and the difference in the individual parameter given in Table 3 is due to compositional or microstructural difference among these fibers. For each fiber type, the tendency of NCS behaved similarly at different temperatures.

Furthermore, since the bend stress relaxation tests were performed in air, the oxidation of surface of SiC fibers would enhance the surface stress relaxation [76]. Especially, for the long time BSR test of fine-diameter SiC fibers, the Silica layer carrying a part of initial applied stress by stress sharing mechanism might be possible, which results in an overestimation of NCS. This was reflected by somewhat high time exponent in high temperature region. According to the oxidation kinetics, the silica thickness obeys a parabolic law for T $\leq$1400 °C, which has been well investigated in literatures [106-107]. SEM observation on cross section of crept fibers at 1400 °C for 25 h, has revealed that fibers have a silica layer of about 1.0 um.

The result in present work is in agreement well with previous work [96], although the fiber types are different. Further comparison was made in BSR test between present and previous work [96]. It can be seen that HNL fiber has an similar BSR creep resistance with that of SCS-6 CVD SiC fiber (higher than 5% excess carbon); the near-stoichiometric HNLS and TySA fiber behaved similarly in BSR creep resistance with that of 2 mil CVD SiC fibers (nearly pure SiC, excess carbon less than 1%). Both CVD SiC fibers appeared that the stress exponent, n, is approximate 1.

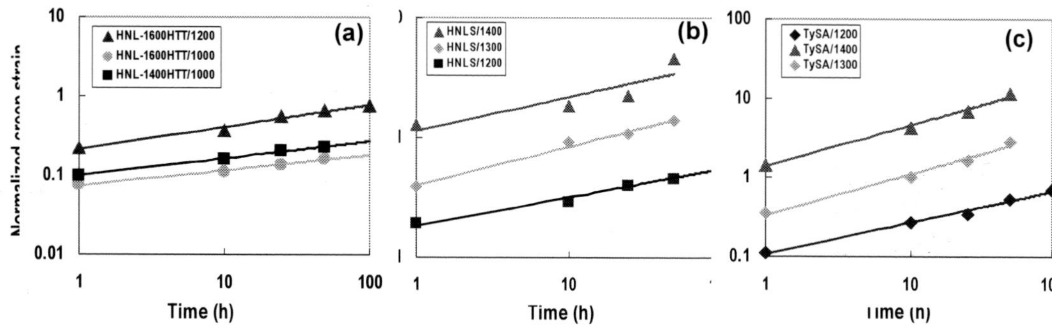

Reprinted by permission of Elsevier from [84].

Figure 35. Normalized creep strain (NCS) predicted from BSR data versus time for HNL fiber, (b) HNLS fiber, (c) TySA fiber. The parameters presented in Table 3 are the best fit of these curves.

Figure 36 showed the tensile creep strain, which was predicted by BSR data. Generally, this prediction showed a similar time and temperature dependence with that of the primary creep stage. As pointed out by other researchers [96], for silicon carbide fibers which have uniform microstructures, the BSR predictions usually are very near the magnitude of tensile creep strain.

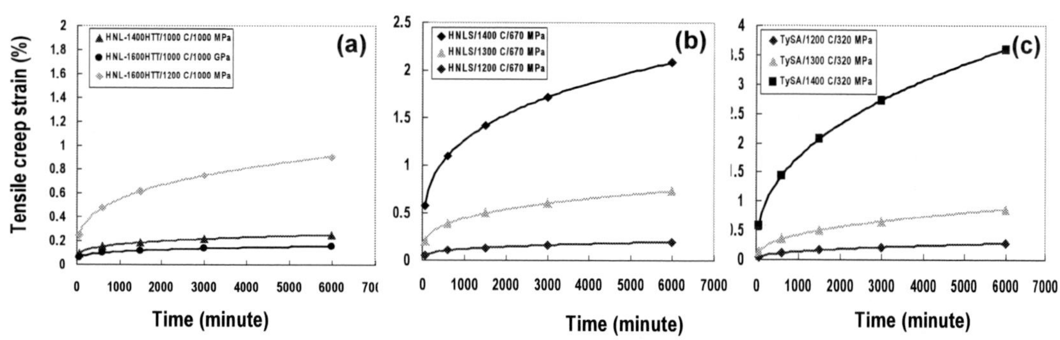

Reprinted by permission of Elsevier from [84].

Figure 36. Showed the tensile creep strain, which was predicted by BSR data parameters using Equation (30): (a) HNL fiber; (b) HNLS fiber; (c) TySA fiber.

In addition, the actual tensile creep strains from Refs. [41] were also listed in Table 3, and compared with that of BSR predictions. It was observed that tensile creep strains from BSR predictions showed same order of magnitude with the actual tensile creep strains. Noting in many tensile creep tests of ceramic fibers, the data scattering in creep strain is significant. On the other hand, because of the lack of tensile creep data on advanced SiC-based fibers, this comparison was made among the studies with different fiber batch. The properties of SiC fibers are different from batch to batch and they are still in developing. Further demonstration from tensile creep test with same fiber batch would be necessary. Nevertheless, in present work, the BSR data predicted the same time and temperature dependence of tensile creep for advanced SiC-based fibers.

Table 3. Tensile creep parameters from the best fit of BSR data with Equation (30)

Fiber Type Experimental	NCS (50h)	A_0 (GPA) Bestfit	p Best Fit	εc (%) Best Fit	εc (%) Actual (from tensile creep test)
HNLS				670 MPa	670 MPa
1200 °C	0.45	0.000435	0.24	0.19 (50h)	
1300 °C	1.38	0.000955	0.32	0.83 (50h)	0.36 (50h)[#1]
1400 °C	4.56	0.002721	0.28	1.72 (50h)	
TySA				320 Mpa	320 Mpa
1200 °C	0.54	0.000269	0.4	0.21 (50h)	
1300 °C	2.85	0.000833	0.4	0.66 (50h)	1.72 (20h)[#1]
1400 °C	11.5	0.000346	0.4	2.72 (50h)	
HNL				1000 Mpa	1000 Mpa
As received					0.3 (1200 °C, 3 5h)[#2]
HNL 1400 C HTT/1000 °C	0.23	0.000366	0.22	0.22 (50h)	
HNL 1600 C HTT/1000 °C	0.16	0.000274	0.20	0.14 (50h)	
HNL 1600 C HTT/1200 °C	0.66	0.000793	0.28	0.75 (50h)	
				.24 (0.83h)	0.13 (0.39h), 0.2 (0.83h)[#2]

Reprinted by permission of Elsevier from [84].

Note: $NCS=1/m-1=A_0 \cdot E \cdot t^p$, $\varepsilon_c=\sigma \cdot A_0 \cdot t^p$, $\sigma=E \bullet \varepsilon_e$, t in minutes, σ in Gpa; HTT: Heat Treatment Temperature; #1: [41]; #2: [107].

Furthermore, Combining the analysis of activation energies with the predicted result (creep rate is linearly dependent on the applied stress) in our work, suggests that the main mechanism responsible for creep of SiC fibers is controlled by GBS accommodated by diffusion (prevalent at low temperatures and small grain size) [103-104]. Since the silicon self-diffusion coefficient is one order of magnitude slower than the carbon self-diffusion coefficient [99-100], so the silicon is the controlling species during the diffusion process. The strain rate for a pure diffusion lattice mechanism is given by Nabarro-Herring creep model [104].

The time exponent p and its slight variation at high temperature indicate that additional mechanism with similar activation energy might be operating besides GBS accommodated by diffusion. Lane et al. [90] found an increase of activation energy with temperature due to the transition from GBS accommodated by grain boundary diffusion and low dislocation activity, to GBS accommodated by bulk diffusion and high dislocation activity. The dislocation activity increased with the coalescence of the precipitate phase, because the interaction of dislocation with the precipitates makes it difficult to gild. This mechanism would be possible in present test, because the SiC fiber used is not so pure.

6. Conlusion

The review of the recent studies on the mechanical durability and microstructure stability of SiC-based fibers presented in this chapter has shown that they are promising reinforcement for CMCs, which have been proposed as potential structural materials for advanced energy systems and propulsion systems. Especially, the development of advanced SiC fibers with near stoichiometry and high crystallinity has improved their mechanical and thermal stabilities significantly. Mechanical and microstructure characterization have led to a deeper knowledge of the relationship between the microstructure and the mechanical behaviour. The main completion can be summarized as follows:

(i) Most of the initial strength for SiC fibers with near stoichiometric composition is retained up to very high temperature in inert atmosphere. The strength of fiber is sensitive to the critical flaws caused during fabrication process or by exposure to service environments. The heat treatment above the processing temperature could improve the creep resistance due to the crystallization, the grain coarsening and composition changes at grain boundaries (GB). The GB composition could affect the stability of GB boundaries.

(ii) The environment-pertinent degradation mechanism is complex for SiC materials subjected to realistic application. When the SiC-based fibers were exposed to the inert atmosphere with different oxygen partial pressures, the SiC-based fibers can be oxidized in passive/active oxidation regimes. The strengths of SiC-based fibers are strongly dependent on the oxygen partial pressures, being decrease with decreasing the oxygen partial pressures when they are oxidized in active oxidation regime. Strength degradation was caused by different oxidation mechanism in different atmospheres. In contrast, no obvious dependence of creep resistance on oxygen partial pressures was observed. On the other hand, the microstructure observation revealed that the oxidation and applied stress/loading can result in the nucleation and growth of new flaw leading to the stress corrosion. The stress corrosion is detrimental to the integrated performance of materials.

(iii) The apparent activation energies were calculated by cross-cut method from the long time bend stress relaxation (BSR) tests. The activation energies calculated from BSR data are in agreement with the activation energies of grain boundary diffusion of carbon and silicon in β-SiC, suggesting the main mechanisms responsible for creep of SiC-based fiber is controlled by GBS accommodated by diffusion. The activation energies are different at low and high temperature regions. This phenomenon could be related to the concurrent microstructure change during BSR test at high temperatures. The tensile creep of SiC-based fibers was predicted from the BSR data by a defined normalized creep strain (NCS). Tensile creep predicted from BSR data could reflect the tendency of primary creep of the SiC-based fibers with similar time and temperature dependence.

Finally, the appropriate understanding of the environmental durability of SiC fibers is essential for the reliable evaluation of CMCs. Although progress in some aspects is achieved, but a number of key issues still remain open. Currently there is much interest in SiC fiber reinforced CMCs. Stress corrosion of CMCs is another interesting point because the stress and corrosion (oxidative and corrosive environments) often act synergistically. The synergistic linkage of mechanical stability and corrosion under stress is an important area of research towards applying SiC fiber reinforced CMCs as structural components.

Acknowledgements:

This work was partially supported by the National Nature Science Foundation of China (grant No. 50871092). The author acknowledges the support of the NSFC.

References

[1] Naslain, R. Design, preparation and properties of non-oxide CMCs for application in engines and nuclear reactors: An overview. *Composite Science and Technology*, 2004, 64, 155-170.

[2] Kohyama, A. Advanced ceramic composite materials for fusion/fission reactor systems. *Ceramics* (in Japanese), 2004, 39, 838-842.

[3] Katoh, Y; Kohyama, A; and Hinoki, T. Progress in SiC-based ceramic composites for fusion applications. *Fusion Science and Technology*, 2003, 44, 155-162.

[4] Spriet, P; and Habarou, G. Applications of CMCs to turbojet engines: overview of the SEP experience. *Key Engineering Materials*, 1997, 127–131, 1267–1276.

[5] Ohnabe, H; Masaki, S; Onozuka, M; Miyahara K; and Sasa, T. *Potential application of ceramic matrix composites to aero-engine components. Composites: Part A*, 1999, 30, 489–496.

[6] Staehler, JM; and Zawada, LP. Performace of four ceramic-matrix composites divergent flap inserts following ground testing on an F110 turbofan engine. *J. Am. Ceram. Soc.*, 2000, 83, 1727-38.

[7] Evans, AG; and Zok, FW. The physics and mechanics of fiber-reinforced brittle-matrix composites. *J. Mater. Sci.*, 1994, 29, 3857-3896.

[8] Donald, LW; and Mcmillan, PW. Review ceramic-matrix composites. *J. Mater. Sci.*, 1976, 11, 949-972.

[9] Yajima, S; Hasegawa, X; Hayashi, J; and Iiuma, M. Synthesis of continuous SiC fibers with high tensile strength and modulus. *J. Mater. Sci.*, 1978, 13, 2569–2576.

[10] Mazdiyasni, KS; West, R; and David, LD. Characterization of organosilicon-infiltrated porous reaction-sintered Si_3N_4. *J. Am. Ceram. Soc.*, 1978, 61, 504-508.

[11] Mah, T; Hecht, NL; McCullum, DE; Hoenigman, JR; Kim, HM; Katz, P; and Lipsitt, HA. Thermal stability of SiC fibers (Nicalon). *J. Mat. Sci.*, 1984, 19, 1191-1201.

[12] Johnson, SM; Brittain, RD; Lamoreaus RH; and Rowcliffe, DJ. Degradation mechanisms of Silicon-carbide fibers. *J. Amer. Ceram. Soc.*, 1988, 71, 132-135.

[13] Okamura, K; Sato, M; Seguchi, T; and Kawanishi, S. Application of electron beam irradiation for preparation of SiC fiber Hi-Nicalon for ultra-high temperature. *Composite Science and Technology*, 1994, 51, 135-144.

[14] Okamura, K; and Seguchi, T. Application of radiation curing in the preparation of polycarbosilane-derived SiC fibers. *J. Inorganic and Organometallic Polymers*, 1992, 2, 171-179.

[15] Takeda, M; Imai, Y; Ichikawa, H; Ishikawa, T; Segushi, T; and Okamura, K. Properties of the low oxygen conent SiC fibers on high temperature treatment. *Ceram. Eng. and Sci. Proc.*, 1991, 12, 1007-1018.

[16] Takeda, M; Imai, Y; Ichikawa, H; Ishikawa, T; Kasai, N; Segushi T; and Okamura, K. *Ceram. Eng. and Sci. Proc.*, 1992, 13, 209-217.

[17] Ichikawa, H. Recent advances in Nicalon ceramic fibers including Hi-Nicalon Type S. Annales De Chimie-Science Des Materiaux, 2000, 52, 523-528.

[18] [18] Ishikawa, T; Kohtoka, Y; Kumagawa, K; Yamamura T; and Nagasawa, T. High-strength alkali-resistant sintered SiC fiber stable to 2200 °C. *Nature*, 1998, 391, 773-774.

[19] [19] Yun, HM; and Dicarlo, JA. Comparison of the tensile, creep and rupture strength properties of stoichiometric SiC fibers. *Ceram. Eng. and Sci. Proc.*, 1999, 20, 259-272.

[20] [20] Takeda, M; Sakamoto, J; Imai, Y; Ichikawa, H. Thermal stability of the low-oxygen-content silicon carbide fiber, Hi-Nicalon^{TM}. *Composites Science and Technology*, 1999, 59, 913-919.

[21] Takeda, M; Urano, A; Sakamoto, J; and Imai, Y. Microstructure and oxidative degradation behaviors of silicon carbide fiber Hi-Nicalon type S. J. Nucl. Mater., 1998, 258-263, 1594-1599.

[22] ASTM D3379-75 (reapproved 1989), Standard test method for tensile strength and young's Modulus for High-Modulus Single-Fiber materials.

[23] Weibull, W. A statistical distribution function of wide applicability. *Journal of Applied Mechanics-Transcations of the ASME*, 1951, 18, 293-297.

[24] Patankar, SN. Weibull distribution as applied to ceramic fibers. *J. Mater. Sci. Lett.*, 1991, 10, 1176-1181.

[25] Morscher, GN; and Dicarlo, JA. A simple test for thermomechanical evaluation of ceramic fibers. *J. Am. Ceram. Soc.*, 1992, 75, 136-140.

[26] Taylor, ST; Zhu, YT; Blumenthal, WR; Stout, MG; Butt, DP; and Lowe, TC. Characterization of Nicalon fibers with varying diameters: Part I strength and fracture studies. *J. Mater. Sci*., 1998, 33, 1465-1473.
[27] Youngblood, GE; Lewinsohn, C; Jones RH; and Kohyama, A. Tensile strength and fracture surface characterization of Hi-Nicalon™ SiC fibers. *J. Nucl. Mater.,* 2001, 289, 1-9.
[28] Serizawa, H; Lewinsohn, CA; Youngblood, GE; Jones, RH; Johnston, DE; and Kohyama, A. High temperature properties of newly developed advanced SiC fibers. *Key Engineering Materials*, 1999, 164-165, 287-290.
[29] Eckel, AJ; and Bradt, RC. Strength distribution of reinforced fibers in a Nicalon fiber/chemically vapor infiltrated silicon carbide matrix composite. *J. Am. Ceram. Soc*., 1989, 72, 455-458.
[30] Wagner, HD. Stochastic concepts in the study of size effects in the mechanical strength of highly oriented polymeric materials. *Journal of Polymer Science Part B-Polymer Physics*, 1989, 27, 115-149.
[31] Zhu, YT; Blumenthal, WR; Taylor, ST; and Lowe, TC. Analysis of size dependence of ceramic fiber and whisker strength. *J. Am. Ceram. Soc*., 1997, 80, 1447-1452.
[32] Sawyer, LC; Jamieson, M; Brikowski, D; Haider, MI; and Chen, RT. Strength, structure, and fracture properties of ceramic fibers produced from polymeric precursors. *Journal of the American Ceramic Society*, 1987, 70, 798-810.
[33] Mecholsky, JJ; Freiman, SW; and Rice, RW. Effect of flow shape and size on strength and fracture in glass. *Report of NRL Progress*, 1975, 11, 7-10.
[34] Griffith, AA. The phenomenon of rupture and flow in solids. *Phil. Trans. Roy. Soc. London A*, 1920, 221, 163-198.
[35] Jones, RH; Lewinsohn, CA; and Youngblood, GE. Properties of advanced fibers for SiC/SiC composite applications in fusion energy systems. *High temperature ceramic matrix composites III,* 1999, 164-1, 405-408.
[36] Thouless, MD; Sbaizero, O; Sigl, LS; and Evans, AG. Effet of interface mechanical-properties on pullout in a SiC-fiber-reinforced lithium aluminum silicate glass-ceramic. *Journal of the American Ceramic Society*, 1989, 72, 525-532.
[37] Dong, SM; Katoh, Y; and Kohyama, A. Preparation of SiC/SiC composites by hot pressing, using Tyranno-SA fiber as reinforcement. *J. Am. Ceram. Soc*., 2003, 86, 26-32.
[38] Lee, SP; Katoh, Y; Park, JS; Dong, SM; Kohyama, A; Suyama, S; and Yoon, HK. Microstructural and mechanical characteristics of SiC/SiC composites with modified-RS process. *J. Nucl. Mater*., 2001, 289, 30-36.
[39] Chollon, G; Pailler, R; Naslain, R; Laanami, F; Monthioux M; and Olry, P. Thermal stability of a PCS-derived SiC fibre with a low oxygen content (Hi-Nicalon). *J. Mater. Sci*., 1997, 32, 327-347.
[40] Ichikawa, H. Advances in Nicalon ceramic fibers including Hi-Nicalon type S. *Annales De Chimie-Science Des Materiaux*, 2000, 25, 523-528.
[41] Bunsell, AR; and Berger, MH. Fine diameter ceramic fibres. *J. Euro. Ceram. Soc*., 2000, 20, 2249-2260.
[42] Dong, SM; Chollon, G; Labrugere, C; Lahaye, M; Guette, A; Bruneel, JL; Couzi, M; Naslain, R; and Jiang, DL. Characterization of nearly stoichiometric SiC ceramic fibers. *J. Mater. Sci*., 2001, 36, 2371-2381.

[43] Ichikawa, H; and Ishikawa, T: Silicon carbide fibers (organometallic Pyrolysis). In: Kelly, A; Zweben, C; Chou, T. Comprehensive composite Materials. Oxford, England: *Elsevier Science Ltd.*; 2000; 107-145.

[44] Takeda, M; Saeki, A; Sakamoto, J; Imai, Y; and Ichikawa, H. Properties of polycarbosilane-derived silicon carbide fibers with various C/SiC compositions. *Composite Science and Technology*, 1999, 59, 787-792.

[45] Yajima, S; Okamura, K; Matsuzawa, T; Hasegawa, Y; and Shishido, T; Anomalous characteristics of the microscrystalline state of SiC fibers. *Nature*, 1979, 279, 706-707.

[46] Sasaki, Y; Nishina, Y; Sato, M; and Okamura, K. Raman-study of SiC fibers mae from polycarbosilane. *J. Mater. Sci.*, 1987, 22, 443-448.

[47] Sacks, MD. Effect of composition and heat treatment conditions on the tensile strength and creep resistance of SiC-based fibers. *Journal of the European ceramic society*, 1999, 19, 2305-2315.

[48] Havel, M; and Colomban, Ph. Skin/bulk nanostructure and corrosion of SiC-based fibers: a surface Rayleigh and Raman Study. *Journal of Raman Spectroscopy*, 2003, 34, 786-794.

[49] Havel, M; and Colomban, Ph. Rayleigh and Raman images of the bulk/surface nanostructure of SiC based fibers. *Compoistes: Part B-Engineering*, 2004, 35, 139-147.

[50] Delhaes, P; and Carmona, F. Physical-properties of noncrystalline carbons. Chemistry and Physics of carbon, 1981, 17, 89-174.

[51] Schneider, B; Guette, A; Naslain, R; Cataldi, M; and Costecalde, A. A theoretical and experimental approach to the active-to-passive transition in the oxidation of silicon carbide: experiments at high temperatures and low total pressures. *J. Mater. Sci.*, 1998, 33, 535-547.

[52] Shimoo, T.; Morisada Y.; and Okamura K. Oxidation behavior of Si-M-C-O fibers under wide range of oxygen partial pressures. *J. Mater. Sci.*, 2002, 37, 4361-4368.

[53] Shimoo, T; Okamura, K; and Morita, T. High temperatue exposure of low-oxygen SiC fibers (Hi-Nicalon) in CO-CO_2 gas mixture. *J. Mater. Sci.*, 2004, 39, 7031-7039.

[54] Kumagawa, K; Yamaoka, H; Shibuya, M; and Yamakura, T. Fabrication and mechanical properties of new improved Si-M-C-(O) tyranno fiber. *Ceram. Eng. Sci. Proc.* 1998, 19, 65-70.

[55] Toreki, W; Batich, CD; Sacks, MD; Saleem, M; Choi, GJ; and Morrone, AA. Polymer-derived silicon-carbide fibers with low-oxygen content and improved thermomechanical stability. *Composites Science and Technology*, 1994, 51, 145-159.

[56] Ramberg, CE; and Worrell, WL. Oxygen transport in silica at high temperatures: implications of oxidation kinetics. *J. Am. Ceram. Soc.*, 2001, 84, 2607-2616.

[57] Mazdiyasni, KS; and Aangvil, A. Effect of impurities on SiC whisker morphology. *J. Am. Ceram. Soc.*, 1985, 68, C142-C144.

[58] Shimoo, T; Takeuchi, H; and Okamura, K. Thermal stability of polycarbosilane-derived silicon carbide fibers under reduced pressures. *J. Am. Ceram. Soc.*, 2001, 84, 566-570.

[59] He, G; Shibayama, T; T and akahashi, H; Microstructural evolution of Hi-Nicalon^{TM} SiC fibers annealed and crept in various oxygen partial pressure atmospheres. *J. Mater. Sci.*, 2000, 35, 1153-1164.

[60] Shimoo, T; Okamura, K; Ito, M; and Takeda, M; High-temperature stability of low-oxygen silicon carbide fiber heat-treated under different atmosphere. *J. Mater. Sci.*,

2000, 35, 3733-3739. [61] Jacobson, NS. Corrosion of Silicon-based ceramics in combustion environments. J. Am. Ceram. Soc., 1993, 76, 3-28.

[61] Maeda, M; Nakamura, K; Ohkubo, T; Ito, J; and Ishii, E. Oxidation behavior of silicon nitride under static and cyclic conditions. *Ceramic International*, 1989, 15, 247-53.

[62] Singhal, SC. Oxidtion-kinetics of hot-pressed silicon-carbide. *J. Mater. Sci.*, 1976, 11, 1246-1253.

[63] Singhal, SC. Oxidation of silicon-based structural ceramics. *Journal of the Electrochemical Society*, 1976, 123, 253-253.

[64] Gallardo-Lopez, A; Munoz, A; Martinez-Fernandez, J; and Dominguez-Rodriguez, A; High-temperature compressive creep of liquid phase sintered silicon carbide. *Acta mate.*, 1999, 47, 2185-2195.

[65] Sha, JJ; Hinoki, T; and Kohyama, A. Mechanical and thermal stabilities of Hi-Nicalon fiber under annealing and creep in various oxygen partial pressure. *Corrosion Science*, in press, 2008.

[66] Shimoo, T; Okamura, K; and Morisada, Y Active-to-passive oxidation transition for polycarbosilane-derived silicon carbide fibers heated in Ar-O_2 gas mixtures. *J. Mater. Sci.*, 2002, 37, 1793-1800.

[67] Shimoo, T; Morisada, Y; and Okamura, K; Oxidation behavior of Si-M-C-O fibers under wide range of oxygen partial pressures. *J. Mater. Sci.*, 2002, 37, 4361-4368.

[68] Sha, JJ; Nozawa, T; Park, JS; Katoh, Y; and Kohyama, A. Effect of heat treatment on the tensile strength and creep resistance of advanced SiC fibers. *J. Nucl. Mater.*, 2004, 329-333, 592-596.

[69] Nickel, KG. Corrosion of Non-oxide ceramics. *Ceramics international*, 1997, 23, 127-133.

[70] Shimoo, T; Toyoda, F; and Okamura, K. Thermal stability of low-oxygen silicon carbide fiber (Hi-Nicalon) subjected to selected oxidation treatment. *J. Am.Ceram. Soc.*, 2000, 83, 1450-1456.

[71] Luthra, KL. Some new perspectives on oxidation of silicon carbide & silicon Nitride. *J. Am. Ceram. Soc.*, 1991, 74, 1095-1103.

[72] Jaskowiak, MH; and Dicarlo, JA. Pressure effects on the thermal-stability of silicon-carbide fibers. *J. Am. Ceram. Soc.*, 1989, 72, 192-197.

[73] Giannuzzi, LA; Lewinsohn, CL; Bakis, CE; T and ressler RE. Microstructural development of SCS-6 SiC fibers during high temperature creep. *Journal of Materials Research*, 1998, 13, 1853-1860.

[74] Dicarlo, JA; Yun, HM; and Hurst, JB. Fracture mechanisms for SiC fibers and SiC/SiC composites under stress-rupture conditions at high temperatures. *Applied mathematics and computation*, 2004, 152, 473-481.

[75] Sha, JJ; Park, JS; Hinoki, T; Kohyama, A; and Yu, J. Tensile properties and creep behavior of SiC-based fibers under various oxygen partial pressures. *Materials Science forum*, 2005, 475-479, 1333-1336.

[76] Pysher, DJ; Goretta, KC; Hodder, RS; and Tressler, RE. Strength of ceramic fibers at elevated-temperatures. *J. Am. Ceram. Soc.*, 1989, 72, 284-288.

[77] Sha, JJ; Park, JS; Hinoki, T; and Kohyama, A. Hot corrosion, oxidation and their effects on the tensile strength of SiC fiber in alkaline melts. *Materials Transactions*, 2005, 46, 1032-1035.

[78] Sha, JJ; Park, JS; Hinoki, T; and Kohyama A. Tensile behavior and microstructural characterization of SiC fibers under loading. *Materials Science and Engineering A*, 2007, 456, 72-77.
[79] Shimoo, T; Tsukata, I; Seguchi, T; and Okamura, K. Effect of firing temperature on the thermal stability of low-oxygen silicon carbide fibers. *J. Am. Ceram. Soc.*, 1998, 81, 2109-2115.
[80] Bodet, R; Bourrat, X; Lamon, J; and Naslain, R. Tensile creep-behavior of a silicon carbide-based fiber with al low-oxygen content. *J. Mater. Sci.*, 1995, 30, 661-677.
[81] Shimoo, T; Tsukata, I; Seguchi, T; Okamura, K. Effect of firing temperature on the thermal stability of low-oxygen silicon carbide fibers. *J. Am. Ceram. Soc.*, 1998, 81, 2109-2115.
[82] Gogotsi, YG; and Grathwohl, G. Stress-enhanced oxidation of silicon nitride ceramics. *J. Am. Ceram. Soc.*, 1993, 76, 3093-3104.
[83] Sha, JJ; Park, JS; Hinoki, T; and Kohyama, A. Bend stress relaxation of advanced SiGbased fibers and its prediction to tensile creep. *Mechanics of Materials*, 2007, 39, 175-182.
[84] Sixta, ME; Zhang, XF; and Jonghe, LC. Flexural creep of an in situ-toughened silicon carbide. *J. Am. Ceram. Soc.*, 2001, 84, 2002-2028.
[85] Hon, MH; and Davis, RF. Self-diffusion of C-14 in polycrystalline beta-SiC. *J. Mater. Sci.*, 1979, 14, 2411-2421.
[86] Sargent, PM; and Ashby, MF. Deformation-mechanism maps for silicon-caride. *Scri. Metall.*, 1983, 17, 951-57.
[87] Besson, JL; Doucey, B; Lucas, S; Bahloul-Hourlier, D; and Goursat, P. SiCN nanocomposite: creep behaviour. *J. Euro. Ceram. Soc.*, 2001, 21, 959-968.
[88] Honda, S; Nagano, T; Kaneko, K; and Kodama, H. Compressive deformation behavior of Al-doped β-SiC at elevated temperatures. *J. Euro. Ceram. Soc.*, 2002, 22, 979-985.
[89] Lane, JE; Carter CH; and Davis, RF. Kinetics and mechanisms of high-temperature creep in silicon-carbide: III, sintered alpha-silicon carbide. *J. Am. Ceram. Soc.*, 1988, 71, 281-295.
[90] Holmes, JW; Park, YH; and Jones, JW. Tensile creep and creep recovery behaviour of a SiC-fiber-Si_3N_4-matrix composite. *J. Am. Ceram. Soc.*, 1993, 76, 1281-1293.
[91] Lamouroux, F; Steen, M; and Valles, JL; Uniaxial tensile and creep-behavior of an alumina fiber-reinforced ceramic-matirx composite: expermental-study. *J. Euro. Ceram. Soc.*, 1994, 14, 529-537.
[92] Evans, AG; and Weber, C; Creep damage in SiC/SiC composites. *Mater. Sci. Eng*. A, 1996, 208, 1-6.
[93] Dicarlo, JA. Creep of chemically vapor-deposited SiC fibers. *J. Mater. Sci.*, 1986, 21, 217-24.
[94] Lewinsohn, CA; Giannuzzi, LA; Bakis, CE; and Tressler, RE. High-temperature creep and microstructural evolution of chemically vapor-deposited silicon carbide fibers. *J. Am. Ceram. Soc.*, 1999, 82, 407-13.
[95] Morscher, GN; Lewinsohn, CA; Bakis, CE; and Tressler, RE. Comparison of bend stress relaxation and tensile creep of SVD SiC fibers. *J. Am. Ceram. Soc.*, 1995, 78, 3244-3252.
[96] Shoji, S; Katase, Y; Okamura, K; Creep behavior of SiC fibers at high temperatures using the BSR method. *Key Eng. Mater.*, 2003, 247, 187-190.

[97] Hong, JD; and Davis, RF. Self-diffusion of C-14 in high-purity and N-doped alpha-Sic single-crystals. *J. Am. Ceram.* Soc., 1980, 63, 546-552.

[98] Hong, JD; Davis, RF; and Newbury, DE. Self-diffusion of Si-30 in alpha-SiC single-crystals. *J. Mater. Sci*., 1981, 16, 2485-2494.

[99] Hong, JD; Hon, MH; and Davis, RF; Si-30 diffusion in alpha-SiC and beta-SiC. *American Ceramic Society Bulletin*, 1979, 58, 348-348.

[100] Shames, H; and Cozzarelli, FA. Elastic and Inelastic Stress Analysis. Prentice Hall, Englewood Cliffs, NJ; 1992.

[101] Ogasawara, T; Ishikawa, T; Ohsawa, Y; Ochi, Y; and Zhu, S; Tensile creep behavior and thermal stability of orthogonal three-dimensional woven Tyranno ™ ZMI fiber/silicon-titanium-carbon-oxygen matrix composites. *J. Am. Ceram. Soc*., 2002, 85, 393-400.

[102] Coble, RL. A model for boundary diffusion controlled creep in polycrystalline materials. *J. Appl. Phys*., 1963, 34, 1679-82.

[103] Herring, C. Diffusional viscosity of a polycrystalline solid. *J. Appl. Phys.,* 1950, 21, 437-445.

[104] Cannon, WR; Langdon, TG. Creep of ceramics: mechanical characteristics. *J. Mater. Sci*., 1983, 18, 1-50.

[105] Shimoo, T; Toyoda, F; and Okamura, K. Oxidation kinetics of low-oxygen silicon carbide fiber. *J. Mater. Sci*., 2000, 35, 3301-3306.

[106] Chollon, G; Railler, R; Naslain, R; and Olry, P. Correlation between microstructure and mechanical behaviour at high temperatures of a SiC fibre with a low oxygen content (Hi-Nicalon). *J. Mater. Sci*., 1997, 32, 1133-47.

In: Strength of Materials
Editors: G. Mendes and B. Lago, pp. 61-96
ISBN: 978-1-60741-500-8

Chapter 2

IONOMERS AS CANDIDATES FOR STRUCTURAL MATERIALS

Daniel J. Klein
Ashland Performance Materials,
5200 Blazer Parkway, Dublin, Ohio, USA, 43017

Abstract

The field of ionomers is an often overlooked and under-utilized branch of polymer research. Although ionomers can be broadly described as a class of polymers that contain any number of ionic groups, from a structural property standpoint only a low percent of ionic groups are necessary to impart significantly improved properties over the nonionic version of the same polymer. Current trends in the field of ionomers are highly focused on the field of fuel cell technology. There appears to be a significant hole remaining in the study of imparting strength to materials using ionic groups. This hole is very significant from an industrial point of view, and has a large commercial potential. There are very few commercially available ionomers, which shows how little this field has been explored to date.

This chapter will focus on several aspects of ionomer research from a physical property standpoint:

1) A history of ionomer research
2) Current trends in ionomer research
    a) stand-alone polymers
    b) nanocomposites
    c) blends
3) A commentary on the immediate needs in the field of ionomer research

Introduction

(1)The field of ionomers is a very broad area of research. There are vast numbers of published works, including books, journal articles, and patents, which account for the wide range of ionomers. Due to the large number, and numerous applications of ionomers, a complete review of the field will not be discussed here. However, there are several good review articles, chapters, and books [1-8] that will give the reader a good understanding of the overall subject matter. However, one of the goals of this chapter is to hit on some key areas of ionomer research as they could possibly relate to potential structural materials. The focus of this chapter will be based on commercially viable options for ionomers.

(2)The word ionomer has many different meanings depending upon the researcher. In this chapter the word ionomer will be defined according to Eisenberg [1]. This is defined as "polymers in which the bulk properties are governed by ionic interactions in discrete regions of the material." From a structural polymer standpoint it is most beneficial to keep the ionomer ion content relatively low. This is due to the fact that polymers containing high ion contents tend to be brittle.

(3)From a microstructural standpoint there are different structures that form depending upon the ion content. It is generally accepted that the ions coordinate into either multiplets or clusters [3]. Both represent sites where the mobility of the polymer chain segments in the proximity of the ions is reduced relative to the regions that are not restricted. Multiplets are usually only small amounts of ionic coordination to tie the chains together (figure 1) [4]. When the regions are of sufficient enough size, and are in close proximity with one another, clusters form. Both act as non-covalent crosslink sites. In the region of restricted mobility the polymer has its own set of properties separate from the rest of the polymer. Obtaining the desired properties of the ionomer is highly dependent upon proper selection of ion concentration and counterion (type, size, charge, etc.). These, along with the polymer structure and polarity, help determine whether the ionomers form multiplets or clusters.

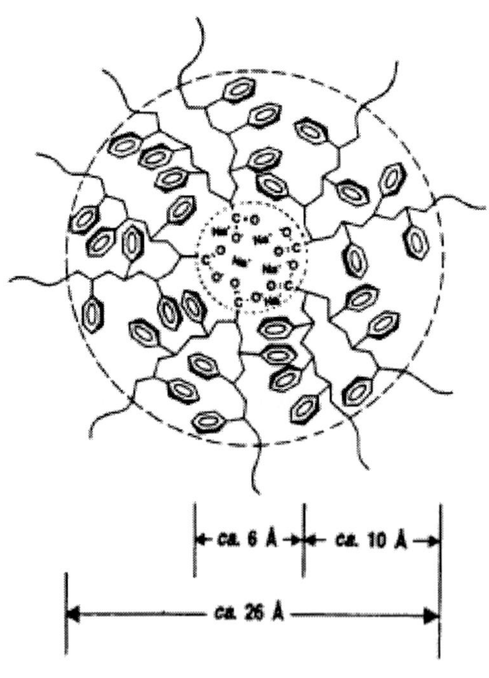

Figure 1. Example of a multiplet based on poly(styrene-co-sodium neutralized methacrylic acid).

(4)An interesting aspect of ionomers is that they can be used in several different applications. Examples include: use as a stand-alone polymer, be used to incorporate various fillers in nanocomposite applications, and can act as a compatibilizer between two dissimilar polymers as part of a blend. Each of these subject areas will be discussed in this chapter.

(5)Although there has been a lot of research in the field of ionomers, and many research groups continue to explore this field to date, the field still remains a largely untapped resource from a commercial polymer standpoint. An extensively thorough review of each topic will not be covered, but rather several potentially important systems will be discussed. The goal of this chapter is to introduce the reader to the field of ionomers as structural materials, and to urge the reader to consider the possibilities that do exist in this field.

Advantages/Disadvantages of Ions in Polymers

(6)There are many ways in which ionic character can be incorporated into polymers. From an industrial point of view, most of these techniques are undesirable since they require an added step in order to convert the polymer into its ionic form. Additional steps add additional cost to the final product. However, the benefits associated with conversion to the ionomeric state, such as increased glass transition temperatures (T_g), tensile properties, fatigue properties, scratch resistance, and optical transparency may be worth the extra effort and cost. And since there are many commercially available ionic or potentially ionic monomers that can be incorporated into polymers during the polymerization reaction, the overall cost can be minimized. This is more cost effective than developing new monomers to achieve the same benefits as can be obtained through ionic forces.

(7)One major benefit of having ionic segments in polymers is that they can form a type of crosslink, either as a multiplet or cluster, in the system. As opposed to most covalent crosslinks, the ionic crosslinks can be broken and reformed. This allows thermoplastic polymers to gain mechanical strength as a result of the ionic crosslinks, but still be able to maintain the thermoplastic nature of the overall polymer. Over time the thermoformed polymer will again reform the ionic crosslinks, and regain the prior properties. Hence, the ionomer has properties of both a thermoset and thermoplastic polymer.

(8)Due to the number of counterions available, the properties can readily be tailored through proper selection of these counterions. Additionally, multiple counterions can be combined to vary the properties relative to the ionomer containing only one type of counterion [9-13]. The polarity and size of the counterion can alter the properties of the polymer. This wide range of counterions adds an easy benefit to changing the properties of the polymer with little effort.

(9)Since ionic segments can cause aggregation, there are segments in the polymer microstructure that have limited mobility. In semi-crystalline polymers this can cause a reduction in the percent crystallinity and crystallization rate due to the limited mobility of segments of the polymer chains. This reduction in crystallinity and crystal size can lead to polymers that have a greater transparency than their non-ionic analogs. This can be beneficial in applications in which optical clarity is an important quality.

(10)The introduction of ionic groups into polymers has a large influence on the melt rheological behavior of the system. This can potentially be a big disadvantage in that the ionic forces can create such a high viscosity that the polymer cannot be processed using traditional thermal methods. Achieving a balance between ionic content and processability helps achieve desirable properties. Due to the potential for the ionic groups to impart ionic crosslinking, the polymer could behave similar to a thermoset at room temperature yet remain a thermoplastic.

(11)A potential disadvantage of ion-containing polymers is that the ionic content typically causes the polymer to absorb more water than if there were no ionic portions present [5-7]. The water absorption is tied to the counterion used in the ionomer. This water absorption can break-up the ionic crosslinks in the polymer, and weaken the polymer. Also, water absorption is very undesirable in environments in which moisture is avoided, such as in electronic applications. Proper selection of the counterion can minimize water absorption, yet still take advantage of the ionic character of the polymer.

(12)Another disadvantage of ionomers is a tendency to exhibit stress relaxation over time, especially in elastomeric ionomers [8-10]. Under a load ionomers tend to display a high level of permanent set. This is believed to be due to an ion-hopping mechanism in which the ions migrate to new sites, which thereby forms new ionic crosslinks. This then prevents the ionomers from regaining their original dimensions, leaving a high degree of permanent set.

Roles of Ions in Properties of Polymers

(13)There are a wide range of polymers that can be used as the base polymer for ion-containing polymers. The incorporation of ionic or potentially ionic units is very amenable to the chemistry used to make them. Structurally the location of the ionic segments in polymers can be varied based upon the chemistry used to make the polymer. The locations of the ionic groups includes: mono or telechelic, block, random, and combinations or variations of such. The main polymer chains are typically linear, star, hyperbranched, or dendrimeric. However, from a commercial standpoint linear polymers are preferred.

(14)There has been significant progress in the past 40-50 years as to determine what is happening in ionomers that causes such a dramatic change in properties with slight changes in the concentration of ionic units. Much of these changes in properties centers on the formation of multiplets and/or clusters. These two are much more complicated than just serving as crosslinks. However, multiplets and clusters are not always possible in polymeric systems. These are highly dependent on the polarity of the polymer structure.

(15)The generally accepted model for ion multiplets and clusters is based upon the Eisenberg-Hird-Moore (EHM) model [4], which is a core-shell structure (Figure 1). In this model the ions coordinate in the center of the shell, and the polymer segments in the immediate vicinity have restricted mobility. As the number of segments away from the center increases the mobility increases. Hence, what are formed are portions of the polymer with distinct properties from one another. As the number and location of these multiplets increases, regions of greater restricted mobility increases. This leads to cluster formation in which properties such as a well-defined cluster T_g can result. These regions essentially act as separate phases within the polymer system. Such cluster formations leads to stronger virtual crosslinks than what is achieved with multiplets. And, like multiplets, cluster formations are able to be broken and reformed.

(16)When the ion content becomes sufficiently high, which is dependent on the polymer structure and counterion, the ionomer can become brittle. The restriction in mobility does not allow the polymer chains to disentangle or release pressure, which leads to a structure that has little toughness. Therefore, controlling the upper limit of ionic content is necessary in order to reap the maximum benefits of its incorporation into the polymer.

Research in Ionic Polymers

(17)As mentioned previously among the many uses of ionomers is that they can act as structural materials by themselves, be used as a matrix to incorporate fillers, or benefit by incorporation as a compatibilizing agent between two incompatible polymers as part of a blend. By far most of the research has been pursued in studying the ionomers by themselves. There has been a lot of research in the past few years in incorporation of fillers, such as clays, to create new nanocomposites. Also, the incorporation of ionomers into two highly incompatible polymers as a polymer blend has garnered much interest of late.

(18)Care must be taken when comparing ionomers vs. non ion-containing polymers. For example, when comparing the properties of ion-containing sulfonated polystyrene it is best to compare it with sulfonated polystyrene rather than comparing it to polystyrene. There is the possibility that simple functionalization of the polymer, rather than the fact that it contains ions, is the reason for the change in properties. Although the ions in these polymers impart changes in the properties of the overall polymer, it is also possible that the non-ionic portions also cause similar changes. This is important due to the fact that an additional step can be avoided should the conversion to an ionomer not be necessary. Simple functionalization may give the desired properties by themselves.

(19)A good example of comparing the acid neutralized vs. ion-functionlized polymer is poly(ethylene-co-methacrylic acid). This polymer is commercialized under the name Nucrel® by DuPont. When Nucrel® is neutralized to various degrees using cations, such as Li^+, Na^+, K^+, or Zn^{2+}, it falls under the name Surlyn®. There has been a tendency to compare Surlyn® directly to polyethylene (either low density or high density). However, when one compares the properties of Nucrel® and Surlyn® there are many similarities between the two. The methacrylic acid segments cause the overall crystallinity of the polymer to be reduced relative to polyethylene. In fact, there is evidence that the methacrylic acid segments interact with one another in a similar fashion as the ionic segments of Surlyn® to form multiplets [14-16]. Therefore, Surlyn® should be compared to other ionomeric polyethylene copolymers, or to the respective acid-neutralized analogs, rather than to polyethylene itself.

(20)A wide range of polymers can be considered as structural materials. Perhaps the most well-known ionomeric structural polymer is Surlyn®. Due to the number of research groups that have studied Surlyn® over the years there has been a wealth of information gathered that can be translated to other ionomeric systems. Selection of the valency of the metal salt counterion plays an important role in the final properties of this polymer. The fact that multiplets and clusters form in this polymer [17-22] shows that there are regions of restricted mobility in the polymer. This leads to a separate phase in the polymer that displays its own separate properties from the bulk polymer.

(21)Surlyn® has applications as golf-ball covers due to the impact/scratch resistance, clarity, and water-resistance of the polymer. The success of Surlyn® shows that ion-containing polymers can be commercialized as long as there is a need for polymers that have those specific properties. This work transitions into other research that has taken place which has potential for commercialization.

Ionomers as Stand-Alone Polymers

(22) Several research groups have synthesized and studied ionomeric polyurethanes. Polyurethanes are a common class of commercial polymers that can have their properties altered through proper selection of the properties of the soft and hard segments, such as molecular weight and chemical structure. Typically the potentially ionic segments are incorporated into polyurethanes through a modified chain-extending unit. Polyurethanes have an advantage in that these polymers can easily be modified to have either cationic or anionic counterions. Selection of the molecular weights of the hard and soft segments allows for an adjustment of regularly spaced ionic sites along the polyurethane backbone. This allows for a control of the concentration of ionic groups in the polymer.

(23)"Anionomeric" polyurethanes were prepared from polypropylene glycol, dimethylolpropionic acid, and isophorone diisocyanate where polymers were neutralized using triethylamine (TEA) or N-methyldiethanolamine (MDEA) [23]. The tensile properties were determined on these two ionomers, and then subsequently compared to the non-neutralized analog (Figure 2). The researchers claimed that there were no significant differences between the polymers. It was found that the ionomer using TEA as the counterion had slightly higher tensile strength and elongation than when MDEA was the counterion. This may have been due to the differences in strengths of the ionic interactions of the two systems. Using DMTA it was found that the ionomers both had greater phase separation than the non-neutralized analog.

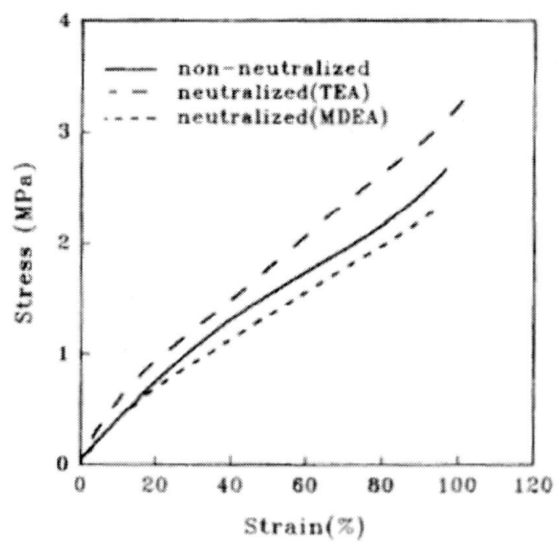

Figure 2. The effect of counterion on the tensile properties of polyurethanes.

(24)In similar work, polyurethanes "cationomers" were synthesized in a similar approach to the previous example. The chain extender for these polyurethanes was methylene diisocyanate, with poly(caprolactone) glycol being the comonomer [24,25]. The resulting polymer was converted to the ion-containing polymer through addition of MDEA. It was stated that the ionization causes a reduction in the order of the hard segments. As a result, it was found that the tensile strengths and tensile moduli increased upon ionization (table 1). It was concluded that ionization increases the hard domain cohesion. Also, it was shown that by increasing the molecular weight between the urethane and the tertiary nitrogen groups that the tensile strength, tensile modulus, and Shore A hardness decreased, and the ultimate elongation increased. This is a reflection of the decrease in the ionic crosslinking upon the decrease in the density of ionic sites. This work also showed that increased ionic content led to increased water uptake. This resulted in a decrease in the tensile properties, but could be regained through removal of the water.

Table 1. The effect of the degree of quaternization on the physical properties of polyurethanes

DQ[1]	Tensile strength (MPa)	Tensile modulus at 100% elongation (MPa)	Tensile modulus at 300% elongation (MPa)	Elongation (%)	Shore A hardness
10	4.64	0.77	1.55	1279	48
20	7.47	1.81	2.46	1082	57
30	19.28	2.21	3.86	998	65
40	20.56	2.64	4.64	836	69
60	21.87	5.04	5.71	788	73
80	24.96	5.38	8.07	770	78
100	29.31	6.49	9.68	714	82

[1] degree of quaternization

Table 2. Molecular weight and molecular weight distribution data of polyurethanes

Chain extender	MW (g/mol)	MWD
ED	27×10^4	1.7
MDA	8×10^4	1.2
BDDS-0.2[1]	21×10^4	1.5
BDDS-0.8[1]	9×10^4	1.3
BDDS-1.4[1]	7×10^4	1.3
BDDS-2.0[1]	4×10^4	1.3
BDDS-1.4-ArgMe[2]	7.4×10^4	1.3
BDDS-1.4-Asp[2]	7.5×10^4	1.3
BDDS-1.4-Gly[2]	7.6×10^4	1.3
BDDS-1.4-Lys[2]	6.1×10^4	1.3

[1] sulfonyl content was 0.2, 0.8, 1.4, and 2.0 wt%
[2] carboxyl content was 1.4 wt%

(25) In another study [26], segmented ion-containing polyurethanes were prepared using different hard segment lengths. The chain-extending unit was the disodium salt of 4,4′-diamino-2,2′-biphenyldisulfonic acid (BDDS). The non-ionic chain extenders that were used were methylenedianiline (MDA) and ethylenediamine (ED). Table 2 lists the molecular

weights and molecular weight distributions of the polymers in this study. It was found that the tensile strength could be enhanced to about double of the polymers containing the non-ionic chain extenders, with a 2-30 fold increase in tensile modulus (table 3). Increasing ionic content led to decreasing ultimate elongations, which was due to increasing levels of ionic crosslinking. It was speculated that the decrease in tensile strength from BDDS-1.4 to BDDS-2.0 was due to molecular weight differences.

Table 3. Mechanical properties of polyurethanes

Chain extender	Tensile Strength (MPa), ± 10%	Tensile modulus at 100% elongation (MPa), ± 15%	Ultimate elongation (%), ± 12%
ED	12.8	3	1340
MDA	10.1	34	365
BDDS-0.2[1]	ND[3]	6	>3900
BDDS-0.8[1]	6.0	8	750
BDDS-1.4[1]	28	60	380
BDDS-2.0[1]	11	100	240
BDDS-1.4-ArgMe[2]	26	18	450
BDDS-1.4-Asp[2]	36	140	390
BDDS-1.4-Gly[2]	34	100	400
BDDS-1.4-Lys[2]	37	32	380

[1] sulfonyl content was 0.2, 0.8, 1.4, and 2.0 wt%
[2] carboxyl content was 1.4 wt%
[3] not determined

(26)In the same study the ionic groups in the ionic polyurethanes were also varied to determine the effect of sulfonic and carboxylic groups in many polymeric systems. The greater polarity of the sulfonyl group vs. the carboxyl groups should show several differences in the properties of the resultant polymers. Four different amino acids were incorporated into the polyurethanes, each of which contains carboxyl groups. It was found that at equivalent ion contents that the ones containing carboxyl groups had higher tensile moduli for the Asp and Gly, but not for the ArgMe and Lys, when compared to the ones containing sulfonyl groups (BDDS-1.4) at equivalent ion contents. The researcherrs concluded that the long hydrocarbon chains of the ArgMe and Lys groups hinder close packing of ion aggregates, thereby lowering the tensile moduli. However, the results of this work show that the tensile properties of the polyurethanes can be significantly improved through incorporation of a small percentage (1.4%) of ionic groups.

(27)To generalize that the same carboxyl vs. sulfonyl ionic trends holds true for all polymers would be incorrect. On the one hand it was found that there are dramatic differences in the properties of carboxylic acid vs. sulfonic acid-containing polystyrene [27]. However, this dramatic effect was not seen in the properties of ionomeric poly(styrene-ethylene-butadiene) [28]. Therefore, it appears that the trends should be determined for each class of ionomers.

(28)Ionomeric polystyrenes have been studied for well over 50 years, with the first patent reported in 1954 [29]. Although these polymers are rarely studied for their stand-alone properties, much insight can be gained through a study of these systems. Sulfonyl groups can be incorporated either through a sulfonation reaction of polystyrene, or through copolymerization with styrene sulfonate. Pure poly(styrene sulfonate) is a brittle polymer. With decreasing levels of ionic content this polymer gains structural strength. In essence, at less than 100% sulfonation of polystyrene the polymer can be referred to as a polystyrene copolymer (a copolymer of styrene and styrene sulfonate).

(29)The addition of small amounts of ionic content into polystyrene has a very significant effect on the physical properties of the polymer, such as increased toughness, fatigue resistance, and improved tensile properties. As mentioned previously the level of crystallinity tends to decrease as ions are incorporated into the polymer. The same holds true for ion-containing polystyrene. Incorporation of sulfonate groups into polystyrene does restrict chain mobility, which causes a reduction in crystallinity [30]. However, the neutralization of the sulfonic acid groups further restricts chain mobility, which causes an even further reduction in the crystallization rate. Also, the crystal morphologies can vary from those of pure polystyrene. This causes a decrease in the melting point of the ion-containing polystyrene relative to the pure polystyrene. The decrease in number and type of crystals in ionomeric polystyrene means that the optical clarity should increase relative to polystyrene

(30)In terms of physical properties of sulfonated polystyrenes, it was found that the tensile strengths of thin films increased up to approximately 7 mol% Na^+ ionic content before its properties started to decrease (Figure 3) [31]. The authors also measured the toughness of the polymer with increasing ion content. The maximum was reached at the same level as the maximum tensile strength (Figure 4). This percentage is approximately the same as found in other studies [32-33]. This level appears to be the critical ion level in which cluster formation starts. The toughness of the 7% ion-containing polystyrene neutralized using Na^+ was approximately two times that of polystyrene. This increase in properties is very attractive in applications using polystyrene in which increased mechanical properties are desirable.

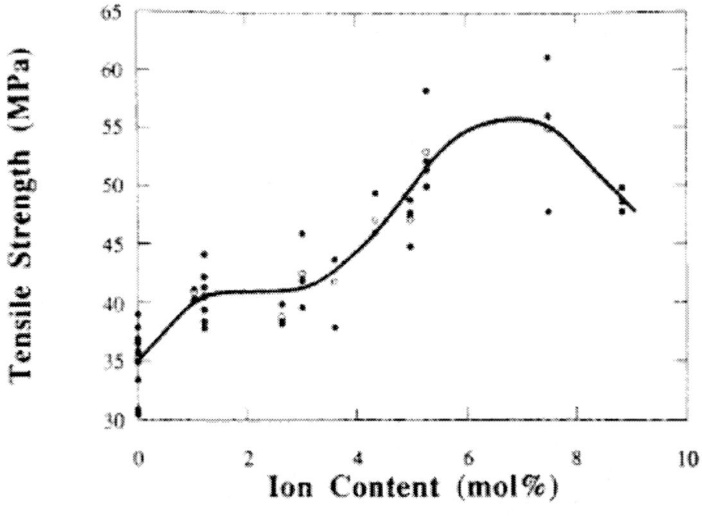

Figure 3. The effect of ion content on the tensile strength of polystyrene ionomers.

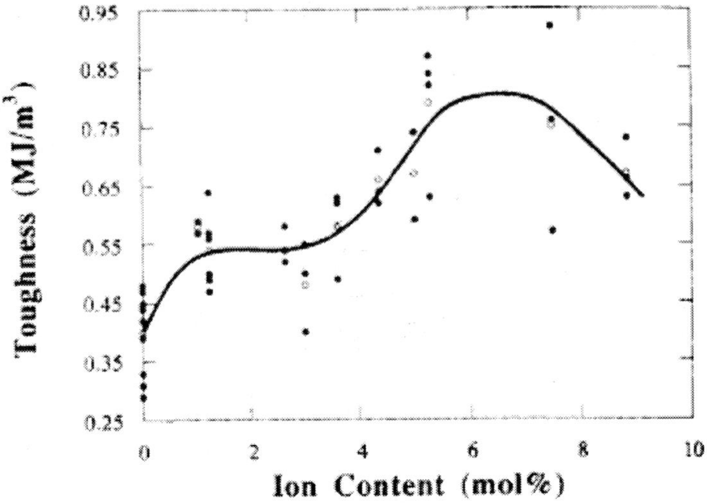

Figure 4. The effect of ion content on the toughness of polystyrene ionomers.

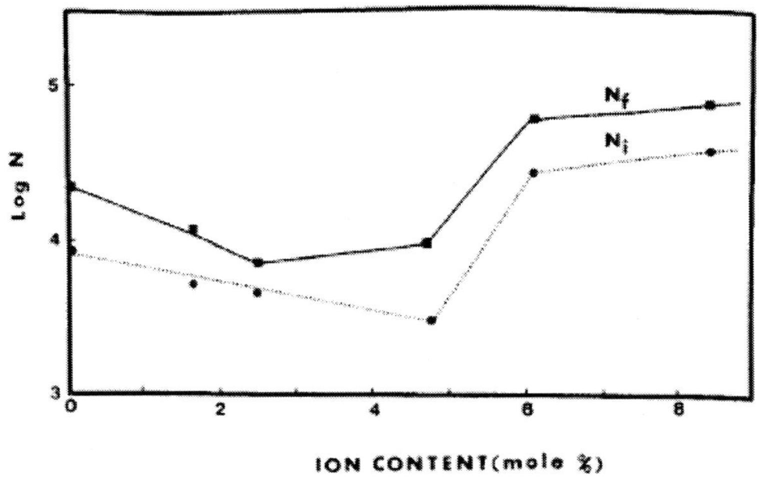

Figure 5. The effect of ion content on the cycles to initiate damage (N_i) and cycles to fracture (N_f) of Na^+ neutralized sulfonated polystyrene ionomers.

(31)Fatigue resistance is an important property in polymers. It was believed that incorporation of ionic groups into polystyrene should increase fatigue resistance. This stems from the ionic groups forming a type of crosslinking that would stabilize crack formation when it occurs. From a fatigue standpoint it was found that the fatigue resistance increases with increasing ion content (Figure 5) [34]. At lower ion contents, in which multiplets are dominant, the fatigue resistance improvement was lower than when clusters were able to form. The clusters were able to form at slightly higher concentrations than when multiplets are formed, which fell at a concentration similar to that mentioned in the previous paragraph. Also, a study of divalent cations revealed that, at equivalent concentrations as the monovalent cations, the properties were significantly improved. DMTA analyses revealed that ionomers based upon Ca^{2+} had a larger rubber plateau modulus, indicative of greater ionic crosslinking, than those from K^+ and Cs^+ (Figure 6) [35]. Thus, it was found that the ionomers using Ca^{2+} as the counterion had three times the fatigue lifetime of polystyrene. Hence, proper selection of ion concentration in polystyrene maximizes fatigue resistance, thereby improving polystyrene as a structural material.

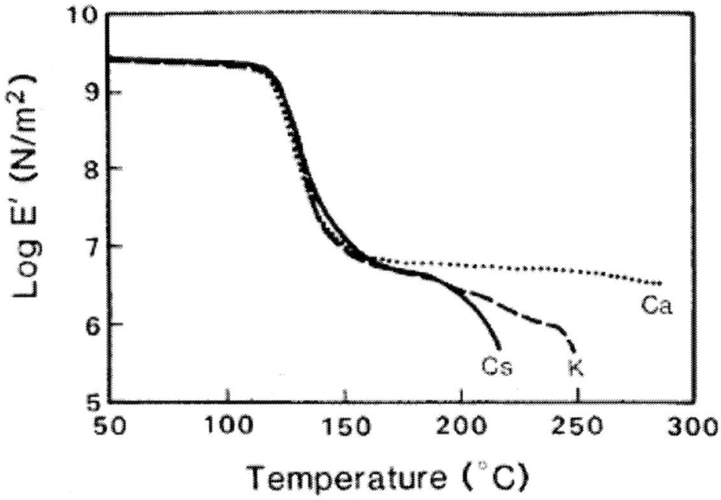

Figure 6. The effect of counterion on the storage modulus of polystyrene ionomers.

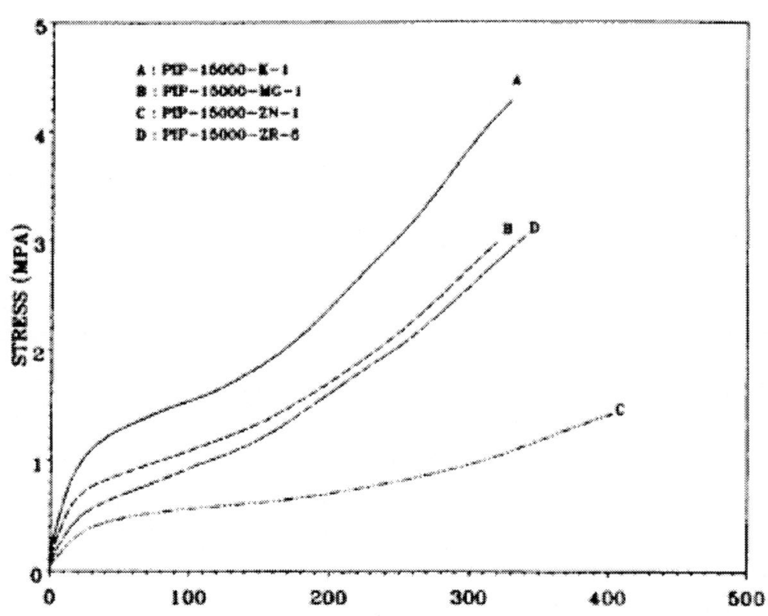

Figure 7. Stress-strain curves for K^+, Mg^{2+}, Zn^{2+}, and Zr^{4+} neutralized sulfonated polyisoprene ionomers.

(32)A study of telechelic ionomers based upon polyisoprene was conducted in which a comparison was made between carboxylate and sulfonate groups [36]. In this work it was found that the polymers based upon sulfonate salts were several orders of magnitude greater than the respective carboxylate salts in terms of tensile stress (Figure 7 and 8). The counterions used in this study were K^{+}, Mg^{2+}, Zn^{2+}, and Zr^{4+}. Polymers based upon the K^{+} counterion led to the greatest tensile strengths in this study. Although the Mg^{2+} counterion is smaller and divalent, whereas the K^{+} is larger and monovalent, polymers based upon the Mg^{2+} counterion had lower tensile strengths at equivalent stresses. It was speculated that the Mg^{2+} counterion would lead to smaller, more stable ionic domains. This was reflected in the longer time creep and stress relaxation experiments. Also, it was found that the sulfonated ionomers exhibited stronger aggregation than the carboxylated analog.

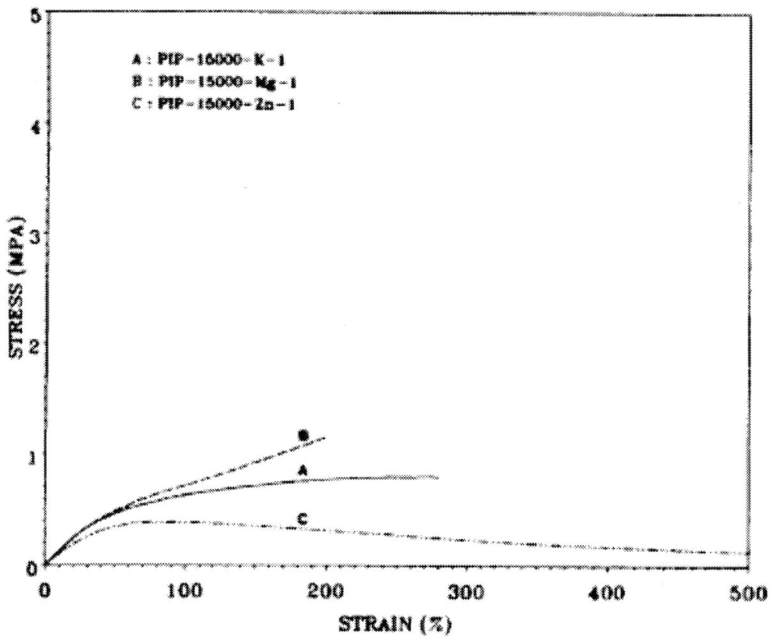

Figure 8. Stress-strain curves of K^{+}, Mg^{2+}, and Zn^{2+} neutralized carboxylated polyisoprene ionomers.

(33)In a study of carboxy-telechelic ionomers based upon polyisoprene, the carboxy groups were neutralized with various divalent cations [37]. Among these were two alkaline earth cations (Ca^{2+} and Sr^{2+}) and three rare earth cations (Ni^{2+}, Zn^{2+}, and Cd^{2+}). The ionomers neutralized using the alkaline earth cations displayed the highest tensile moduli and tensile strengths, and the lowest percent elongations in the series (Figure 9a and 9b). Through SAXS studies it was concluded that the alkaline earth cations lead to polymers with larger aggregates than the rare earth cations. Researchers speculated that the polymer chains form loops when the two aggregates merge. Because the alkaline earth cations can accommodate larger numbers of cations due to the larger core radii of both cations compared to the rare earth cations, this leads to larger numbers of entanglements [38]. This translates into the higher tensile moduli for the ionomers from the alkaline earth cations compared to the rare earth cations. EXAFS confirmed that the alkaline earth cations formed more cohesive ionic microdomains. This caused strain hardening of these two polymers, whereas the polymers using the rare earth cations could relax through ion-hopping. Hence, the authors showed a significant difference in tensile properties through proper selection of the counterion.

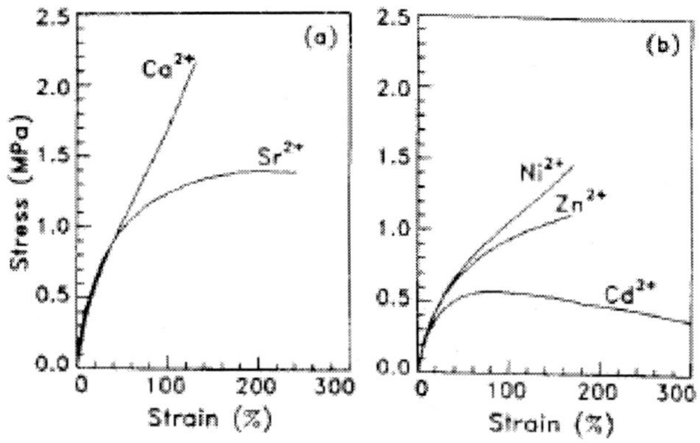

Figure 9. Stress-strain curves of carboxy-telechelic polyisoprenes; (a) neutralized using Ca^{2+} and Sr^{2+}; (b) neutralized using Ni^{2+}, Zn^{2+}, and Cd^{2+}

(34)Polyesters are a commercially important class of polymers that have also been studied as ionomers. Research on polyester ionomers has focused on the random and telechelic incorporation of ionic groups. One study involved the incorporation of ionic segments into poly(ethylene terephthalate) (PET). The use of 5-(sodiosulfo)isophthalate (5-SSI) led to direct incorporation of ionic groups without further modification of the polymer [39]. In this study the ion content was varied from 0 to 9 mol% 5-SSI groups. The T_gs of the ionic and the acid-neutralized analogs were identical, and the increasing concentration of the 5-SSI groups had little effect on the T_gs. Variation in the T_gs ranged only ± 4.5 °C, but did show a minimum at 4 mol% 5-SSI groups (Figure 10), after which point the T_g increased. Incorporation of the 5-SSI groups did decrease the crystallinity and crystallization rate relative to unmodified PET, but increased the crystallization temperature. Also, the melting points decreased with increasing 5-SSI content. Although no physical properties were reported, one would expect beneficial properties typical of ionomers to be extended to this system, which could bring valuable benefits to PET, such as improved tensile properties and optical clarity.

(35)Ionomeric polyester liquid crystalline polymers have also been synthesized in order to increase the tensile and compressive properties relative to the nonionic analog. In this system the polymer was based upon 4-hydroxybenzoic acid, 6-hydroxy-2-naphthoic acid, and 5-SSI. The divalent cations Mg^{2+}, Ca^{2+}, Ba^{2+}, and Zn^{2+} were used as the counterions [40]. The tensile properties of the resulting ionomers and the acid-neutralized analog are listed in table 4. It was speculated that the poor properties of the ionomers based on Zn^{2+} and Mg^{2+} were due to the low molecular weight, as evidenced by the low intrinsic viscosity values. At equivalent concentrations the Ca^{2+} ion is more effective than the Na^{+} ion in increasing the physical properties. Such behavior has been documented by several researchers [32,41]. The increase in the physical properties was explained through SEM analysis, which showed that the non-ionic polymer failed at localized locations, while the ionic analogs showed no localized failure. These marked improvements in the properties of the polymer relative to the acid-neutralized analog are good incentives to benefit from a change to the ionomeric form of this polymer system.

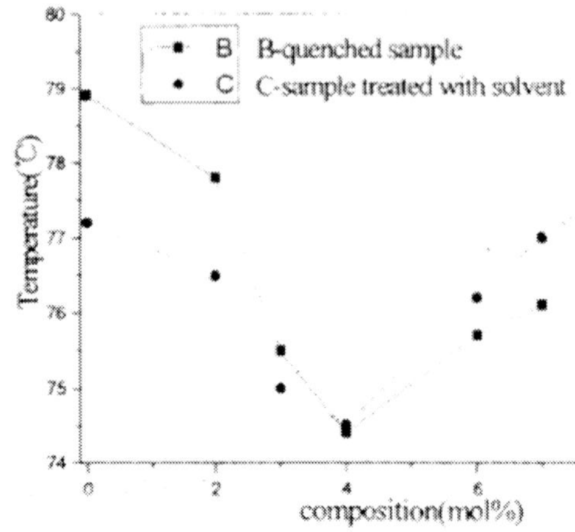

Figure 10. The effect of ion content on the melting point of Na^+ neutralized sulfonated PET.

Table 4. Mechanical properties and intrinsic viscosities of ionomeric liquid crystalline polymers

Cation	Tensile Strength (MPa)	Tensile Modulus (GPa)	Ultimate elongation (%)	Intrinsic viscosity (dL/g)
None	142	13.3	1.27	5.12
Na^+	162	14.9	1.22	5.66
Mg^{2+}	106	16.4	0.61	2.50
Ca^{2+}	351	23.4	1.59	4.83
Ba^{2+}	ND[1]	ND[1]	ND[1]	2.24
Zn^{2+}	86	10.0	0.85	3.88

[1] ND = not determined because films were brittle

Ionomers in Nanocomposites

(36)The field of nanocomposites has gathered much interest in the use of ionomers as compatibilizing agents. Typically non ion-containing polymers have been used in dispersing the fillers such as layered silicates. These layered silicates are typically of the monmorillonite (MMT) variety. The key to obtaining a good dispersion is to maximize exfoliation of the filler particles. Although techniques do exist in which this can be attained without using any addition of ionomeric compatibilizing agents, these techniques tend to be complex. Such other techniques include solution intercalation and *in situ* polymerization techniques, along with traditional melt intercalation.

(37)One approach that can be used to get good exfoliation of the filler is to use ionomers. The use of such polymers should take advantage of the charged surfaces of many different types of fillers. The interactions between the ionomers and the charged filler surfaces are reasons why ionomers are good candidates for exfoliation of these fillers. Along these lines, the use of crystalline polymers is also beneficial to take advantage of the thermal stability of such structures.

(37')One such ionomer is Na^+ neutralized sulfonated poly(butylene terephthalate) (PBT). Studies on this ionomer has shown that the ionic groups aggregate to form ionic domains [42]. The incorporation of sulfonate groups was easily obtained through incorporation of the dimethyl-5-sodioisophthalate monomer [43]. Significant increases in the tensile moduli were observed when the MMT contained alkylammonium ions, and the polymer contained various levels of sodium sulfonate groups (table 5). This modulus increased with increasing sodium sulfonate concentration. However, this increase in moduli was not seen when the clay contained sodium ions. It was determined that the alkylammonium-modified clays have smaller particle sizes than the sodium-modified clays. This leads to greater interactions with the polymer, which translates into a higher amount of energy required to deform the nanocomposite. The high degree of exfoliation was determined through TEM analysis.

Table 5. Young's Moduli of non-ionic and ionomeric PBT

Clay	PBT (psi)	PBT-3% $SO_3^-Na^+$ (psi)	PBT-5% $SO_3^-Na^+$ (psi)
None	169000 ± 7500	165000 ± 7400	157000 ± 5800
Na^+ MMT	182000 ± 50	178000 ± 5100	178000 ± 2000
R_4N^+ MMT	196000 ± 3800	210000 ± 7300	215000 ± 7600

(37a)Ionomeric sulfonated PET/MMT nanocomposites were also recently prepared to take advantage of the benefits of incorporation of MMT [44]. The ionic segments were incorporated through addition of dimethyl-5-sodiosulfoisophthalate during the synthesis of the polymer. The levels of sulfonation were controlled from 0-8 mol%. Nanocomposites were prepared using 5 wt% MMT. Increasing levels of sulfonate groups led to better dispersion of MMT. Interestingly it was found that the ionomer alone showed no evidence of crystallization, but with the addition of MMT the ability to crystallize returned. The researchers speculated that the MMT could be acting as a nucleation site in the system, or due to a combination of the MMT with the polymer to form the nucleation site. Also, it was found that addition of MMT increased the thermal stability of the polymer relative to the unfilled analog (Table 6).

Table 6. Decomposition temperatures of ionomeric PET and ionomeric PET/MMT nanocomposites

Sample	T_d, 5% (°C)	T_d, 10% (°C)	Wt. at 600 °C
SPET2[a]	299	327	0.0
SPET6[b]	258	288	1.4
SPET2M5[c]	324	344	6.9
SPET6M5[d]	343	357	11.8

[a] sulfonated PET, 2 mol%
[b] sulfonated PET, 6 mol%
[c] sulfonated PET, 2 mol%; 5% MMT
[d] sulfonated PET, 6 mol%; 5% MMT

(38)A recent study of polyurethane ionomers based on poly(tetramethylene oxide)/4,4′-diphenylmethylene diisocyanate which contained 3 mol% quaternary ammonium groups was used to incorporate MMT [45]. Simple blending of the non-ionomeric polyurethane (PU) did not lead to exfoliated nanocomposites, as evidenced by WAXS. Incoroporation of the polyurethane ionomer (PUC) led to apparent exfoliation of the filler, as determined using WAXS and TEM. This led to nanocomposites that displayed an increase in moduli with increasing MMT content. As is typical with most nanocomposites of this nature, there was a subsequent decrease in the ultimate elongation (table 7). The authors concluded that the changes in the tensile properties of the nanocomposites upon use of PUC was due to its strong interaction with the exfoliated MMT layers, which resulted in a less phase-separated morphology.

Table 7. Mechanical properties of ionomeric polyurethanes/MMT nanocomposites

Sample	Maximum stress (kgf/mm^2)	Elongation at break (%)	Young's modulus (kgf/mm^2)
PU	4.88	806	4.11
PUC	4.90	824	4.15
1 wt% MMT/PUC	4.93	760	5.60
3 wt% MMT/PUC	3.86	568	6.26
5 wt% MMT/PUC	3.66	495	7.20
7 wt% MMT/PUC	3.45	487	9.43

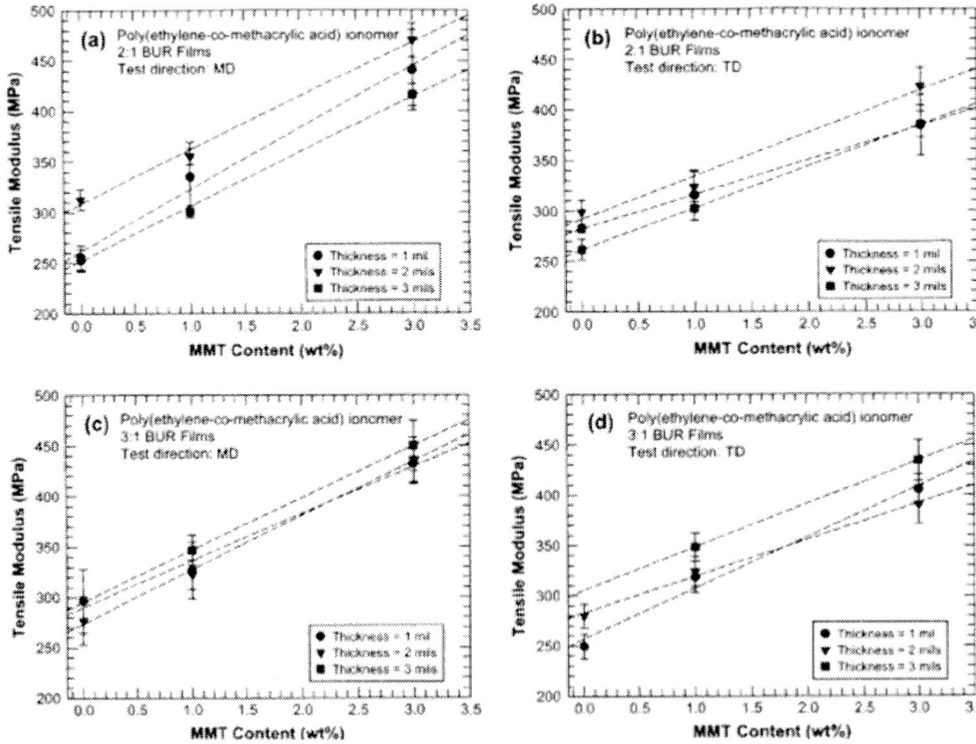

Figure 11. Continued on next page.

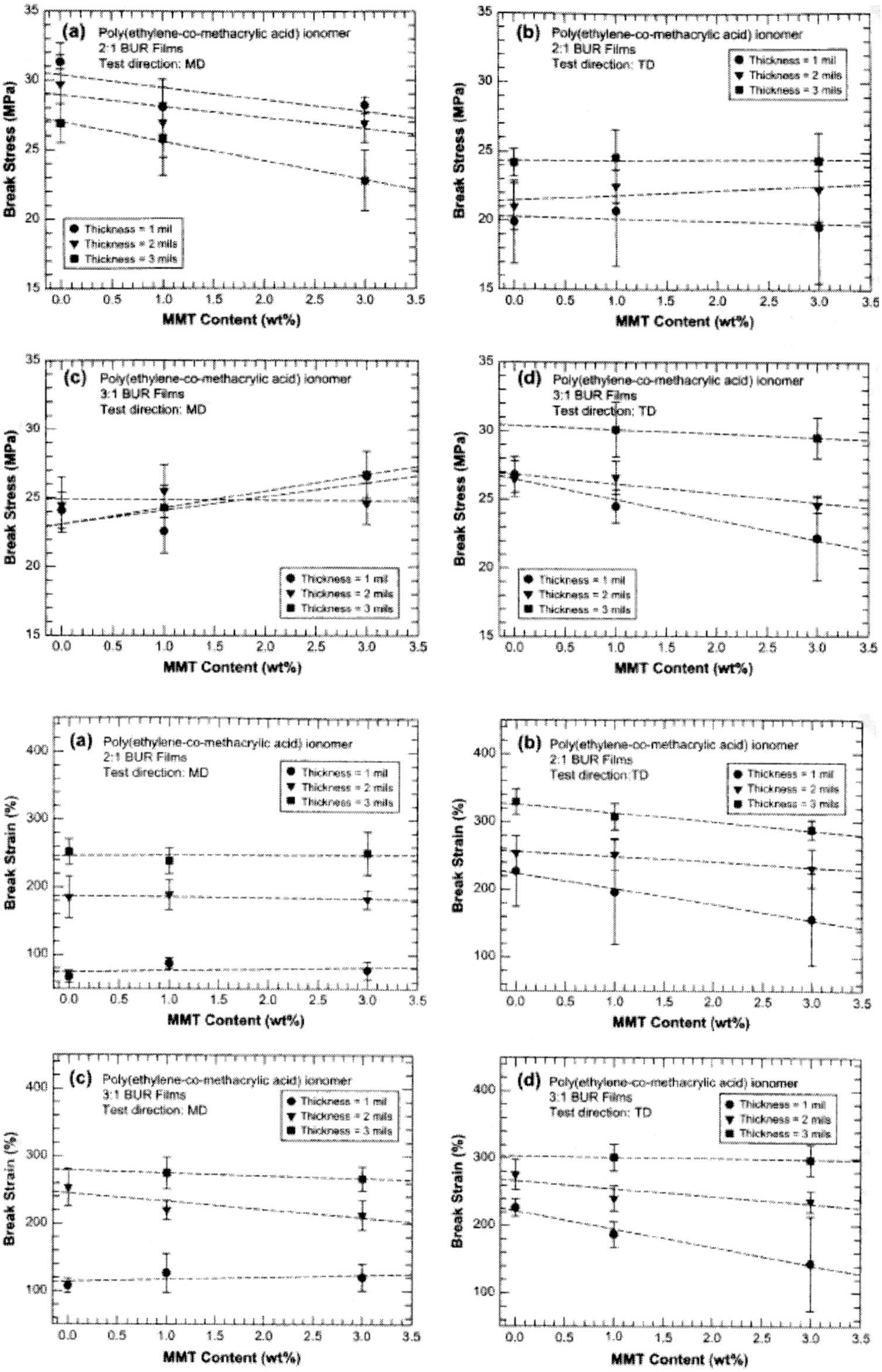

Figure 11. The effect of MMT(HT_2) concentration on the mechanical properties of Surlyn®/MMT nanocomposites. (BUR = ratio of the diameter of the film to the diameter of the die. MD = machine direction. TD = transverse direction.)

Table 8. Mechanical properties of LDPE, Nucrel®, and Surlyn®/MMT nanocomposites

Polymer	Organoclay type	MMT (wt%)	Tensile modulus (MPa)	Relative modulus[1]	Tensile strength (MPa)	Elongation (%)
LDPE	None	0.0	114	1.00	13.6	108
	HT_1	2.5	155	1.36	14.4	87
		5.0	172	1.51	14.4	80
		7.5	194	1.70	14.1	73
		10.0	218	1.91	14.0	67
	HT_2	2.5	178	1.56	14.2	83
		5.0	227	1.99	14.3	77
		7.5	280	2.46	14.3	70
		10.0	375	3.29	14.2	62
Nucrel®[2]	None	0.0	118	1.00	13.9	136
	HT_1	2.5	151	1.31	14.3	120
		5.0	180	1.52	14.5	108
		7.5	220	1.86	14.9	99
		10.0	260	2.20	15.2	90
	HT_2	2.5	189	1.60	14.7	111
		5.0	259	2.20	16.5	99
		7.5	328	2.78	17.5	91
		10.0	425	3.60	18.0	82
Nucrel®[3]	None	0.0	73	1.00	15.4	185
	HT_1	2.5	112	1.53	16.1	176
		5.0	133	1.82	16.7	165
		7.5	178	2.44	17.4	148
		10.0	220	3.01	18.5	133
	HT_2	2.5	147	2.01	17.8	156
		5.0	203	2.78	19.1	143
		7.5	254	3.48	20.6	134
		10.0	353	4.83	22.2	120
Surlyn®[4]	None	0.0	262	1.00	21.3	194
	HT_1	2.5	349	1.33	21.2	117
		5.0	410	1.56	21.0	111
		7.5	465	1.77	22.1	119
		10.0	563	2.15	23.2	116
	HT_2	2.5	403	1.54	22.3	127
		5.0	560	2.14	23.8	111
		7.5	732	2.79	26.6	72
		10.0	919	3.51	29.4	65

[1] Relative modulus = modulus with organoclay/modulus without organoclay
[2] Contains 3.9 wt% methacrylic acid
[3] Contains 8.9 wt% methacrylic acid
[4] Contains 15.2 wt% methacrylic acid

(39)Surlyn® has also been used as the matrix polymer for dispersing clays. In this specific study the authors tested the properties of various blow-molded Surlyn®/MMT nanocomposites at different clay concentrations [46]. The clays that were used in this study were both single (HT_1) and dual tail hydrogenated tallow oil (HT_2). It was found that the tensile moduli increase as the MMT concentration increases. The tensile strength and elongation do not change much in the concentration range tested in this study (Figure 11). The authors compare the Surlyn®/MMT nanocomposites vs. LDPE/MMT nanocomposites. Due to the structural differences between these polymers these systems should not be directly compared to one another. The high crystalline content of LDPE is broken up by the incorporation of methacrylic acid monomeric units. A more direct comparison should be made between Surlyn®/MMT nanocomposites vs. Nucrel®/MMT nanocomposites. However, favorable results showing increased tensile moduli, tensile strengths, and improved impact resistance bode well as candidates for structural materials.

(40)Subsequent work by the same researchers did compare two grades of Nucrel® vs. Surlyn® and LDPE as matrices for nanocomposites [47]. The Nucrels® used in this study contained 3.9 and 8.9 wt% methacrylic acid groups, while Surlyn® contained 15.2% methacrylic acid groups. The clays that were used in this study were both single and dual tail hydrogenated tallow oil. It was found that exfoliation increased when a dual tail HT was used vs. single tail HT clay. The tensile strengths and tensile moduli increased, and elongation decreased, with increasing levels of clay (table 8). In the two Nucrels® that were studied the tensile modulus increased with increasing methacrylic acid content. It was suggested that Surlyn® exfoliated the clays the most due to the largest particle aspect ratios in the series. Conversion of this Surlyn® to the acid-neutralized analog, and subsequent comparison to Surlyn®, would have shown whether the ions were responsible for the greater exfoliation. However, the lesson learned was that the tensile strengths and moduli could markedly be increased through proper selection of polymer matrix, and selection of methacrylic acid content.

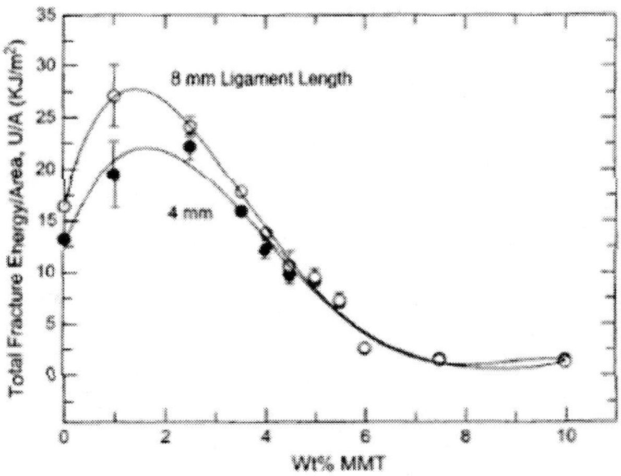

Figure 12. The effect of MMT on the total fracture energy per unit area.

(41)Nanocomposites using Surlyn® 8945, a Na^+ neutralized version of Surlyn®, as the matrix were also studied to determine the improvement in fracture toughness relative to the unfilled analog [48]. In this work the fracture toughness was maximized through proper selection of the clay loading. What was found was that initially the fracture toughness increases quickly with the addition of clay, but then decreases slowly as the concentration continues to increase (Figure 12). At MMT concentrations of greater of 2.5% the fracture energy starts to decrease. At MMT concentrations of 5% and greater the nanocomposites display brittle failure. In applications where an increase in stiffness is required, these nanocomposites, with low MMT concentrations, would be good candidates. The negative associated with this system is that the addition of the clay causes a reduction in the overall ultimate elongation and increase in brittleness relative to the unfilled system.

(42)In terms of tensile properties of Surlyn® nanocomposites, a series of three different Surlyns® neutralized using three different cations were studied [49]. All three of the Surlyns® contained similar methacrylic acid contents. The three cations that were studied were Na^+, Li^+, and Zn^{2+}. Analysis of the Surlyn®/clay nanocomposites revealed that the Surlyn® using Li^+ as the counterion had the worst exfoliation in the series. In each series the tensile moduli and tensile strengths increased with increasing clay content, but the elongation decreased in the same series (table 9). Because the Li^+ neutralized Surlyn® did not exfoliate the clay particles completely, the tensile properties reflected numbers closer to the matrix resin than the other two polymers in the series. The Surlyn® neutralized using the Zn^{2+} counterion yielded the nanocomposite with the highest relative modulus. Although no clear conclusion was reached as to why this is the case, the authors do provide some possible reasons as to why this occurs. It was speculated that the Zn^{2+} neutralized ionomer causes anhydride formation, but not those using Na^+ and Li^+ cations.

Table 9. Mechanical properties of Surlyn®/MMT nanocomposites

Cation	MMT (wt%)	Tensile modulus (MPa)	Relative modulus[a]	Elongation (%)	Tensile strength (MPa)
Zn^{2+}	0.0	176	1.00	172	19.3
Zn^{2+}	2.5	314	1.78	116	22.2
Zn^{2+}	5.0	447	2.50	86	24.2
Zn^{2+}	10.0	795	4.51	59	29.2
Na^+	0.0	260	1.00	194	21.0
Na^+	2.5	412	1.58	130	22.6
Na^+	5.0	568	2.18	119	25.9
Na^+	10.0	908	3.49	66	28.8
Li^+	0.0	292	1.00	136	21.0
Li^+	2.5	407	1.38	116	24.2
Li^+	5.0	491	1.68	104	24.8
Li^+	10.0	676	2.32	98	27.2

[a] Relative modulus = (modulus at MMT wt% > 0%)/(modulus without MMT)

(43)Rubbers have also been used as matrices in nanocomposites. One such polymeric system is based on poly(isobutylene-co-isoprene) (BIIR). In this work the BIIR matrix was used to incorporate ion-exchanged MMT (NR^+-MMT) and precipitated silica [50]. The BIIR polymers were modified to make the triphenylphosphonium bromide salts (IIR-PPh_3Br). The researchers hypothesized that the quaternary phosphonium cations could displace the ammonium ions of the modified clays. This could then lead to an interaction with the particles, which would lead to exfoliation of the NR^+-MMT. Incorporation of NR^+-MMT into the non ion-containing polymer that was crosslinked using ZnO did not lead to any significant reinforcement of the system. Use of the ionomeric analog led to an increase in both the tensile strengths and tensile moduli of the system (table 10). As noted by these researchers, and which is common throughout ion-containing polymeric systems, this system exhibited stress relaxation. It was also noted that this stress relaxation was not improved through incorporation of NR^+-MMT.

(44)As a continuation of this work the ionomer was used to incorporate precipitated silica. The researchers indicated that there were no signs of silica agglomeration at any of the tested concentrations. Based upon the tensile properties it appeared that the ion-containing analog had a greater degree of exfoliation than the non ion-containing polymer. Strain relaxation analysis, which involves the determination of the storage modulus as a function of amplitude, revealed that the non ion-containing polymer containing silica had agglomeration of the silica particles, while that ion-containing polymer did not.

Table 10. Mechanical properties of cured BIIR/NR^+-MMT and IIR-PPh3Br/NR^+-MMT nanocomposites

Polymer/filler	Filler level (wt%)	Tensile strength (MPa)	Tensile Modulus (MPa)	Elongation (%)
IIR-PPh_3Br/NR^+-MMT	0	5.0	0.55	320
	3	4.4	0.64	280
	5	8.1	0.79	235
	15	5.4	2.15	78
BIIR-ZnO/ NR^+-MMT	0	3.1	0.33	375
	5	1.8	0.43	285
	15	1.2	0.65	285
IIR-PPh_3Br/silica	0	5.0	0.55	320
	15	3.6	1.18	270
	30	3.9	3.60	60
BIIR-ZnO/silica	0	3.1	0.33	330
	15	2.3	0.64	280
	30	3.7	2.90	200

(45)An ion-containing poly(ethylene-graft-maleic anhydride) was used to incorporate silica and MMT [51]. TEM analysis clearly showed the exfoliation of the particles. This is very unlike standard HDPE, which led to silica aggregates rather than platelets. In both versions of the nanocomposites the tensile strengths and tensile moduli increased with increasing filler levels (table 11). A reduction in the ultimate elongation was seen with increasing levels of filler, which was expected. A comparison of the nanocomposites using MMT vs. silica fillers showed that those from silica displayed greater ductility than those from MMT. The authors concluded that the nanocomposites using silica provided better overall melt viscosity, stiffness, and mode of failure than those using MMT.

(46)Wood flour, a common filler used in the wood plastics industry, has been used as a filler in a HDPE matrix to form nanocomposites. Four different grades of Surlyn® were used as the compatibilizing agents [52]. Two of the Surlyns® were Zn^{2+} neutralized, and two were Na^{+} neutralized. Use of the Na^{+} neutralized ionomers led to an improvement in the tensile properties at concentrations above 8%. At this point it was found that the mode of failure occurred along the shear plane. This is in contrast to lower concentrations of this ionomers in which the mode of failure was similar to that of HDPE/wood flour blends in which the polymers cracked. The ionomers neutralized using Zn^{2+} displayed significant improvement in the mechanical properties at concentrations of 2%. Whereas the modulus of elasticity decreased with increasing ionomer content, the modulus of rupture increased. This implies that there is an increase in toughness of the composite due to increased compatibility. In terms of toughness it was found that the Na^{+} neutralized ionomers displayed a better improvement in toughness compared to the Zn^{2+} neutralized ionomers. In the Zn^{2+} neutralized series there was no evidence of an improvement or deterioration of the toughness with a change in the concentration.

Table 11. Mechanical properties of ion-containing poly(ethylene-graft-maleic anhydride)/MMR-NR_4^+ and SiO_2 nanocomposites

Filler	Filler level (wt%)	Tensile modulus (MPa)	Yield Stress (MPa)	Elongation (%)
MMT-NR_4^+	0	290 ± 20	26 ± 1	2190 ± 340
MMT-NR_4^+	1	320 ± 10	25 ± 1	1570 ± 230
MMT-NR_4^+	5	320 ± 60	27 ± 1	690 ± 310
MMT-NR_4^+	9	410 ± 30	30 ± 2	740 ± 300
MMT-NR_4^+	13	420 ± 60	27 ± 3	12 ± 1
MMT-NR_4^+	29	470 ± 125	29 ± 7	11 ± 5
SiO_2	0	30 ± 10	26 ± 1	2190 ± 340
SiO_2	1	330 ± 40	27 ± 1	1150 ± 380
SiO_2	5	340 ± 20	28 ± 1	1250 ± 290
SiO_2	9	360 ± 50	30 ± 1	2010 ± 360
SiO_2	13	400 ± 50	31 ± 2	1220 ± 360

(47)Polypropylene (PP)/Vectra B composites were achieved through addition of Zn^{2+} neutralized Surlyn® [53]. Blending Vectra B with ionomer led to an immiscible blend, as evidenced by two T_gs using DMTA. Blending PP with ionomer led to a miscible blend, as determined by the single T_g using DMTA. Tensile testing revealed that the binary blends had yielding, while the compatibilized ternary blends did not. The higher modulus values of the ternary blends it indicated that Vectra B was a large contributor to the mechanical response.

Ionomers as Blend Compatibilizers

(48)The field of polymer blends is one in which the work is limited by the compatibility of the polymers. The low entropy of mixing causes many polymer blends to be thermodynamically unstable, which leads to low interfacial adhesion [54]. In order to compatibilize blends, the use of additives that reduce the interfacial tensions of the polymers are used [55]. In recent years there has been an interest in using various ionomers as the compatibilizing agents. Due to the commercial availability of several ionomers, this portion of the field of polymer blends can be pursued relatively quickly. As such, there has been a significant amount of work published in the past decade using ionomers in this application.

(49)From an industrial standpoint polymer blends are of high interest due to the availability of the starting polymers. Through proper selection of the polymers used in the blends, specific properties can be targeted that benefit from the advantages of the polymers used in the blend. Also, rather than investing in research and development of new polymers, blending of several commercially available polymers may accomplish the same goals. Hence, blending of polymers offers an attractive alternative to the synthesis of new polymers.

(50)Polyurethane (PU) was blended with a Zn^{2+} neutralized poly(ethylene-co-methacrylic acid-co-isobutyl acrylate) (EMI-Zn) ionomer [56]. Most of the tensile properties of the blends were lower than that of pure PU (table 12). In one case the ionomer was neutralized with H^+ to determine the effect of the ions on the tensile properties. What was found was that that there was a significant difference between the propeties of the Zn^{2+} neutralized ionomer vs. the H^+ neutralized analog. This implies that the blend had better compatibility when using an ionomer vs. the non-ionomeric version. The thermal properties of the blends were not much different than that of PU, but there were differences in crystallinity. In the case of the 90/10, 70/30, and 50/50 blends the crystallinity increased relative to PU.

Table 12. Mechanical and crystallinity data of PU/EMI-Zn blends

PU/EMI-Zn	Tensile strength (MPa)	Elongation at break (%)	Tensile modulus (J/cm^3)	χ_c, PU (%)	χ_c, EMI-Zn (%)
100/0	57 ± 10	731 ± 57	220 ± 20	33.8	-----
90/10	42 ± 7	639 ± 84	142 ± 37	39.3	6.7
70/30	28 ± 3	624 ± 75	98 ± 23	41.1	7.4
70/30[a]	24 ± 3	518 ± 69	74 ± 15	ND[c]	ND
50/50	32 ± 4	657 ± 66	128 ± 21	39.2	12.2
30/70	27 ± 2	743 ± 39	117 ± 13	18.7	11.7
10/90	25 ± 3	747 ± 55	106 ± 18	-----	11.1
0/100	31 ± 3	562 ± 48	109 ± 15	-----	8.8

[a] The Zn^{2+} was replaced with H^+

[b] χ_c = percent crystallinity

[c] ND = not determined

(51)PU was also blended with HDPE using EMI-Zn as the ionomeric compatibilizing agent [57]. Optimization of the properties was achieved by melt mixing all of the components for 15 minutes at the same time, except for the 28.3/56.7/15 blend. Longer mixing times led to a decrease in the properties, which may be due to ionomer agglomeration [58]. The properties of the blends increased as the PU content increased (table 13). Use of Na^+ neutralized EMI led to blends with inferior properties to that containing EMI-Zn. This was thought to be due to the greater ionic character, which would lead to a decreased amount of dispersion [59].

Table 13. Mechanical properties of PU/HDPE/EMI-Zn blends

PU/HDPE/EMI-Zn	Mixing time (min)	Tensile strength (MPa)	Elongation at break (%)	Tensile modulus (J/cm^3)
66.7/33.3/0	15	9 ± 1	28 ± 6	2.0 ± 0.1
56.7/28.3/15	15	14 ± 1	280 ± 6	46 ± 1
	15[1]	13 ± 1	52 ± 16	7 ± 3
	20	14 ± 4	209 ± 41	38 ± 7
50/50/0	15	11 ± 2	9 ± 1	1.0 ± 0.1
47.5/47.5/5	15	13 ± 2	93 ± 20	7 ± 3
45/45/10	15	14 ± 1	95 ± 30	11 ± 1
42.5/42.5/15	10	14 ± 1	103 ± 20	13 ± 4
	15	14 ± 1	152 ± 30	21 ± 3
	20	12 ± 1	144 ± 30	20 ± 4
33.3/66.7/0	15	9 ± 1	8 ± 2	0.4 ± 0.1
28.3/56.7/15	15	15 ± 1	19 ± 2	3 ± 0
	20	14 ± 2	185 ± 35	23 ± 7
	25	11 ± 1	11 ± 4	2.5 ± 1

[1] HDPE added to premixed PU/EMI-Zn

Table 14. Mechanical properties of HDPE/EVOH/Surlyn® blends

HDPE/EVOH/ Ionomer[a]	Tensile strength at yield (MPa)	Tensile strength at break (MPa)	Elongation at break (%)	Tensile modulus (J/cm^3)
66.7/33.3/0	-----	22 ± 1	7 ± 0	1
60/30/10	-----	17 ± 1	7 ± 2	1
53.3/26.7/20	20 ± 1	17 ± 1	78 ± 21	13
50/50/0	-----	29± 3	7 ± 0	1
45/45/10	29 ± 2	24 ± 2	37 ± 13	8
40/40/20	25 ± 3	26 ± 4	346 ± 36	78

[a] Polymers were conditioned at 60% relative humidity, and quenched at 0 °C

(52)Several blends using poly(ethylene-co-vinyl alcohol) (EVOH) have been pursued with several polymers. EVOH offers beneficial gas barrier properties, processability, optical clarity, and oil resistance [60, 61]. A largely detrimental property of EVOH is its absorption of water, which causes a reduction in its gas barrier properties. One such blend system that was attempted to overcome the water absorption issue was EVOH/HDPE [62]. It was believed that the HDPE portion of the blend would lead to a decrease in the water absorption ability compared to pure EVOH. In this study the ionomer compatibilizer was Zn^{2+} neutralized Surlyn®. The ratio of EVOH to HDPE was varied in order to determine the concentration of ionomer necessary to compatibilize the blend. It was found that at a 2/1 ratio of HDPE to EVOH that a larger concentration of ionomer was required to compatibilize the blend than when the ratio was 1/1. However, even at the 1/1 ratio the 20% ionomer had a dramatic effect on the tensile moduli and elongation values (table 14). Blending EVOH with ionomer was enhanced through treatment in a humid environment. It is believed that the water bridges the components to aid in the compatibilization of the two polymers, which is why the ternary blends were also treated in a humid environment.

(54)EVOH/polypropylene (PP) blends were compatibilized using Surlyn® neutralized using Na^+ [63]. The reasoning behind this polymer blend was for the same reasons listed for the EVOH/HDPE blends. The ratios of the PP to EVOH of 90/10 and 80/20 yielded the best ionomer compatilization, which was proven in previous work [64,65]. It was determined that increasing levels of ionomer did not lead to better tensile properties (table 15). In fact, increasing levels of ionomer had a detrimental effect, which may be due to the tensile strength and modulus of Surlyn®. PP/EVOH blends with ionomer displayed a better fracture parameter than those without ionomer. The EVOH/PP blends, both with and without ionomer, did increase the ductile-type of fracture, as opposed to PP that is not stable to crack propagation.

Table 15. Mechanical properties of PP/EVOH/Surlyn® blends

Polymer (PP/EVOH/ Surlyn®)	Young's modulus (GPa)	Yield strength (MPa)	Deformation at yield (%)	Break strength (MPa)	Deformation at break (%)
100/0/0	1.39 ± 0.11	24.73 ± 0.31	10.99± 0.46	21.74 ± 0.77	619.20 ± 4.70
90/10/0	1.60 ± 0.15	27.07 ± 0.55	8.34 ± 0.40	13.80 ± 1.83	113.80 ± 5.07
90/10/2	1.84 ± 0.29	27.19 ± 0.53	8.01 ± 0.45	10.63 ± 0.58	60.25 ± 13.90
90/10/5	1.62 ± 0.10	25.75 ± 0.82	8.29 ± 0.52	12.55 ± 1.99	92.58 ± 19.36
90/10/10	1.46 ± 0.11	25.71 ± 0.28	8.26 ± 0.18	9.62 ± 0.48	282.00 ± 33.77
80/20/0	1.85 ± 0.07	27.13 ± 0.55	5.48 ± 0.61	29.26 ± 0.68	8.50 ± 0.40
80/20/2	1.82 ± 0.09	27.10 ± 0.37	5.98 ± 0.19	29.17 ± 0.62	8.82 ± 0.17
80/20/5	2.08 ± 0.37	27.52 ± 0.50	6.54 ± 0.28	11.88 ± 1.87	34.25 ± 5.92
80/20/10	1.58 ± 0.19	27.88 ± 0.29	6.68 ± 0.59	12.92 ± 1.49	27.42 ± 5.63
0/100/0	5.03 ± 0.83	67.80 ± 0.77	5.09 ± 0.14	28.26 ± 2.02	20.69 ± 2.45
0/0/100	(30.33 ± 0.08) x 10^{-3}	11.66 ± 0.11	154.70 ± 0.86	11.50 ± 1.11	174.40 ± 5.95

(53)EVOH was blended with poly(ethylene-co-cyclohexane-1,4-dimethanol) (PETG) using two different versions of Surlyn® [66]. One of version of Surlyn® was neutralized using Zn^{2+} and the other using Na^+. As in the previous work a certain percent ionomer was required based upon the ratio of EVOH to PETG. It was found that the ionomer neutralized using Na^+ was able to compatibilize the EVOH/PETG blends at a lower concentration (5%) than the analogous ionomer neutralized using Zn^{2+} (15%). These findings were supported using tensile testing (table 16) and DMTA. It was also found that the ionomer neutralized using Na^{+} had better barrier properties than that neutralized using Zn^{2+}. The combination of the benefits of EVOH (oxygen and carbon dioxide gas barrier properties and processing ease) and PETG (toughness and clarity), along with the inherent properties of the ionomer lead to this high strength material that targets a specific application.

Table 16. Mechanical properties of PETG/EVOH/Surlyn® blends

PETG/EVOH/ Surlyn®	Tensile strength at yield (MPa)	Tensile strength at break (MPa)	Elongation at break (%)	Tensile modulus (J/cm^3)
66.7/33.3/0	-----	34 ± 4	7 ± 2	1 ± 0
63.3/31.7/5	37 ± 3	36 ± 3	305 ± 60	89 ± 7
60/30/10	28 ± 1	32 ± 3	324 ± 45	78 ± 9
56.7/28.3/15	34 ± 4	33 ± 3	205 ± 45	59 ± 5
53.3/26.7/20	27 ± 1	30 ± 2	265 ± 21	62 ± 4
67/33/0	-----	35 ± 3	8 ± 2	2 ± 1
63.3/31.7/5	34 ± 2	33 ± 3	311 ± 39	75 ± 12
60/30/10	32 ± 1	31 ± 4	331 ± 39	80 ± 11
56.7/28.3/15	27 ± 1	26 ± 4	287 ± 38	59 ± 10
53.3/26.7/20	26 ± 2	25 ± 4	289 ± 58	62 ± 11

Table 17. Mechanical properties of PET/nylon-6/Surlyn® blends

PET/nylon-6/Surlyn®	Tensile stress (MPa)	Elongation (%)	Tensile modulus (J/cm^3)
100/0/0	46 ± 2	587 ± 41	156 ± 11
0/100/0	36 ± 2	423 ± 31	135 ± 4
50/50/0	42 ± 4	396 ± 43	128 ± 14
48.5/48.5/3	26 ± 2	279 ± 27	65 ± 6
47.5/47.5/5	32 ± 4	357 ± 60	88 ± 3
45/45/10	33 ± 3	333 ± 48	86 ± 6
42.5/42.5/15	33 ± 3	420 ± 38	101 ± 5
37.5/37.5/25	34 ± 2	452 ± 22	99 ± 11
33.3/33.3/33.3	35 ± 5	481 ± 54	117 ± 9
66.7/33.3/0	37 ± 3	413 ± 31	108 ± 6
56.7/28.3/15	38 ± 3	486 ± 65	127 ± 9
33.3/66.7/0	47 ± 5	505 ± 53	149 ± 19
28.3/56.7/15	30 ± 2	347 ± 46	72 ± 7

(55)Two common and well known commercial thermoplastics are PET and nylon-6. Surlyn® neutralized using Zn^{2+} was used as the compatibilizing agent when blending these two polymers [67]. Table 17 shows the mechanical properties of the blends from this work. Aging studies were conducted on blends of the PET/nylon-6/ionomer in ratios of 50/50/5, 47.5/47.5/5, and 42.5/42.5/15. It was found that at the 5% ionomer level that the tensile properties decreased over a month (tables 18-20). However, when 15% ionomer was used the properties remained fairly constant. Also, the thermal properties and percent crystallinity did not change over a month. These results show that the ionomer can be used as a compatibilizing agent, and also as a stabilizer of the physical properties of the blends.

Table 18. Tensile stress aging studies of PET/nylon-6/Surlyn® blends

PET/nylon-6/Surlyn®	1 day	1 week	2 weeks	1 month
50/50/5	35 ± 1	36 ± 9	41 ± 1	40 ± 3
47.5/47.5/5	31 ± 3	22 ± 5	24 ± 2	20 ± 4
42.5/42.5/15	33 ± 3	31 ± 7	32 ± 2	33 ± 2

Table 19. % Elongation aging studies of PET/nylon-6/Surlyn® blends

PET/nylon-6/Surlyn®	1 day	1 week	2 weeks	1 month
50/50/5	396 ± 43	373 ± 34	327 ± 28	17 ± 3
47.5/47.5/5	323 ± 56	25 ± 9	25 ± 14	15 ± 7
42.5/42.5/15	420 ± 38	373 ± 31	402 ± 47	420 ± 30

Table 20. Tensile modulus aging studies of PET/nylon-6/Surlyn® blends

PET/nylon-6/Surlyn®	1 day	1 week	2 weeks	1 month
50/50/5	128 ± 14	103 ± 10	94 ± 8	3 ± 1
47.5/47.5/5	88 ± 3	2 ± 1	1 ± 0	1 ± 0
42.5/42.5/15	101 ± 5	82 ± 6	95 ± 7	118 ± 6

(56)Nylon-6 was blended with LDPE using Surlyn® neutralized using Na^+ [68]. Based on the successes of blending maleic anhydride-grafted polyethylene with nylon-6, this work sought to use Surlyn® instead of a modified polyethylene. Although no tensile data was conducted on these blends, the blends using the ionomer as the compatibilizing agent had a very significant effect using only a small concentration. For example, the size of the dispersed phase was decreased five-fold through addition of only 0.5 phr ionomer. Also, it was found that the Na^+ neutralized ionomer was more effective than the Zn^{2+} neutralized ionomer as a compatibilizer. This was based on the lower dispersed phase size and higher thermal stability of the resulting blend. The major drawback is that the Na^+ neutralized Surlyn® absorbs more water than the Zn^{2+} neutralized analog. There was no reported change in the crystallization rate of the ternary blend relative to pure nylon-6.

(57)Nylon-6/poly(ethylene-co-vinyl acetate) (EVAc) blends were compatibilized using Na^+ neutralized Surlyn® [69]. Marked improvements in the tensile (table 21) and impact (figure 13) properties were observed through the compatibilization of the blends using ionomer. At the ionomer levels tested, all exhibited about a 3 times improvement in the notched impact strength relative to the blend not containing ionomer. Analysis of the impact-fracture area indicated that the energy was dissipated effectively, which can be attributed to the good interfacial adhesion between the blend components.

Table 21. Mechanical properties of Nylon-6/EVAc/Surlyn® blends

Nylon-6/EVAc/Surlyn®	Tensile stress (MPa)	Elongation (%)	Tensile modulus (MPa)
80/20/0	38	264	820
80/19.6/0.4	42	350	862
80/19.2/0.8	40	365	848
80/18.8/1.2	42	382	830
80/18.4/1.6	40	325	825

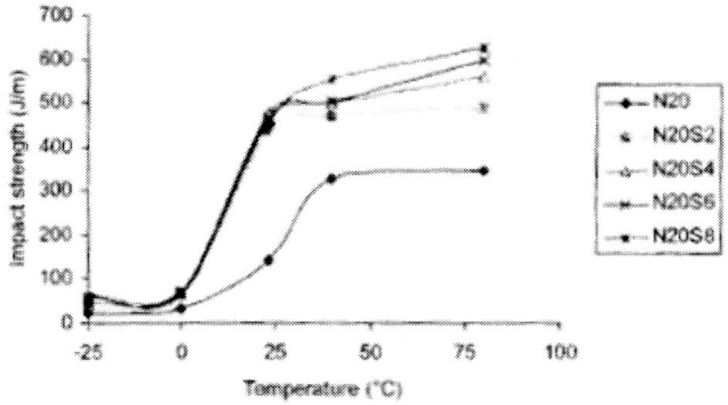

Figure 13. The effect of concentration of Surlyn® on the impact strength of blends; N20=20% nylon-6; S=Surlyn® concentration.

(58)Binary blends of poly(ethylene-co-acrylic acid) (PEA) and Zn^{2+} neutralized carboxylated nitrile rubber (XNBR) were prepared in an attempt to incorporate rubber and plastic in a single compatible blend (table 22) [70]. It was believed that the ionic segments would help compatibilize the typically difficult blending of these two polymers with high interfacial tension. Conversion of XNBR to the ionomeric form was accomplished through melt mixing with ZnO and stearic acid. Physical properties of the ionomeric blends were compared to that of the non-ionomeric blend (table 23). The optimum properties appeared to be in blends **2** and **3**. X-ray analysis determined that the percent crystallinity of these two blends (5% and 11%, respectively) were less than of the starting polymers XNBR and PEA (13% and 61%, respectively). The conclusion was that the improvement in properties was due to the compatible blend rather than an increase in crystallinity.

Table 22. XNBR/PEA blend compositions

Component	1	2	3	4	5	6	7	8
XNBR	100	90	80	70	60	50	0	80
PEA	0	10	20	30	40	50	100	20
ZnO	20	20	20	20	20	20	20	0
Stearic acid	1	1	1	1	1	1	1	1

Table 23. Physical properties of XNBR/PEA blends

Property	1	2	3	4	5	6	7	8
Modulus at 50% Elongation (MPa)	1.6	5.3	6.4	8.8	10.3	-----	-----	1.1
Modulus at 200% elongation (MPa)	2.9	10.9	10.8	11.3	-----	-----	-----	1.7
Tensile strength (MPa)	13.8	35.7	26.5	13.3	12.4	12.5	16.2	1.8
Elongation at break (%)	765	650	555	260	76	45	24	670
Tear strength (KN/m)	50.0	88.8	101.6	88.0	70.0	45.6	65.4	24.5
Shore A hardness	57	70	75	85	87	90	90	44
Tension at 100% elongation (%)	10	10	20	45	-----	-----	-----	50

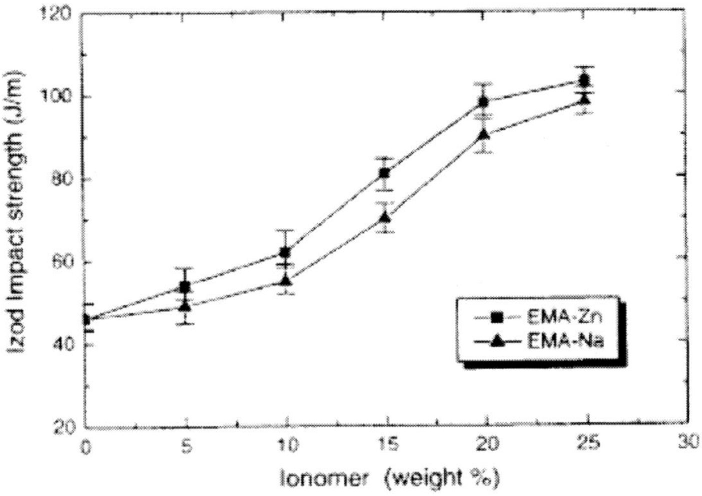

Figure 14. The effect of ionomer concentration on impact strength of POM/Surlyn® blends.

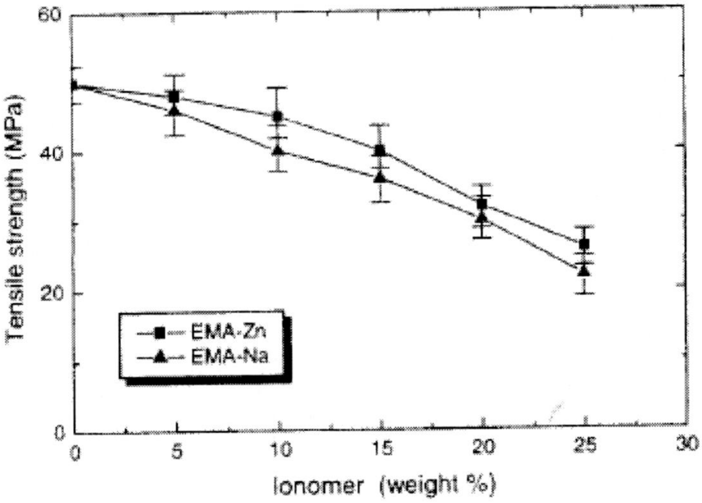

Figure 15. The effect of ionomer concentration on tensile strength of POM/Surlyn® blends.

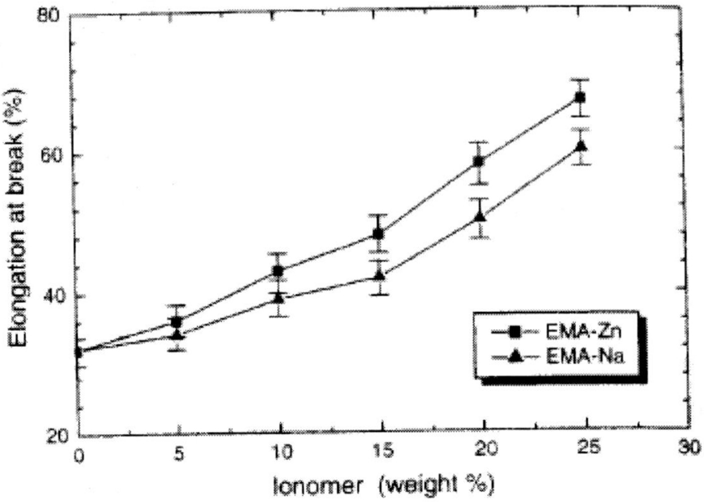

Figure 16. The effect of ionomer concentration on the elongation of POM/Surlyn® blends.

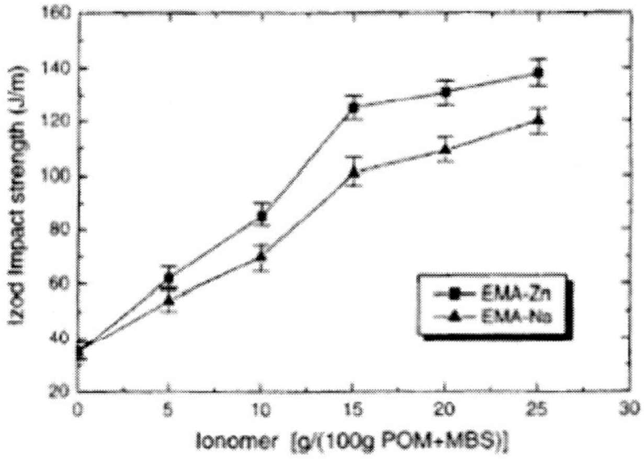

Figure 17. The effect of ionomer concentration on the impact strength of POM/MBS/Surlyn® blends (POM/MBS = 80/20).

(59)An example of tailoring a blend for an application involved the blending of poly(oxymethylene) (POM) with poly(methyl methacrylate-styrene-butadiene) (MBS) using Na^+ and Zn^{2+} neutralized Surlyn® [71]. The goals of this work were to reduce the crystallinity of the POM while maintaining the high creep resistance, fatigue resistance, heat resistance, and solvent resistance. In the binary blend of POM and Surlyn® it was found that the Zn^{2+} neutralized Surlyn® was the better compatibilizing agent, leading to better toughening (Figure 14-16). In the ternary blends the Zn^{2+} neutralized Surlyn® also led to better compatibility between the components than without Surlyn® (Figure 17-19). From a tensile strength standpoint the maximum was reached at 15% ionomer. The incorporation of MBS and ionomer also led to a decrease in the crystallinity of the blend relative to pure POM.

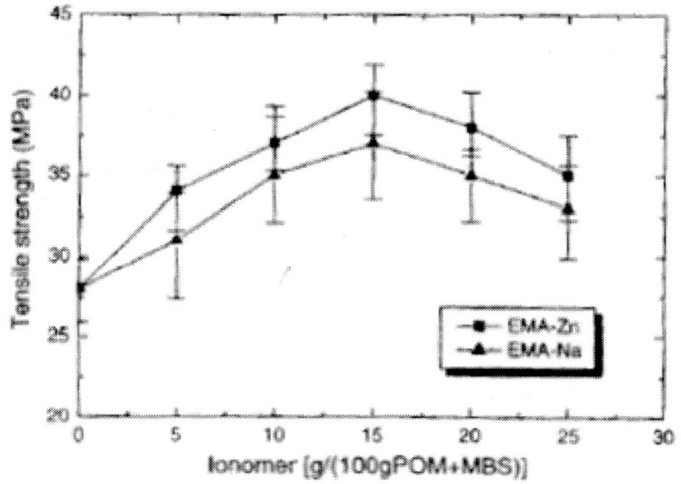

Figure 18. The effect of ionomer concentration on the tensile strength of POM/MBS/Surlyn® blends (POM/MBS = 80/20).

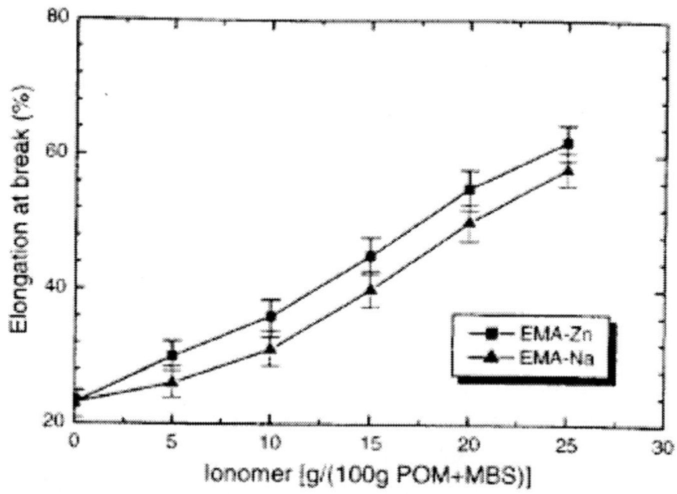

Figure 19. The effect of ionomer concentration on the elongation of POM/MBS/Surlyn® blends (POM/MBS = 80/20).

Commentary and Current and Future Directions of Ionomer Research in the Field of Structural Materials

Much of the current focus on ionomer research has been towards obtaining proton exchange membranes (PEM) for fuel cell applications. On the one hand this is good for the field of ionomer research. This potentially will lead to advancements in learning more about the fundamentals of structure-property relationships in ionomeric structures. Also, new ionomers may be formed which are not applicable for PEMs, but can be used for other applications. However, when these new ionomers are formed one must be careful to not give up on them due to their lack of success in the area of PEMs. Rather, there is a potential that it can be used in other applications.

There continues to be a lack of commercially available ionomers. Many ionomers exist that can be used in membrane applications, but cannot be used in dry environments due to their brittle nature when dry. There continues to be a large interest in ionomers in the field of dentistry. However, from a high volume industrial point of view there has not been a significant contribution in the field of ionomers in a very long time.

The use of ionomers in nanocomposites has shown a large degree of success. This can open the door to future ionomer research into making new ionomers rather than relying on the standard few commercially available ionomers. Earlier in this chapter there was mention of a few of polymers that were modified to make them ionic in order to incorporate MMT. In these cases the polymer was easily amenable to functionalization in order to achieve this result. Similar techniques can be used to alter other polymeric structures in order to achieve a high degree of exfoliation of filler particles.

There appears to be a lot of room for advancement in the field of nanocomposites using ionomers, including the incorporation of nanotubes. There has been significant progress over the past few years, but it appears that there is a lot of room to grow and advance. The large volume of research in basic ionomer structures should be helpful in the advancement of the use of ionomers as matrices for nanocomposites.

Blending two or more incompatible polymers to make them compatible has been an art form over the years. However, the use of ionomers to compatibilize these incompatible polymers appears to have large commercial importance. From an economical viewpoint the field of blending is very attractive compared to the development of new polymers. Commercially available polymers have established themselves, and their long term use is well-known. Taking advantage of these known properties and availability would allow for making several new potentially important polymer blends.

A potential exists in combining several aspects of the field of ionomers. For example, the use of ionomers as compatibilizing agents in blends may also be used in conjunction with incorporation of fillers. Hence, there could be an opportunity in making polymer blend nanocomposites. Blending targets specific applications based upon taking advantage of the benefits of the properties of the parent polymers in the blend. However, addition of fillers to these to make nanocomposites may be able to make stronger blends without losing any of the beneficial properties of the blend.

There is a large volume of ionomer research that is constantly being published and patented for the reader to consider. This chapter only touched on a few potentially commercially important systems from a structural polymer standpoint. Creativity and imagination are important in the use of ionomers as structural materials. Simple functionalization of existing polymers may lead to the next great breakthrough in the field. The reader is urged to look beyond what has been published and patented, and consider new possibilities. There is a large potential market for the use of ionomers in structural materials, and the field of ionomers still appears to be more of a niche field of study rather than a major focus.

References

[1] Eisenberg, A. & King, M. *Ion Containing Polymers: Physical Properties and Structure,* vol. 2. New York: Halsted-Wiley; 1977.

[2] Tant, M.R., Mauritz, K.A., & Wilkes, G.L. *Ionomers: Synthesis, Structure, Properties and Applications.* New York: Van Nostrand Reinhold Company Inc.; 1997.

[3] Eisenberg, A. & Kim J-S Introduction to Ionomers. New York: John Wiley & Sons, Inc.; 1998.

[4] MacKnight, W.J. & Earnest, T.R., Jr. (1981) *J. Polym.. Sci. Macromol. Rev.,***16**, 41-122.

[5] Eisenberg, A. & Bailey, F, Eds. (1986) Coulombic Interactions in Macromolecular Systems. *ACS Symposium Series* No. **302**. American Chemical Society, Washington, DC.

[6] Tant, M.R. & Wilkes, G.L. (1988) An Overview of the Viscous and Viscoelastic Behavior of Ionomers in Bulk and Solution. *J. Macromol. Sci. Part C: Polym. Rev.*, **28**, 1-63.

[7] Lantman, C.W., MacKnight, W.J., & Lundberg, R.D. (1989) Structural Properties of Ionomers. *Annu. Rev. Mater. Sci.,* **19**, 295-317.

[8] Fitzgerald, J.J. & Weiss, R.A. (1988) Synthesis, Properties, and Structure of Sulfonate Ionomers. *J. Macromol. Sci. Part C: Polym. Rev.,* **28**, 99-185.

[9] Tachino, H., Hara, H., Hirasawa, E., Kutsumizu, S., & Yano, S. (1994) Structure and properties of ethylene ionomers neutralized with binary metal cations. *Macromolecules,* **27**, 372-378.

[10] Kutsumizu, S., Hara, H., Tachino, H., Shimabayashi, K., & Yano, S. (1999) Infrared spectroscopic study of the binary blends of sodium and zinc salt ionomers produced from poly(ethylene-co-methacrylic acid). *Macromolecules,* **32**, 6340-6347.

[11] Jeon, H.S. & Kim, J-S (2003) Dynamic mechanical properties and morphology of sulfonated polystyrene ionomers neutralized with mixtures of various cations. *Polym. Bull.,* **49**, 457-464.

[12] Phillips, A.K. & Moore, R.B. (2006) Mechanical and transport property modifications of perfluorosulfonate ionomer membranes prepared with mixed organic and inorganic counterions. *J. Polym. Sci. Part B: Polym. Phys.,* **44**, 2267-2277.

[13] Nishio, M., Nichioka, T.T., & Koyama, K. (2005) Rheological properties of ethylene ionomer neutralized with binary metal cation. *Polymer,* **46**, 261-266.

[14] Tadano, K., Hirasawa, E., Yamamoto, H., & Yano, S. (1989) Order-disorder transition of ionic clusters in ionomers. *Macromolecules*, **22**, 226-233.

[15] Hirasawa, E., Yamamoto, Y., Tadano, K., & Yano, S. (1989) Formation of ionic crystallites and its effect on the modulus of ethylene ionomers. *Macromolecules,* **22**, 2776-2780.

[16] Hirasawa, E., Yamamoto, Y., Tadano, K., Yano, S. (1991) Effect of metal cation type on the structure and properties of ethylene ionomers. *J. Appl. Polym. Sci.,* **42**, 351-362.

[17] MacKnight, W.J., McKenna, L.W., & Read, B.E. (1967) Properties of ethylene-methacrylic acid copolymers and their sodium salts: Mechanical relaxations. *J. Appl. Phys.,* **38**, 4208-4212.

[18] McKenna, L.W., Kajiyama, T., & MacKnight, W.J. (1969) The γ relaxation of a poly(ethylene-co-methacrylic acid) polymer and its salts. *Macromolecules,* **2**, 58-61.

[19] Register, R.A. & Cooper, S.L. (1990) Anomalous small-angle x-ray scattering from nickel-neutralized ionomers. 2. Semicrystalline polymer matrices. *Macromolecules,* **23**, 318-323.

[20] Quiram, D.J., Register, R.A., & Ryan, A.J. (1998) Crystallization and ionic associations in semicrystalline ionomers. *Macromolecules,* **31**, 1432-1435.

[21] Sauer, B.B. & McLean, R.S. (2000) AFM and x-ray studies of crystal and ionic domain morphology in poly(ethylene-co-methacrylic acid) ionomers. *Macromolecules,* **33**, 7939-7949.

[22] Walters, R.M., Sohn, K.E., Winey, K.I., & Composto, R.J. (2002) Local acid environment in poly(ethylene-ran-methacrylic acid) ionomers. *J. Polym. Sci. Part B: Polym. Phys.,* **41**, 2833-2841.

[23] Lee, K.H. & Kim, B.K. (1996) Structure-property relationships of polyurethane anionomer acrylates. *Polymer,* **37**, 2251-2257.

[24] Al-Salah, H.A., Xiao, H.X., McLean, J.A., Jr., & Frisch, K.C. (1988) Polyurethane cationomers. I. Structure-properties relationships. *J. Polym. Sci., Part A: Polym. Chem.,* **26**, 1609-1620.

[25] Hsu, S.L., Xiao, H.X., Szmant, H.H., & Frisch, K.C. (1984) Polyurethane ionomers. I. Structure-properties relationships of polyurethane ionomers. *J. Appl. Polym. Sci*., **29**, 2467-2479.

[26] Santerre, J.P. & Brasch, J.L. (1997) Physical properties of nonionomeric and ionomeric segmented polyurethanes: Effect of sulfonated, carboxylate, and quaternary ammonium ions in the hard segment *Ind. Eng. Chem. Res*., **36**, 1352-1359.

[27] Lundberg, RD; Makowski, HS *Ions in Polymers*, ACS Adv. Chem. Ser. 187; Eisenberg, A Ed; American Chemical Society: Washington, DC, 1980, p. 21.

[28] Tierney, N.K., Trzaska, S.T., & Register, R.A. (2004) Matched random ionomers: Carboxylate vs sulfonate. *Macromolecules*, **37**, 10205-10207.

[29] Roth, H.H. *Water-dispersible molding compositions and articles prepared therefrom.* U.S. Patent 2,691,639, Oct. 12, 1954.

[30] Orler, E.B. & Moore, R.B. (1994) Influence of ionic interactions on the crystallization of lightly sulfonated syndiotactic polystyrene ionomers. *Macromolecules*, **27**, 4774-4780.

[31] Bellinger, M., Sauer, J.A., & Hara, M. (1994) Tensile fracture properties of sulfonated polystyrene ionomers. 1. Effect of ion content. *Macromolecules,* **27**, 1407-1412.

[32] Eisenberg, A; King, M. *Ion Containing Polymers*, New York, Halsted-Wiley, 1975.

[33] Hird, ,B. & Eisenberg, A. (1990) Phase dominance, multiple glass transitions, and apparent activation energies in some poly(styrene-co-alkali methacrylate) ionomers. *J. Polym. Sci., Polym. Sci. Polym. Phys. Ed.,* **28**, 1665-1675.

[34] Hara, M., Jar, P-Y, & Sauer, J.A. (1988) Fatigue behavior of ionomers. 1. Ion content effect on sulfonated polystyrene ionomers. *Macromolecules,* **21**, 3183-3186.

[35] Hara, M., Jar, P., & Sauer, J.A. (1990) Fatigue behavior of ionomers. 2. Effect of counterion on sulfonated polystyrene ionomers. *Macromolecules*, **23**, 4465-4469.

[36] Venkateshwaran, L.N., Tant, M.R., Wilkes, G.L., Charlier, P., & Jérôme, R. (1992) Structure-property comparison of sulfonated and carboxylated telechelic ionomers based on polyisoprene. *Macromolecules,* **25**, 3996-4001.

[37] Tant, M.R., Song, J.H., Wilkes, G.L., Horrion, J., & Jérôme, R. (1986) Mechanical properties of carboxylato-telechelic polyisoprene. *Polymer, 27*, 1815-1818.

[38] Register, R.A., Foucart, M., Jérôme, R., Ding, S., & Cooper, S.L. (1988) Structure-property relationships in elastomeric carboxy-telechelic polyisoprene ionomers neutralized with divalent cations. *Macromolecules,* **21**, 1009-1015.

[39] Guo, X-Y, Gu, L-X, & Feng, X-X (2002) The glass transition, crystallization, and melting characteristics of a class of polyester ionomers. *J. Appl. Polym. Sci.,*, **86**, 3660-3666.
[40] Xue, Y., Hara, M., & Yoon, H.N. (1998) Ionic naphthalene thermotropic copolyesters: Divalent salts and tensile mechanical properties. *Macromolecules,* **31**, 7806-7813.
[41] Eisenberg, A; King, M. *Ion Containing Polymers*, New York, Halsted-Wiley, 1975.
[42] Holliday, L. *Ionic Polymers. London: Applied Science*; 1975.
[43] Ma, X; Sauer, JA; Hara, M. (1995) Poly(methyl methacrylate) based ionomers. 1. Dynamic mechanical properties and morphology. *Macromolecules*, **28**, 3953-3962.
[44] Chisholm, B.J., Moore, R.B., Barber, G., Khouri, F., Hempstead, A., Larsen, M., Olson, E., Kelley, J., Balch, G., & Caraher, J. (2002) Nanocomposites derived from sulfonated poly(butylene terephthalate). *Macromolecules,* **35**, 5508-5516.
[45] Gorda, KR; Peiffer, D.G. (1992) Properties of sulfonated poly(butylene terephthalate). *J. Polym. Sci., Polym. Phys. Ed.,* **30**, 281-292.
[46] Li, Y., Ma, J., Wang, Y., & Liang, B. (2005) Synthesis and characterization of sulfonated poly(ethylene terephthalate)/montmorillonite nanocomposites. *J. Appl. Polym. Sci.,* **98**, 1150-1156.
[47] Jeong, E.H., Yang, J., Hong, J.H., Kim, T.G., Kim, J.H., & Youk, J.H. (2007) Effective preparation of montmorillonite/polyurethane nanocomposites by introducing cationic groups into the polyurethane main chain. *Eur. Polym. J,* **43**, 2286-2291.
[48] Shah, R.K., Krishnaswamy, R.K., Takahasi, S., & Paul, D.R. (2006) Blown films of nanocomposites prepared from low density polyethylene and a sodium ionomer of poly(ethylene-co-methacrylic acid). *Polymer,* **47**, 6187-6201.
[49] Shah, R.K., Kim, D.H., & Paul, D.R. (2007) Morphology and properties of nanocomposites formed from ethylene.methacrylic acid copolymers and organoclays. *Polymer,* **48**, 1047-1057.
[50] Yoo, Y., Shah, R.K., & Paul, D.R. (2007) Fracture behavior of nanocomposites based on poly(ethylene-co-methacrylic acid) ionomers. *Polymer,* **48**, 4867-4873.
[51] Shah, R.K. & Paul, D.R. (2006) Comparison of nanocomposites prepared from sodium, zinc, and lithium ionomers of ethylene/methacrylic acid copolymers. *Macromolecules,* **39**, 3327-3336.
[52] Parent, J.S., Liskova, A., & Resendes, R. (2004) Isobutylene-based ionomer composites: siliceous filler reinforcement. *Polymer,* **45**, 8091-8096.
[53] Lee, J.A., Kontopoulou, M., & Parent, J.S. (2005) Synthesis and characterization of polyethylene-based ionomer nanocomposites. *Polymer,* **46**, 5040-5049.
[54] Li, T. & Yan, N. (2007) Mechanical properties of wood flour/HDPE/ionomer composites. *Composites Part A,* **38**, 1-12.
[55] Vallejo, F.J., Eguiazábal, J.I., & Nazábal, J. (2000) Compatibilization of PP/Vectra B "in situ" composites by means of an ionomer. *Polymer,* **41**, 6311-6321.
[56] Paul, D.R. & Newman, S. *Polymer Blends*, Vol. 1. New York: Academic Press; 1979.
[57] Utracki, L.A. & Favis, B.D. *Polymer alloys and Blends.* In: Cheremisinoff, N.P., editor Encyclopedia of Engineering Materials. New Jersey: Marcel Dekker; 1988.
[58] Papadopoulou, C.P. & Kalfoglou, N.K. (1998) Polyurethane/olefinic ionomer blends. 1: Compatibility characterization. *Polymer,* **39**, 7015-7021.
[59] Papadopoulou, C.P. & Kalfoglou, N.K. (1999) Polyurethane/HDPE blends: 2. Compatibilization with an olefinic ionomer. *Polymer,* **40**, 905-911.

[60] Willis, J.M. & Favis, B.D. (1988) Processing-morphology relationships of compatibilized polyolefin/polyamide blends. Part I: The effect of an ionomer compatibilizer on blend morphology. *Polym. Eng. Sci.,* **28**, 1416-1426.

[61] Bazuin, C.G., Rancourt, L., Villeneuve, S., & Soldera, A. (1993) Miscibility of blends of poly(2,6-dimethyl 1,4-phenylene oxide) with polystyrene-based ionomers and copolymers. *J. Polym. Sci., Part B: Polym. Phys.,* **31**, 1431-1440.

[62] Kit, K.M., Schultz, J.M., & Gohil, R.M. (1995) Morphology and barrier properties of oriented blends of poly(ethylene terephthalate) and poly(ethylene 2,6-naphthalate) with poly(ethylene-co-vinyl alcohol). *Polym. Eng. Sci.,* **35**, 680-692.

[63] Ahn, T.O., Kim, C.K., Kim, B.K., Jeong, H.M., & Huh, J.D. (1990) Binary blends of nylons with ethylene vinyl alcohol copolymers: Morphological, thermal, reheological, and mechanical behavior. *Polym. Eng. Sci.,* **35**, 341-349.

[64] Samios, C.K. & Kalfoglou, N.K. (1998) Compatibilization of poly(ethylene-co-vinyl alcohol) (EVOH) and EVOH/HDPE blends with ionomers. Structure and properties. *Polymer,* **39**, 3863-3870.

[65] Montoya, M., Abad, M.J., Barral, L., & Bernal, C. (2006) Mechanical and fracture behavior of polypropylene/poly(ethylene-co-vinyl alcohol) blends compatibilized with ionomer Na^{+}. *Eur. Polym. J.,* **42**, 265-273.

[66] Abad, M.J., Ares, A., Barral, L., Cano, J., Díez, F.J., García-Garabal, S., López, J., & Ramírez, C. (2004) Use of a sodium ionomer as a compatibilizer in polypropylene/high-barrier ethylene-vinyl alcohol copolymer blends: The processability of the blends and their physical properties. *J. Appl. Polym. Sci.,* **94**, 1763-1770.

[67] Abad, M.J., Barral, L., & Eguiazabal, J.I. (2005) Influence of the ethylene-(methacrylic acid)-Zn^{2+} ionomer on the thermal and mechanical properties of blends of poly(propylene) (PP)/ethylene-(vinyl alcohol) copolymer (EVOH). *Polym. Int.,* **54**, 673-678.

[68] Samios, C.K. & Kalfoglou, N.K. (2001) Acrylic-modified polyolefin ionomers as compatibilizers for poly(ethylene-co-vinyl alcohol)/aromatic copolyester blends. *Polymer,* **42**, 3687-3696.

[69] Samios, C. & Kalfoglou, N.K. (1999) Compatibilization of poly(ethylene terephthalate)/polyamide-6 alloys: Mechanical, thermal, and morphological characterization. *Polymer,* **40**, 4811-4819

[70] Lahor, A., Nithitanakul, M., & Grady, B.P. (2004) Blends of low-density polyethylene with nylon compatibilized with a sodium-neutralized caboxylate ionomer. *Eur. Polym. J.,* **40**, 2409-2420.

[71] Bhattacharyya, A.R., Ghosh, A.K., & Misra, A. (2003) Ionomer compatibilized PA6/EVA blends: mechanical properties and morphological characterization. *Polymer,* **44**, 1725-1733.

[72] Antony, P., Bandyopadhyay, S., & De, S.K. (2000) Synergism is properties of ionomeric polyblends based on zinc salts of carboxylated nitrile rubber and poly(ethylene-co-acrylic acid). *Polymer,* **41**, 787-793.

[73] Wang, X. & Cui, X. (2005) Effect of ionomers on mechanical properties, morphology, and rheology of polyoxymethylene and its blends with methyl methacrylate-styrene-butadiene copolymer. *Eur. Polym. J.,* **41**, 871-880.

In: Strength of Materials
Editors: G. Mendes and B. Lago, pp. 97-131

ISBN 978-1-60741-500-8

Chapter 3

FAILURE OF LAYERED COMPOSITES SUBJECT TO IMPACTS: CONSTITUTIVE MODELING AND PARAMETER IDENTIFICATION ISSUES

Stefano Mariani*
Politecnico di Milano, Dipartimento di Ingegneria Strutturale
Piazza L. da Vinci 32, 20133 - Milano (Italy)

Abstract

Layered composites subject to impacts can fail by delamination, i.e. by debonding between laminae, if the stress waves cause damaging phenomena to take place mainly within the resin-enriched interlaminar phases. To simulate delamination at the structural level, processes dissipating energy are lumped onto fictitious zero-thickness interlaminar surfaces, and softening interface constitutive laws are adopted to describe the progressive failure of the interlaminar phases.

Since delamination occurs inside very narrow regions, results of experimental testing on whole composites need to be accurately and reliably filtered to calibrate the interface constitutive laws. To this aim, here we propose a sigma-point Kalman filter approach. The performances of the proposed methodology, in terms of constitutive parameter estimations and dynamic delamination tracking, are assessed through pseudo-experimental testings on a two-layer composite, and real testings on multi-layer glass fiber reinforced plastic composites.

Keywords: composites, delamination, impact loading, interface constitutive modeling, parameter identification, sigma-point Kalman filter.

1. Introduction

Foreign objects striking the outer surface of composite structures may cause permanent damage, or even sudden failure [1]. On the basis of the velocity of the striker, impacts can be roughly distinguished into two main classes: low-velocity impacts, characterized by small or negligible effects of inertial forces on the damage/failure mode; high-velocity

*E-mail address: stefano.mariani@polimi.it

impacts, whereby inertial forces strongly affects the damage/failure event. In case of a high-velocity impact, the eventual failure of layered composites can be due to the propagation of interlaminar cracks only (delamination), or to a local perforation accompanied also by intralaminar damage [2].

In this work we focus on impact-induced delamination and provide a review of a classical approach to debonding in layered bodies [3–8]. Within this approach it is assumed that laminae always behave elastically (i.e. intralaminar damaging phenomena are disregarded, or thought to be negligible), and interlaminar resin-enriched phases are lumped onto zero-thickness surfaces, along which debonding can occur because of the impinging shock waves. To model delamination, softening interface constitutive laws are adopted to link the tractions acting upon each interlaminar surface with the displacement jumps occurring across it. If these laws are able to phenomenologically describe the micromechanical processes leading to debonding, the above approach furnishes accurate results at the structural level, see e.g. [4]. However, calibration of the interface laws is still an open issue: since damaging and cracking phenomena linked to delamination take place inside very narrow regions, direct testing on the interlaminar phases can not be devised. Instead, inverse analysis procedures can be adopted to efficiently manage experimental data in order to estimate uncertain constitutive parameters [6, 9–11]; needles to say, the aforementioned experimental data have to be informative, i.e. they have to carry information on the current response of interphases to the impact loading.

As for model calibration purposes, standard filtering procedures have been proven accurate enough in case of static loadings [12, 13]; in case of impacts the inverse problem becomes stiff since composite failure, once incepted, usually occurs almost instantaneously [2, 14]. To deal with this issue, we recently adopted the extended Kalman filter (EKF) [11, 15]; when compared to alternative approaches (like, e.g., neural networks and least squares), the EKF has the great advantages of being able to work in real-time and of being explicitly linked to the physics of the ongoing delamination processes. Moreover, the EKF exploits the evolution in time of the measured fields in order to continuously enhance model calibration.

As pointed out by several authors (see, e.g., [16, 17]), the EKF looses stability when nonlinearities become dominant in the equations governing the inverse problem. This is due to the fact that the EKF replaces nonlinear functions with their relevant tangent surfaces, leading to possibly biased or even divergent parameter estimates. An alternative approach to treat nonlinearities in a stochastic framework recently led to the proposal of the so-called sigma-point Kalman filter (SPKF), also termed unscented Kalman filter [16]. Instead of linearizing the governing equations, thereby introducing approximations, the SPKF samples the statistics of the current state of the system and of model parameters to draw a set of sigma-points. These sigma-points are then let to evolve according to the actual, nonlinear dynamics of the problem. The filter estimates are eventually computed by averaging the information conveyed by all the evolved sigma-points through an ad-hoc numerical scheme [16, 18]. When compared to the EKF, the SPKF can achieve a much higher accuracy in the model calibration task [19], often preventing the aforementioned bias and divergence occurrences.

If used to deal with the nonlinearities accompanying delamination in layered composites, the SPKF has to furnish accurate estimates of uncertain interface model parameters

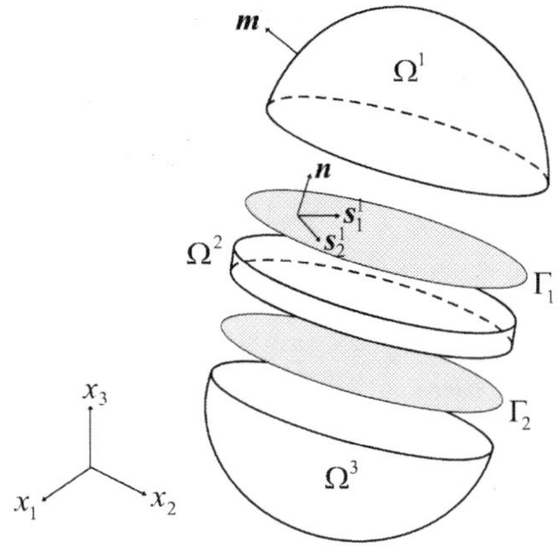

Figure 1. geometry of a three-dimensional layered continuum (here $n_\Gamma = 2$), and notation.

while tracking the state of the laminate. In this Chapter we analyze this framework in details, trying to point out the main strengths of the SPKF. To this purpose, in Section 2. we present the equations governing the dynamics of a layered, possibly delaminating body. Constitutive models for laminae and resin-enriched interphases are then discussed so as to recognize the basic parameters, in need of an accurate estimation. An explicit time integration scheme for the equations of motion of the composite, and an explanation on how to handle model calibration via the SPKF are eventually offered. Section 3. deals with sigma-point Kalman filtering: physical arguments are used to propose a slightly new scheme for the drawing of the sigma-points. In Section 4. the performances of the SPKF are assessed: pseudo-experimental, i.e. fictitious impact testings on a two-layer composite are first considered; hence, real experiments on multi-layer glass fiber reinforced plastic (GRP) composites are used to testify the robustness of the SPKF in promptly detecting delamination.

Throughout the whole Chapter, a matrix notation is adopted, with uppercase and lowercase bold symbols respectively denoting matrices and vectors; a superscript T stands for transpose, while a superposed dot represents time rates.

2. Dynamics of Layered Composites

2.1. Governing Relations

Let Ω be a three-dimensional, layered body; its smooth outer boundary, with unit outward normal $\boldsymbol{m}$, be constituted by the two disjoint parts Γ_u and Γ_τ, where displacement and traction fields are respectively assigned. Ω is assumed to be crossed by n_Γ non-intersecting surfaces, or interfaces Γ_j, $j = 1, ..., n_\Gamma$, see Figure 1. Each resulting portion Ω^j, $j = 1, ..., n_\Gamma + 1$, of the bulk will be here referred to as layer, or lamina. Interfaces in laminates are usually flat and parallel to each other, with a common orientation defined by the unit vector $\boldsymbol{n}$; opening (mode I) and sliding (mode II and mode III) displacement discontinuities along each Γ_j take place along the $\boldsymbol{n}$ direction and in the $\boldsymbol{s}_1^j - \boldsymbol{s}_2^j$ plane, respectively.

The equilibrium of Ω at time t is governed by the following equations:

$$\boldsymbol{\mathcal{C}}^{\mathrm{T}}\boldsymbol{\sigma} + \bar{\boldsymbol{b}} = \varrho\ddot{\boldsymbol{u}} \qquad \text{in } \Omega\backslash\Gamma \tag{1}$$

$$\boldsymbol{\mathcal{M}}\boldsymbol{\sigma} = \bar{\boldsymbol{\tau}} \qquad \text{on } \Gamma_\tau \tag{2}$$

$$\begin{cases} \boldsymbol{\mathcal{N}}\boldsymbol{\sigma} = -\boldsymbol{\tau} & \text{on } \Gamma_j^+ \\ \boldsymbol{\mathcal{N}}\boldsymbol{\sigma} = \boldsymbol{\tau} & \text{on } \Gamma_j^- \end{cases} \qquad j = 1, ..., n_\Gamma \tag{3}$$

where: $\Omega = \cup_{j=1}^{n_\Gamma+1}\Omega^j$ and $\Gamma = \cup_{j=1}^{n_\Gamma}\Gamma_j$; Γ_j^+ and Γ_j^- are the sides of Γ_j respectively belonging to layers Ω^j and Ω^{j+1}, according to the notation of Figure 1. The two sides can not be distinguished in the initial configuration (at $t = 0$) but, as soon as delamination is incepted, they can experience a relative movement. In the equations above, adopting a standard Voigt notation: $\boldsymbol{\sigma}$ is the stress vector, which gathers the independent components of the stress tensor; $\bar{\boldsymbol{b}}$ and $\bar{\boldsymbol{\tau}}$ are the assigned loads in the bulk $\Omega\backslash\Gamma$ and along Γ_τ, respectively; ϱ is the bulk mass density; $\ddot{\boldsymbol{u}}$ is the acceleration field in the bulk $\Omega\backslash\Gamma$; $\boldsymbol{\mathcal{C}}$ is the differential compatibility operator; $\boldsymbol{\mathcal{M}}$ and $\boldsymbol{\mathcal{N}}$ are the matrices respectively collecting the components of the unit vectors $\boldsymbol{m}$ and $\boldsymbol{n}$. To locally ensure equilibrium along each surface, sides Γ_j^+ and Γ_j^- are respectively acted upon by the traction vectors $-\boldsymbol{\tau}$ and $\boldsymbol{\tau}$.

Since we aim at modeling phenomena occurring within very short time intervals after the impact, linearized kinematics proves sufficient. Compatibility thus reads:

$$\boldsymbol{\varepsilon} = \boldsymbol{\mathcal{C}}\boldsymbol{u} \qquad \text{in } \Omega\backslash\Gamma \tag{4}$$

$$\boldsymbol{u} = \bar{\boldsymbol{u}} \qquad \text{on } \Gamma_u \tag{5}$$

where: $\boldsymbol{\varepsilon}$ is the strain vector, which gathers the independent components of the strain tensor; $\boldsymbol{u}$ is the displacement field in the bulk $\Omega\backslash\Gamma$; $\bar{\boldsymbol{u}}$ is the assigned displacement along Γ_u. Across each interface, the displacement discontinuity $[\boldsymbol{u}]$ is defined as:

$$[\boldsymbol{u}] = \boldsymbol{u}\Big|_{\Gamma_j^+} - \boldsymbol{u}\Big|_{\Gamma_j^-} \qquad j = 1, ..., n_\Gamma \tag{6}$$

The body is assumed to be initially at rest, in an undeformed and unstressed state, such that:

$$\begin{cases} \boldsymbol{u}_0 = \boldsymbol{0} \\ \dot{\boldsymbol{u}}_0 = \boldsymbol{0} \end{cases} \qquad \text{in } \Omega \tag{7}$$

2.2. Constitutive Modeling

Energy dissipation in layered composites subject to impacts arises from: the spreading of damage and the subsequent propagation of cracks within the interlaminar, resin-enriched phases; the spreading of damage inside laminae. This latter dissipation mechanism can be disregarded, as done here, if the impact energy is not high enough to cause a penetration of the impactor inside the laminate. In such a case, laminae can be assumed to behave elastically, according to:

$$\boldsymbol{\sigma} = \boldsymbol{E}_{\Omega}\boldsymbol{\varepsilon} \qquad \text{in } \Omega \backslash \Gamma \tag{8}$$

where $\boldsymbol{E}_{\Omega}$ is the elasticity matrix of the bulk. Each lamina is usually an orthotropic body; though not explicitly shown in (8), $\boldsymbol{E}_{\Omega}$ can change from lamina to lamina because of a different orientation of the axes of elastic symmetry.

To simulate strength reduction in the interlaminar phases, eventually leading to debonding, softening interface constitutive laws prove efficient [3–8, 20, 21]. A detailed, micromechanics-based representation of the damaging processes caused by the constrained deformation field inside each interphase is not looked for; instead, a phenomenological relationship is adopted to link the tractions $\boldsymbol{\tau}$ acting upon the interface sides Γ_j^+ and Γ_j^- to the displacement jump $[\boldsymbol{u}]$ occurring across Γ_j. This constitutive law might be conceived as the macroscopic (homogenized through the interphase thickness) behavior of the interphase. Since the ratio between the thickness of each interlaminar phase and the thickness of the whole laminate is usually quite small, this approach can furnish accurate results at the laminate length scale; on the contrary, an accurate representation of the micromechanical phenomena preceding delamination is in need of a multi-scale approach (see, e.g., [22–24]).

Along each Γ_j, both opening/closing (along $\boldsymbol{n}$) and sliding (in the $\boldsymbol{s}_1^j - \boldsymbol{s}_2^j$ plane) discontinuities take place under general loading conditions. By way of a simplified scheme adopted in [25–28], the local kinematics of an interface is governed by the effective displacement discontinuity $[u]$, defined as:

$$[u] = \sqrt{[u]_n^2 + \kappa^2 [u]_s^2} \tag{9}$$

where: $[u]_n = [\boldsymbol{u}]^{\mathrm{T}}\boldsymbol{n}$ and $[u]_s = |[\boldsymbol{u}] - [u]_n\boldsymbol{n}|$ are, respectively, the opening and sliding displacement discontinuities; κ is a coupling coefficient, which accounts for the interaction between stretching and shearing deformation modes inside the interphase. Through an incremental work equivalence the effective traction τ, work-conjugate to $[u]$, turns out to be (see [25, 27]):

$$\tau = \sqrt{\tau_n^2 + \frac{\tau_s^2}{\kappa^2}} \tag{10}$$

where τ_n and τ_s are defined like $[u]_n$ and $[u]_s$. The mechanical behavior of the interface can now be furnished in terms of an effective $\tau - [u]$ relationship.

For quasi-brittle materials, like interphase resins, the tensile strength is typically much smaller than the compressive one. Therefore, when dynamic delamination occurs without intralaminar damage, the response of the interface under compressive stress states along $\boldsymbol{n}$ can be assumed linear elastic, according to:

$$\tau = K[u] \qquad \text{if } [u] < 0 \tag{11}$$

where K is the stiffness of the interface. On the other hand, the response of the interface under tensile stress states along $\boldsymbol{n}$ is characterized by strength reduction, leading to softening, beyond the attainment of a peak traction. A simple way to model the transition from the

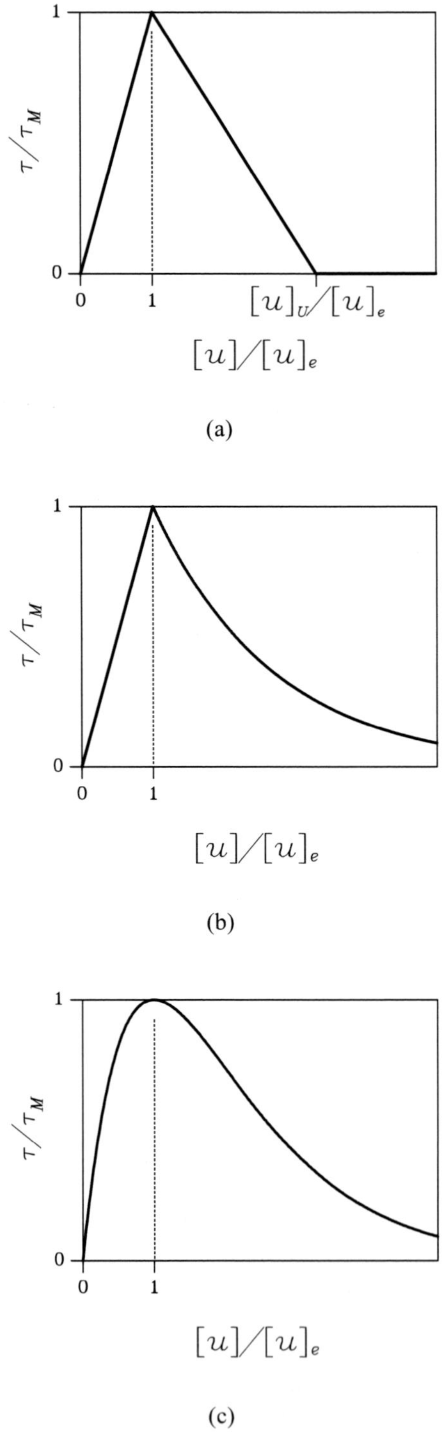

Figure 2. Effective traction-displacement discontinuity relationships under tensile loading. (a) piecewise linear law (12); (b) linear-exponential law (13); (c) exponential law (14).

initial elastic regime to the subsequent softening one, is through a piecewise linear (PWL) law (see Figure 2(a)):

$$\begin{cases} \tau = K[u] & \text{if } [u] \le [u]_e \\ \tau = \tau_M + Q\left([u] - [u]_e\right) & \text{if } [u]_e < [u] \le [u]_U \\ \tau = 0 & \text{if } [u] > [u]_U \end{cases} \tag{12}$$

where: $[u]_e$ is the effective displacement discontinuity corresponding to the peak traction $\tau_M = K[u]_e$; Q is the (negative) slope of the softening branch; $[u]_U = \left(1 - \frac{K}{Q}\right)[u]_e$ is the effective displacement discontinuity at which the interaction between the crack faces ceases.

The linear softening in (12) can be a too crude approximation of the post-peak regime for some materials, whose response features an initial steep descent followed by a long, much less steep tail (see, e.g., [29]). In such a case, the softening regime can be modeled through an exponential law (hereafter referred to as linear-exponential, L-E law because of the pre-peak linear elastic phase) according to (see Figure 2(b)):

$$\begin{cases} \tau = K[u] & \text{if } [u] \le [u]_e \\ \tau = \tau_M \exp\left(-\varsigma([u] - [u]_e)\right) & \text{if } [u] > [u]_e \end{cases} \tag{13}$$

where ς is a model parameter that allows to match the slope of the softening branch just beyond the attainment of the peak traction τ_M.

Sometimes, a smooth transition from the elastic regime to the softening one turns out to be more representative of the actual interphase response. The nonlinear binding model, originally proposed in [30, 31] for metals and bimetallic compounds and later adopted also in nonlinear fracture mechanics [20, 21, 32], allows to describe such smooth transition through the following exponential (EXP) law (see Figure 2(c)):

$$\tau = K[u]\exp\left(-\frac{[u]}{[u]_e}\right) \tag{14}$$

Besides the effective stiffness K and strength τ_M, a full characterization of the nonlinear behavior of the interface has to match the fracture energy, or work of separation G. In terms of effective quantities, G is defined as the amount of energy required to annihilate the interaction between the opening/sliding crack faces, i.e.:

$$G = \int_0^\infty \tau \, d[u] \tag{15}$$

From a model calibration perspective, parameters Q and ς in (12) and (13) can be tuned to accurately match the actual G, since they do affect only the softening branch of the interface law. On the other hand, after having assigned K in (14), only $[u]_e$ can be adjusted: therefore, both τ_M and G can not be accurately matched. In the exponential law, in fact, the following constraint holds:

$$\sqrt{\frac{KG}{\tau_M^2}} = e \tag{16}$$

e being the Nepero number. To avoid problems related to this fictitious constraint, a modified exponential law is here formulated as follows (see Figure 3):

$$\tau = K[u] \exp\left(-\left(\frac{[u]}{[u]_e}\right)^q\right) \tag{17}$$

where q shows up as an additional constitutive parameter. In law (17): K still represents the initial elastic stiffness; $[u]_e$ is a reference effective displacement discontinuity, while peak traction τ_M is attained at $[\bar{u}] \equiv (1/q)^{1/q}\,[u]_e$. The effective peak traction and fracture energy are affected by q, according to:

$$\begin{aligned} \tau_M =& K[u]_e \left(\frac{1}{q}\right)^{1/q} \exp\left(-\frac{1}{q}\right) \\ G =& \frac{K}{q}[u]_e^2\, \gamma_f\left(\frac{2}{q}\right) \end{aligned} \tag{18}$$

γ_f being the gamma function. The dependence of τ_M and G on the parameter q is depicted in Figure 4: it can be seen that G is a monotonically decreasing function of q, whereas τ_M is lower-bounded by the value corresponding to $q = 1$. Having tuned K, this law thus allows the calibration of both τ_M and G.

All the above laws but the piecewise linear one, assume that the interaction between the opening interface sides continues up to $[u] \rightarrow \infty$, which seems not physical at the macroscale. To simulate delamination growth a breakdown threshold therefore needs to be introduced [33, 34]: as soon as the current traction τ reduces to a small fraction (say 5-10 %) of the peak value τ_M, the interaction is suddenly assumed to vanish.

When unloading from the tensile envelope occurs, i.e. when $[\dot{u}] < 0$, the above interface models can be viewed as either reversible, if τ always belongs to the envelope (leading to interface healing if softening has already started), or irreversible, if τ decreases following a radial path to the origin of the $\tau - [u]$ plane. These two alternative constitutive assumptions lead to different entries in the interface tangent stiffness matrix $\boldsymbol{E}_\Gamma$, linking rates of $\boldsymbol{\tau}$ and $[\boldsymbol{u}]$ in the local $\boldsymbol{n} - \boldsymbol{s}_1^j - \boldsymbol{s}_2^j$ reference frame according to:

$$\dot{\boldsymbol{\tau}} = \boldsymbol{E}_\Gamma[\dot{\boldsymbol{u}}] \tag{19}$$

For additional details, readers are referred to [27, 28].

2.3. Finite Element Formulation

The weak form of the equilibrium equations (1)-(3) reads:

$$\int_{\Omega\backslash\Gamma} \boldsymbol{\varepsilon_v}^{\mathrm{T}} \boldsymbol{\sigma}\, d\Omega = \int_{\Omega\backslash\Gamma} \boldsymbol{v}^{\mathrm{T}}(\bar{\boldsymbol{b}} - \varrho\ddot{\boldsymbol{u}})\, d\Omega + \int_{\Gamma_\tau} \boldsymbol{v}^{\mathrm{T}}\bar{\boldsymbol{\tau}}\, d\Gamma_\tau - \sum_{j=1}^{n_\Gamma}\int_{\Gamma_j} [\boldsymbol{v}]^{\mathrm{T}}\boldsymbol{\tau}\, d\Gamma_j \qquad \forall \boldsymbol{v} \in \mathcal{U}_0 \tag{20}$$

where: $\boldsymbol{v}$ is the test function; $\boldsymbol{\varepsilon_v} = \boldsymbol{C}\boldsymbol{v}$; $\mathcal{U}$ is the trial solution space, collecting displacement fields $\boldsymbol{u}$ continuous in $\Omega\backslash\Gamma$, possibly discontinuous along each Γ_j and fulfilling the boundary condition (5) on Γ_u; $\mathcal{U}_0$ is the relevant variation space, with zero prescribed

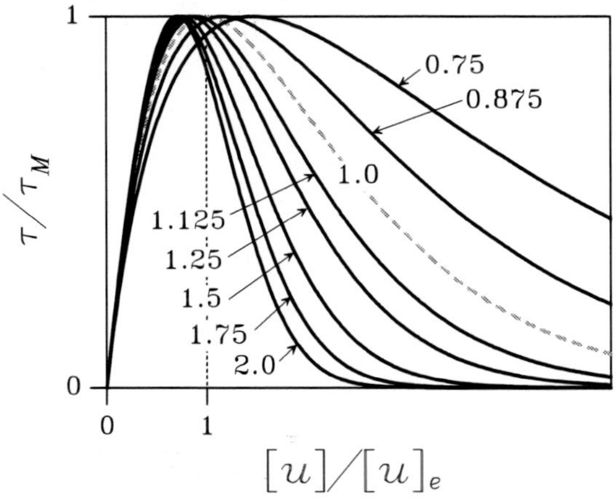

Figure 3. effect of q on the modified exponential traction-displacement discontinuity law (17).

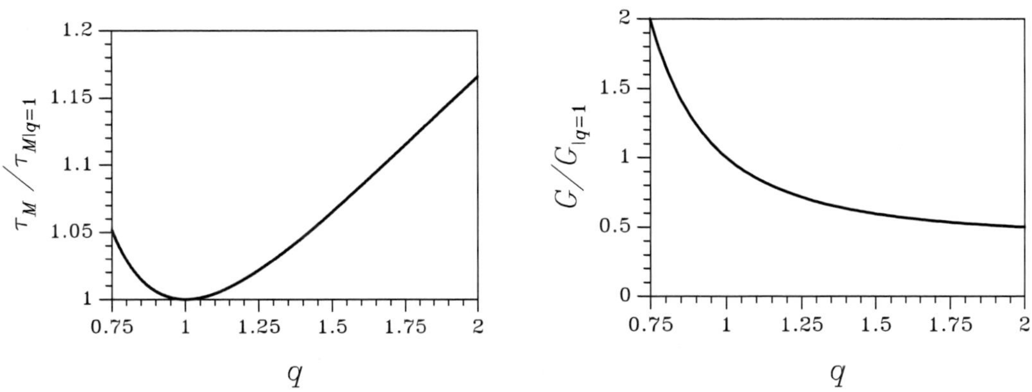

(a) (b)

Figure 4. modified exponential law (17). Effects of q on (a) the effective peak traction τ_M and on (b) the effective fracture energy G.

displacements on Γ_u. In (20), in view of the assumed linearized kinematics, the relation $\Gamma_j \equiv \Gamma_j^+ \equiv \Gamma_j^-$ for each interface has been exploited.

Allowing for the elastic bulk constitutive law (8), the following variational statement is arrived at:

$$\text{find } \boldsymbol{u} \in \mathcal{U} : \int_{\Omega\backslash\Gamma} \varrho\, \boldsymbol{v}^{\mathrm{T}} \ddot{\boldsymbol{u}}\, d\Omega + \int_{\Omega\backslash\Gamma} \boldsymbol{\varepsilon_v}^{\mathrm{T}} \boldsymbol{E}_{\Omega} \boldsymbol{\varepsilon}\, d\Omega + \sum_{j=1}^{n_\Gamma} \int_{\Gamma_j} [\boldsymbol{v}]^{\mathrm{T}} \boldsymbol{\tau}\, d\Gamma_j$$
$$= \int_{\Omega\backslash\Gamma} \boldsymbol{v}^{\mathrm{T}} \bar{\boldsymbol{b}}\, d\Omega + \int_{\Gamma_\tau} \boldsymbol{v}^{\mathrm{T}} \bar{\boldsymbol{\tau}}\, d\Gamma_\tau \quad \forall \boldsymbol{v} \in \mathcal{U}_0 \tag{21}$$

Now, let the finite element approximation of the displacement and deformation fields in $\Omega\backslash\Gamma$ be (see [35] for the notation):

$$\boldsymbol{u} \cong \boldsymbol{\Phi}\boldsymbol{u}^h \tag{22}$$

$$\boldsymbol{\varepsilon} \cong \boldsymbol{C}\boldsymbol{\Phi}\boldsymbol{u}^h = \boldsymbol{B}_\Omega\boldsymbol{u}^h \tag{23}$$

where matrix $\boldsymbol{\Phi}$ gathers the nodal shape functions, and vector $\boldsymbol{u}^h$ collects the nodal displacements.

If delamination is allowed to occur only along element boundaries, the discrete displacement jump field can be written:

$$[\boldsymbol{u}]\Big|_{\Gamma_j} \cong \boldsymbol{B}_{\Gamma_j}\boldsymbol{u}^h \qquad j = 1, ..., n_\Gamma \tag{24}$$

Owing to the discrete interpolation fields defined above, the semi-discretized equations of motion of the composite turn out to be:

$$\boldsymbol{M}\ddot{\boldsymbol{u}}^h + \boldsymbol{K}_\Omega\boldsymbol{u}^h + \sum_{j=1}^{n_\Gamma}\boldsymbol{R}^j = \boldsymbol{F} \tag{25}$$

where the mass matrix $\boldsymbol{M}$, the bulk stiffness matrix $\boldsymbol{K}_\Omega$, the internal force vectors $\boldsymbol{R}^j$ and the external load vector $\boldsymbol{F}$ are, respectively:

$$\begin{aligned}
\boldsymbol{M} &= \int_{\Omega\backslash\Gamma} \varrho\,\boldsymbol{\Phi}^{\mathrm{T}}\boldsymbol{\Phi}\,d\Omega \\
\boldsymbol{K}_\Omega &= \int_{\Omega\backslash\Gamma} \boldsymbol{B}_\Omega^{\mathrm{T}}\boldsymbol{E}_\Omega\boldsymbol{B}_\Omega\,d\Omega \\
\boldsymbol{R}^j &= \int_{\Gamma_j} \boldsymbol{B}_{\Gamma_j}^{\mathrm{T}}\boldsymbol{\tau}\,d\Gamma_j \\
\boldsymbol{F} &= \int_{\Omega\backslash\Gamma} \boldsymbol{\Phi}^{\mathrm{T}}\bar{\boldsymbol{b}}\,d\Omega + \int_{\Gamma_\tau} \boldsymbol{\Phi}^{\mathrm{T}}\bar{\boldsymbol{\tau}}\,d\Gamma_\tau
\end{aligned} \tag{26}$$

Smarter finite element formulations, like the extended or generalized ones [36,37], have been recently formulated to simulate mixed-mode crack growth in homogeneous solids, see e.g. [27,38,39]. These methodologies allow cracks to propagate not only along inter-element edges, but also inside the elements; possible constraints imposed by the mesh layout on crack trajectories, evidenced e.g. in [40], can be therefore alleviated. When dealing with delamination in layered continua, where debonding occurs only along the a-priori known interlaminar surfaces, crack description looks simple and the aforementioned feature of the extended finite element method looses much of its advantages.

2.4. Time Integration

In our previous works [11, 15] it was shown that the time integration scheme can strongly affect the stability of the filtering procedure and, therefore, the accuracy of model parameter estimates.

In case of impact loadings, which cause the propagation of shock waves inside the composite, the time marching algorithm has to dump the spurious high frequency oscillations linked to space discretization. Otherwise, the numerically computed displacement and velocity fields do not prove reliable enough to be compared to the experimental data.

We adopt here the explicit $\alpha-$method [41, 42] to advance in time the solution of the equations of motion (25), see also [19]. After having partitioned the time interval of interest according to $[t_0 \quad t_N] = \cup_{i=0}^{N_t-1}[t_i \quad t_{i+1}]$, at the end of the generic time step $[t_i \quad t_{i+1}]$ the solution to (25) is obtained according to the following predictor-integrator-corrector splitting:

- predictor:

$$\tilde{\boldsymbol{u}}_{i+1} = \boldsymbol{u}_i + \Delta t\, \dot{\boldsymbol{u}}_i + \Delta t^2 (\frac{1}{2} - \beta) \ddot{\boldsymbol{u}}_i \tag{27}$$

$$\tilde{\dot{\boldsymbol{u}}}_{i+1} = \dot{\boldsymbol{u}}_i + \Delta t (1 - \gamma) \ddot{\boldsymbol{u}}_i \tag{28}$$

where $\Delta t = t_{i+1} - t_i$;

- explicit integrator:

$$\ddot{\boldsymbol{u}}_{i+1} = \boldsymbol{M}^{-1} \left(\boldsymbol{F}_{i+1+\alpha} - (1+\alpha) \left(\boldsymbol{K} \tilde{\boldsymbol{u}}_{i+1} + \sum_j \tilde{\boldsymbol{R}}^j_{i+1} \right) + \alpha \left(\boldsymbol{K} \boldsymbol{u}_i + \sum_j \boldsymbol{R}^j_i \right) \right) \tag{29}$$

where: $\boldsymbol{F}_{i+1+\alpha} = \boldsymbol{F}(t_i + (1+\alpha)\Delta t)$; $\boldsymbol{R}^j_i = \int_{\Gamma_j} \boldsymbol{B}^{\mathrm{T}}_{\Gamma_j} \boldsymbol{\tau}_i \, d\Gamma_j$ and $\tilde{\boldsymbol{R}}^j_{i+1} = \int_{\Gamma_j} \boldsymbol{B}^{\mathrm{T}}_{\Gamma_j} \tilde{\boldsymbol{\tau}}_{i+1} \, d\Gamma_j$;

- corrector:

$$\boldsymbol{u}_{i+1} = \tilde{\boldsymbol{u}}_{i+1} + \Delta t^2 \beta \, \ddot{\boldsymbol{u}}_{i+1} \tag{30}$$

$$\dot{\boldsymbol{u}}_{i+1} = \tilde{\dot{\boldsymbol{u}}}_{i+1} + \Delta t \gamma \, \ddot{\boldsymbol{u}}_{i+1} \tag{31}$$

In the above equations α, β and γ are algorithmic coefficients. To get a non-oscillatory velocity field, $\alpha = -0.3$ has been adopted in all the forthcoming simulations; furthermore, β and γ have been finely tuned around the values allowing second-order accuracy in linear elasto-dynamics.

To ensure accuracy of the filtering procedure, the time step size Δt has been always set so as to fulfill the Courant condition in the bulk of the composite. Moreover, to speed up the explicit integrator phase (29), the mass matrix $\boldsymbol{M}$ has been diagonalized by means of a standard row-sum lumping procedure [42].

Account taken of the explicit format of the integrator stage, the space-time discretized equations of motion of the laminate (state equations) can be formally written:

$$\boldsymbol{z}_{i+1} = \begin{Bmatrix} \boldsymbol{u}_{i+1} \\ \dot{\boldsymbol{u}}_{i+1} \\ \ddot{\boldsymbol{u}}_{i+1} \end{Bmatrix} = \boldsymbol{f}_i^z(\boldsymbol{z}_i) \tag{32}$$

where $\boldsymbol{z}$ is the structural state vector, and mapping $\boldsymbol{f}^z$ turns out to be nonlinear because of the softening interface behavior.

3. Constrained Sigma-point Kalman Filtering

3.1. Parameter Identification via Joint Kalman Filtering

According to a standard methodology [43], the calibration of constitutive laws can be pursued by Kalman filtering if a state vector $\boldsymbol{x}$ is obtained by joining the structural state vector $\boldsymbol{z}$ (see Eq. 32) with a vector $\boldsymbol{\vartheta}$ gathering all the model parameters to be tuned. At time t_i this can be written:

$$\boldsymbol{x}_i = \begin{Bmatrix} \boldsymbol{z}_i \\ \boldsymbol{\vartheta}_i \end{Bmatrix} \tag{33}$$

While the current structural state $\boldsymbol{z}$ is always at least partially observed, model parameters to be identified can not be directly measured; by joining $\boldsymbol{z}$ and $\boldsymbol{\vartheta}$, state tracking can consistently improve model calibration.

In case of irreversible constitutive laws, internal state variables must be gathered by $\boldsymbol{x}$ too, see e.g. [12, 19].

Allowing for model and measurement errors, the state-space model describing the evolution within the time interval $[t_i \quad t_{i+1}]$ of the joint state vector and its link with observations turns out to be:

$$\begin{cases} \boldsymbol{x}_{i+1} & = \boldsymbol{f}_i(\boldsymbol{x}_i) + \boldsymbol{v}_i \\ \boldsymbol{y}_i & = \boldsymbol{H}\boldsymbol{x}_i + \boldsymbol{w}_i \end{cases} \tag{34}$$

where: $\boldsymbol{y}$ is the observation vector, which collects the measured components of the state vector; $\boldsymbol{v}$ is the process noises; $\boldsymbol{w}$ is the measurement noise. $\boldsymbol{v}$, $\boldsymbol{w}$ are assumed to be additive, uncorrelated white and Gaussian processes, with zero mean and covariances $\boldsymbol{V}$ and $\boldsymbol{W}$ [44, 45]. Since $\boldsymbol{z}$ is defined according to Eq. (32), the observation equation in (34) shows up as a linear relation between $\boldsymbol{y}$ and $\boldsymbol{x}$. On the contrary, the interface behavior renders the evolution equation $\boldsymbol{f}$ nonlinear.

By way of the EKF [12, 46], within the time step the nonlinear mapping $\boldsymbol{f}$ is expanded in Taylor series, up to the first order, around the current estimates of the state vector and of model parameters. Bounds on the required accuracy of the initialization of $\boldsymbol{x}$, and on the statistics of noises $\boldsymbol{v}$ and $\boldsymbol{w}$ to assure filter stability were provided for linear systems in [47] and, more recently, for nonlinear systems in [48]. Even in the absence of filter instabilities, the softening response of the interlaminar surfaces does not always guarantee the achievement of an accurate model calibration, see [15, 19].

Table 1. Sigma-point Kalman filter.

- Initialization at t_0:

$$\hat{\boldsymbol{x}}_0 = \mathbb{E}[\boldsymbol{x}_0]$$
$$\boldsymbol{P}_0 = \mathbb{E}[(\boldsymbol{x}_0 - \hat{\boldsymbol{x}}_0)(\boldsymbol{x}_0 - \hat{\boldsymbol{x}}_0)^{\mathrm{T}}]$$

- At t_i, for $i = 0, ..., N$

  1. Predictor phase:

$$\hat{\boldsymbol{\chi}}_{i,\mathrm{j}} = \hat{\boldsymbol{x}}_i + \boldsymbol{\Delta\chi}_{i,\mathrm{j}} \qquad \mathrm{j} = 0, ..., N_\chi$$
$$\hat{\boldsymbol{\chi}}^-_{i+1,\mathrm{j}} = \boldsymbol{f}_i(\hat{\boldsymbol{\chi}}_{i,\mathrm{j}})$$

$$\hat{\boldsymbol{x}}^-_{i+1} = \sum_{\mathrm{j}=0}^{N_\chi} \omega_{\mathrm{j}} \, \hat{\boldsymbol{\chi}}^-_{i+1,\mathrm{j}}$$
$$\boldsymbol{P}^-_{i+1} = \boldsymbol{R}^-_{i+1} + \boldsymbol{V}$$

  where

$$\boldsymbol{R}^-_{i+1} = \sum_{\mathrm{j}=0}^{N_\chi} \omega_{\mathrm{j}} \left(\hat{\boldsymbol{\chi}}^-_{i+1,\mathrm{j}} - \hat{\boldsymbol{x}}^-_{i+1}\right)\left(\hat{\boldsymbol{\chi}}^-_{i+1,\mathrm{j}} - \hat{\boldsymbol{x}}^-_{i+1}\right)^{\mathrm{T}}$$

  2. Corrector phase:

$$\hat{\boldsymbol{x}}_i = \hat{\boldsymbol{x}}^-_i + \boldsymbol{G}^U_i \left(\boldsymbol{y}_i - \boldsymbol{H}\hat{\boldsymbol{x}}^-_i\right)$$
$$\boldsymbol{P}_i = \boldsymbol{P}^-_i - \boldsymbol{G}^U_i \boldsymbol{H}\boldsymbol{R}^-_i$$

  where

$$\boldsymbol{G}^U_i = \boldsymbol{R}^-_i \boldsymbol{H}^{\mathrm{T}} \left(\boldsymbol{H}\boldsymbol{R}^-_i\boldsymbol{H}^{\mathrm{T}} + \boldsymbol{W}\right)^{-1}$$

To improve the results when nonlinearities become dominant, the SPKF has been recently proposed [16, 49–51]. At the beginning of the time step, the probability distribution of $\boldsymbol{x}$ is deterministically sampled through a set of sigma-points $\hat{\boldsymbol{\chi}}_{\mathrm{j}}$, $\mathrm{j} = 0, ..., N_\chi$. These sigma-points are then allowed to evolve according to the nonlinear mapping $\boldsymbol{f}$. The statistics of $\boldsymbol{x}$ at the end of the time step are finally obtained through a proper weighted averaging scheme [18]. This filtering procedure is detailed in Table 1, where $\mathbb{E}[\square]$ represents the expected value of $\square$.

The number of sigma-points and their location in the state vector space are accurately chosen, so as to achieve high accuracy in the estimated probability distribution of $\boldsymbol{x}$ at the end of each time step; when compared to the EKF, a better performance of the SPKF, also in terms of model calibration, is therefore expected [16]. The enhanced accuracy of the

SPKF is discussed next; even though these results have been already presented elsewhere, they are here collected to show how possible constraints on parameter estimates, not dealt with by the standard SPKF, can be managed.

3.2. Accuracy of a Constrained Sigma-point Transformation

In this Section we focus on the time interval $[t_i \quad t_{i+1}]$, but we avoid using indexes i and $i+1$ to simplify the notation.

Let $\boldsymbol{x}$ be a random vector, featuring at the beginning of the time step mean $\hat{\boldsymbol{x}}$ and covariance $\boldsymbol{P}$. We conceive $\boldsymbol{x}$ as the sum of the mean $\hat{\boldsymbol{x}}$ and a zero-mean disturbance $\boldsymbol{\Delta x} = \boldsymbol{x} - \hat{\boldsymbol{x}}$. If $\boldsymbol{x}$ undergoes a nonlinear transformation, governed by a mapping $\boldsymbol{f}$ analytic everywhere so that it can be expanded in Taylor series about $\hat{\boldsymbol{x}}$, at the end of the time step we get:

$$\boldsymbol{x}^- = \boldsymbol{f}(\boldsymbol{x}) = \boldsymbol{f}(\hat{\boldsymbol{x}} + \boldsymbol{\Delta x}) = \boldsymbol{f}(\hat{\boldsymbol{x}}) + \sum_{n=1}^{\infty} \frac{1}{n!} \boldsymbol{D}^{\boldsymbol{n}}_{\boldsymbol{\Delta x}} \tag{35}$$

where, with a slight abuse in notation, the $n-$th order term $\boldsymbol{D}^{\boldsymbol{n}}_{\boldsymbol{\Delta x}}$ in the series expansion is:

$$\boldsymbol{D}^{\boldsymbol{n}}_{\boldsymbol{\Delta x}} \equiv \left(\sum_{\ell=1}^{N_x} \left. \frac{\partial \boldsymbol{f}}{\partial x_\ell} \right|_{\boldsymbol{x}=\hat{\boldsymbol{x}}} \Delta x_\ell \right)^n \tag{36}$$

N_x being the number of components of the state vector $\boldsymbol{x}$. Since the derivatives of $\boldsymbol{f}$ in (36) are evaluated at $\boldsymbol{x} = \hat{\boldsymbol{x}}$, they are not random variables. The expected value of $\boldsymbol{x}^-$ therefore reads:

$$\begin{aligned} \hat{\boldsymbol{x}}^- = \mathbb{E}[\boldsymbol{x}^-] &= \mathbb{E}\left[\boldsymbol{f}(\hat{\boldsymbol{x}})\right] + \mathbb{E}\left[\sum_{n=1}^{\infty} \frac{1}{n!} \boldsymbol{D}^{\boldsymbol{n}}_{\boldsymbol{\Delta x}} \right] \\ &= \boldsymbol{f}(\hat{\boldsymbol{x}}) + \sum_{n=1}^{\infty} \frac{1}{n!} \mathbb{E}\left[\boldsymbol{D}^{\boldsymbol{n}}_{\boldsymbol{\Delta x}}\right] \end{aligned} \tag{37}$$

The relevant error covariance matrix is:

$$\begin{aligned} \boldsymbol{P}^- &= \mathbb{E}\left[\left(\boldsymbol{x}^- - \hat{\boldsymbol{x}}^-\right) \left(\boldsymbol{x}^- - \hat{\boldsymbol{x}}^-\right)^{\mathrm{T}} \right] \\ &= \mathbb{E}\left[\left(\sum_{n=1}^{\infty} \frac{1}{n!} \left(\boldsymbol{D}^{\boldsymbol{n}}_{\boldsymbol{\Delta x}} - \mathbb{E}\left[\boldsymbol{D}^{\boldsymbol{n}}_{\boldsymbol{\Delta x}}\right]\right) \right) \left(\sum_{m=1}^{\infty} \frac{1}{m!} \left(\boldsymbol{D}^{\boldsymbol{m}}_{\boldsymbol{\Delta x}} - \mathbb{E}\left[\boldsymbol{D}^{\boldsymbol{m}}_{\boldsymbol{\Delta x}}\right]\right) \right)^{\mathrm{T}} \right] \\ &= \sum_{n=1}^{\infty} \sum_{m=1}^{\infty} \frac{1}{n!} \frac{1}{m!} \left(\mathbb{E}\left[\boldsymbol{D}^{\boldsymbol{n}}_{\boldsymbol{\Delta x}} \boldsymbol{D}^{\boldsymbol{m}}_{\boldsymbol{\Delta x}}{}^{\mathrm{T}} \right] - \mathbb{E}\left[\boldsymbol{D}^{\boldsymbol{n}}_{\boldsymbol{\Delta x}}\right] \mathbb{E}\left[\boldsymbol{D}^{\boldsymbol{m}}_{\boldsymbol{\Delta x}}{}^{\mathrm{T}} \right] \right) \end{aligned} \tag{38}$$

Now, let us suppose to sample the probability distribution of $\boldsymbol{x}$ through a set of sigma-points $\hat{\boldsymbol{\chi}}_{\mathsf{j}}$, $\mathsf{j} = 0, \ldots, N_\chi$, chosen around the current mean $\hat{\boldsymbol{x}}$ according to:

$$\hat{\boldsymbol{\chi}}_{\mathsf{j}} = \hat{\boldsymbol{x}} + \boldsymbol{\Delta\chi}_{\mathsf{j}}, \qquad \mathsf{j} = 0, \ldots, N_\chi \tag{39}$$

where the terms $\boldsymbol{\Delta\chi}_{\mathsf{j}}$ need to be determined.

Similarly to $\boldsymbol{x}$, within the time step each sigma-point undergoes the transformation:

$$\hat{\boldsymbol{\chi}}_{\mathsf{j}}^{-} = \boldsymbol{f}(\hat{\boldsymbol{\chi}}_{\mathsf{j}}) = \boldsymbol{f}(\hat{\boldsymbol{x}}) + \sum_{n=1}^{\infty} \frac{1}{n!} \boldsymbol{D}_{\boldsymbol{\Delta\chi}_{\mathsf{j}}}^{\boldsymbol{n}} \tag{40}$$

where:

$$\boldsymbol{D}_{\boldsymbol{\Delta\chi}_{\mathsf{j}}}^{\boldsymbol{n}} \equiv \left(\sum_{\ell=1}^{N_x} \left. \frac{\partial \boldsymbol{f}}{\partial x_\ell} \right|_{\boldsymbol{x}=\hat{\boldsymbol{x}}} \Delta\chi_{\mathsf{j}\ell} \right)^n \tag{41}$$

At the end of the time step, the information in the evolved sigma-points are collected via a weighted averaging procedure to obtain:

$$\begin{aligned}\hat{\boldsymbol{x}}_{\mathrm{SPT}}^{-} &= \sum_{\mathsf{j}=0}^{N_\chi} \omega_{\mathsf{j}} \, \hat{\boldsymbol{\chi}}_{\mathsf{j}}^{-} \\ &= \left(\sum_{\mathsf{j}=0}^{N_\chi} \omega_{\mathsf{j}} \right) \boldsymbol{f}(\hat{\boldsymbol{x}}) + \sum_{n=1}^{\infty} \frac{1}{n!} \left(\sum_{\mathsf{j}=0}^{N_\chi} \omega_{\mathsf{j}} \, \boldsymbol{D}_{\boldsymbol{\Delta\chi}_{\mathsf{j}}}^{\boldsymbol{n}} \right)\end{aligned} \tag{42}$$

where ω_{j} are the weights of the sigma-point transformation relevant to the mean of $\boldsymbol{x}$. The corresponding error covariance matrix is given by:

$$\begin{aligned}\boldsymbol{P}_{\mathrm{SPT}}^{-} &= \sum_{\mathsf{j}=0}^{N_\chi} \omega_{\mathsf{j}}^{\star} \left(\hat{\boldsymbol{\chi}}_{\mathsf{j}}^{-} - \hat{\boldsymbol{x}}_{\mathrm{SPT}}^{-} \right) \left(\hat{\boldsymbol{\chi}}_{\mathsf{j}}^{-} - \hat{\boldsymbol{x}}_{\mathrm{SPT}}^{-} \right)^{\mathrm{T}} \\ &= \sum_{\mathsf{j}=0}^{N_\chi} \omega_{\mathsf{j}}^{\star} \sum_{n=1}^{\infty} \sum_{m=1}^{\infty} \frac{1}{n!} \frac{1}{m!} \left(\boldsymbol{D}_{\boldsymbol{\Delta\chi}_{\mathsf{j}}}^{\boldsymbol{n}} - \sum_{\mathsf{r}=0}^{N_\chi} \omega_{\mathsf{r}} \, \boldsymbol{D}_{\boldsymbol{\Delta\chi}_{\mathsf{r}}}^{\boldsymbol{n}} \right) \left(\boldsymbol{D}_{\boldsymbol{\Delta\chi}_{\mathsf{j}}}^{\boldsymbol{m}} - \sum_{\mathsf{s}=0}^{N_\chi} \omega_{\mathsf{s}} \, \boldsymbol{D}_{\boldsymbol{\Delta\chi}_{\mathsf{s}}}^{\boldsymbol{m}} \right)^{\mathrm{T}}\end{aligned} \tag{43}$$

where $\omega_{\mathsf{j}}^{\star}$ are the weights of the sigma-point transformation relevant to the covariance of $\boldsymbol{x}$.

If $\boldsymbol{x}$ is a Gaussian random vector, its probability distribution is symmetric with respect to the mean $\hat{\boldsymbol{x}}$; therefore, all the odd central moments $\boldsymbol{D}_{\boldsymbol{\Delta x}}^{\boldsymbol{n}}$, $n = 1, 3, 5, ...$, are zero. To be compliant with this condition, couples of sigma-points are symmetrically placed around $\hat{\boldsymbol{x}}$, according to [16]:

$$\begin{cases} \boldsymbol{\Delta\chi}_0 &= \boldsymbol{0} \\ \boldsymbol{\Delta\chi}_{\mathsf{k}} &= +\psi\sqrt{\boldsymbol{P}}\,\boldsymbol{1}_{\mathsf{k}} \qquad \mathsf{k} = 1, ..., \frac{N_\chi}{2} \\ \boldsymbol{\Delta\chi}_{\frac{N_\chi}{2}+\mathsf{k}} &= -\psi\sqrt{\boldsymbol{P}}\,\boldsymbol{1}_{\mathsf{k}} \end{cases} \tag{44}$$

Here: $\sqrt{\boldsymbol{P}}$ represents the square root of matrix $\boldsymbol{P}$, computed e.g. through a Cholesky factorization; ψ is a scaling parameter; $\boldsymbol{1}_{\mathsf{k}}$ is a unit vector aligned with component k in the state vector space. The series expansions (37) and (42) agree up to third order if:

$$\begin{cases} \sum_{\mathsf{j}=0}^{N_\chi} \omega_\mathsf{j} = 1 \\ \sum_{\mathsf{j}=0}^{N_\chi} \omega_\mathsf{j}\, \boldsymbol{D}^{\mathbf{1}}_{\boldsymbol{\Delta}\boldsymbol{\chi}_\mathsf{j}} = \mathbf{0} \\ \sum_{\mathsf{j}=0}^{N_\chi} \omega_\mathsf{j}\, \boldsymbol{D}^{\mathbf{2}}_{\boldsymbol{\Delta}\boldsymbol{\chi}_\mathsf{j}} = \mathbb{E}\left[\boldsymbol{D}^{\mathbf{2}}_{\boldsymbol{\Delta x}}\right] \\ \sum_{\mathsf{j}=0}^{N_\chi} \omega_\mathsf{j}\, \boldsymbol{D}^{\mathbf{3}}_{\boldsymbol{\Delta}\boldsymbol{\chi}_\mathsf{j}} = \mathbf{0} \end{cases} \tag{45}$$

To simplify the matter, let us assume $\omega_\mathsf{j} = \omega$ for $\mathsf{j} = 1, ..., N_\chi$; relations involving the first and third order terms in (45) are then automatically satisfied. Relations involving the zeroth and second order terms in (45) then furnish:

$$\begin{cases} \omega_0 + N_\chi \omega = 1 \\ 2\psi^2 \omega = 1 \end{cases} \tag{46}$$

A further condition to set ω_0, ω and ψ can be furnished by matching the diagonal entries of the fourth order terms (kurtoses) in (37) and (42). This leads to [19]:

$$\omega_0 = 1 - \frac{N_\chi}{6}, \qquad \omega = \frac{1}{6}, \qquad \psi = \sqrt{3} \tag{47}$$

Here we propose an alternative condition to determine ω_0, ω and ψ, partially exploiting the features of the so-called scaled unscented transformation [52]. Let us assume that model parameters have to satisfy the constraints:

$$\boldsymbol{\vartheta}^m \leq \boldsymbol{\vartheta} \leq \boldsymbol{\vartheta}^M \tag{48}$$

where $\boldsymbol{\vartheta}^m$ and $\boldsymbol{\vartheta}^M$ respectively gather the minimum and maximum (if any) allowed values of model parameters. This requirement must be fulfilled by each sigma-point $\hat{\boldsymbol{\chi}}_\mathsf{j}^-$, $\mathsf{j} = 0, ..., N_\chi$. For $\mathsf{j} = 0$ the conditions (48) are automatically satisfied, since $\hat{\boldsymbol{x}}$ (and, therefore, $\hat{\boldsymbol{\vartheta}}$) is computed at the end of the previous time step by averaging sigma-points all fulfilling the constraints. Further, if $\boldsymbol{\vartheta} = \boldsymbol{B}_\vartheta \boldsymbol{x}$, $\boldsymbol{B}_\vartheta$ being a Boolean matrix, conditions (48) are satisfied by all the sigma-points if:

$$\psi \leq \min\left\{ \frac{\hat{\vartheta}_\ell - \vartheta_\ell^m}{a_\mathsf{k}^\ell}, \frac{\vartheta_\ell^M - \hat{\vartheta}_\ell}{a_\mathsf{k}^\ell} \right\}, \qquad \mathsf{k} = 1, ..., \frac{N_\chi}{2}; \ell = 1, ..., N_\vartheta \tag{49}$$

where $\boldsymbol{a}_\mathsf{k} = \boldsymbol{B}_\vartheta \sqrt{\boldsymbol{P}}\, \mathbf{1}_\mathsf{k}$, and N_ϑ is the number of model parameters in $\boldsymbol{\vartheta}$. In the forthcoming examples, we initially assume $\psi = \sqrt{3}$ (according to what reported in 47) and reduce its value if necessary, according to relation (49).

As for the error covariance matrix $\boldsymbol{P}_{\text{SPT}}^-$, by letting $\omega_\mathsf{j}^\star = \omega_\mathsf{j} = \omega$ for $\mathsf{j} = 1, ..., N_\chi$, we get:

$$\boldsymbol{P}_{\text{SPT}}^- = \sum_{n=1}^{\infty}\sum_{m=1}^{\infty} \frac{1}{n!}\frac{1}{m!} \left(\sum_{\mathsf{j}=0}^{N_\chi} \omega_\mathsf{j} \boldsymbol{D}^{\boldsymbol{n}}_{\boldsymbol{\Delta}\boldsymbol{\chi}_\mathsf{j}} \boldsymbol{D}^{\boldsymbol{m}}_{\boldsymbol{\Delta}\boldsymbol{\chi}_\mathsf{j}}{}^{\mathrm{T}} + (\omega_0^\star + N_\chi \omega - 2) \sum_{\mathsf{j}=0}^{N_\chi} \omega_\mathsf{j} \boldsymbol{D}^{\boldsymbol{n}}_{\boldsymbol{\Delta}\boldsymbol{\chi}_\mathsf{j}} \sum_{\mathsf{k}=0}^{N_\chi} \omega_\mathsf{k} \boldsymbol{D}^{\boldsymbol{m}}_{\boldsymbol{\Delta}\boldsymbol{\chi}_\mathsf{k}}{}^{\mathrm{T}} \right) \tag{50}$$

In case of Gaussian random variables, independently of the value of $\omega_0^\star$, (38) and (50) agree up to the third order. Weight $\omega_0^\star$ can be set by matching part of the fourth order terms (specifically those involving $\boldsymbol{D}^2_{\Delta\chi_j}\boldsymbol{D}^2_{\Delta\chi_k}{}^{\mathrm{T}}$ in 50), thereby obtaining (see also [52]):

$$\omega_0^\star = 4 - \frac{N_\chi}{\psi^2} - \psi^2 \tag{51}$$

In what precedes we have assumed the mapping $\boldsymbol{f}$ to be analytic everywhere in the $\boldsymbol{x}$ space. If $\boldsymbol{f}$ is not differentiable, low order terms of the Taylor series expansions of mean and covariance get affected. It is difficult to quantify the discrepancies with respect to the analytic case, because they depend on whether the sigma-points sample the loci of non-differentiability. However, it can be generally said that the order of accuracy is detrimentally affected.

4. Results

To assess the performances of the proposed filtering approach in calibrating the interlaminar constitutive law while detecting impact-induced delamination, we first study a simple problem consisting of a two-layer composite stricken by a homogeneous impactor. Hence, two different impact tests on GRP composites [14, 53] are considered to mainly show the accuracy in detecting delamination in real-time.

In all the cases, it is assumed that the contact between specimen and impactor is perfect (i.e. distributed all over their approaching surfaces) and that failure of the laminate occurs because of the propagation of dilatational plane waves in the through-the-thickness direction: the interlaminar surfaces are therefore subject to pure mode I loading.

4.1. Pseudo-experimental Testings

As a starting benchmark, a pseudo-experimental testing condition is conceived. The pseudo-experimental response of the laminate to the impact loading has been computed by adding a white noise (of assigned variance) to sampled outcomes of finite element analyses. Even though this approach is sometimes criticized, being not clear whether one is testing with it the filtering approach or its implementation, it helps in getting insights into the performance of the filter in terms of stability and convergence rate. Indeed, calibration of interlaminar constitutive models may become difficult if delamination occurs almost instantaneously: the filter has to be highly sensitive to model parameters to promptly react to the information conveyed by measurements. This requires a careful setting of filter parameters, like $\boldsymbol{P}_0$ and $\boldsymbol{V}$. It is worth mentioning that, to further complicate the problem, the effects of the shape of the tensile envelope of interlaminar laws (at assigned strength and toughness) on the overall response of a laminate subject to impacts have not been thoroughly understood yet: depending on the loading and boundary conditions, on the composite geometry and on the stacking sequence, in some circumstances the shape affects the response, whereas in others it does not [54].

The capabilities of the SPKF are therefore first assessed through a simple test: a two-layer composite is stricken by a homogeneous impactor. The laminae and the impactor are

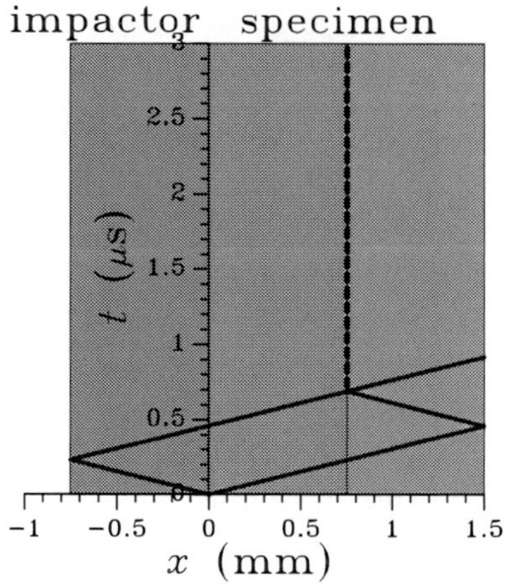

Figure 5. impact on a two-layer composite. Space-time diagram (the vertical dashed line here represents the possibly debonding surface when a brittle, homogeneous material is subject to the same impact).

assumed to be isotropic and elastic, featuring Young's modulus $E = 10$ GPa, Poisson's ratio $\nu = 0.35$ and mass density $\varrho = 1500$ kg/m^3 (see also [11]). Each lamina and the impactor are 0.75 mm in thickness. Target mechanical properties of the interlaminar surfaces are assumed:

$$\begin{aligned} 2K &= 277.09 && (\text{N/mm}^3) \\ \tau_M &= 75 && (\text{MPa}) \\ G &= 0.15 && (\text{N/mm}) \end{aligned} \tag{52}$$

According to the space-time diagram of Figure 5, failure can occur only along the interlaminar surface because of the interaction of the two release waves propagating inwards from the free surfaces of impactor and specimen.

Two different values of the velocity v of the impactor are considered. In a first case $v = 10.19$ m/s leads to the propagation in the through-the-thickness direction of a compressive/tensile wave of amplitude $\bar{\tau} = 50$ MPa, which does not cause interface failure. In a second case $v = 20.38$ m/s causes laminate failure, i.e. whole delamination, because of the propagation of a compressive/tensile wave of amplitude $\bar{\tau} = 100$ MPa. Outcomes of the two tests are respectively reported in Figures 6 and 7, in terms of time evolution of the free surface velocity velocity $\dot{u}_r$ at the rear laminate surface, of the opening displacement discontinuity $[u]$ and of the normal traction τ (here and in what follows the subscript n has been dropped to simplify the notation). Results are shown for all the constitutive models described in Section 2.2., having assumed $q = 1$ for the modified exponential interface law (17). For comparison purposes, the response of an interface-free specimen is reported too; in such a case, the purely elastic behavior of the material leads to the propagation of sharp fronts of a shock wave.

If delamination is not incepted ($\bar{\tau} = 50$ MPa), the response is almost independent of the shape of the interface law. Only in the presence of an interface that behaves according to the exponential model, the pre-peak nonlinearity of the constitutive law (see Figure 2(a))

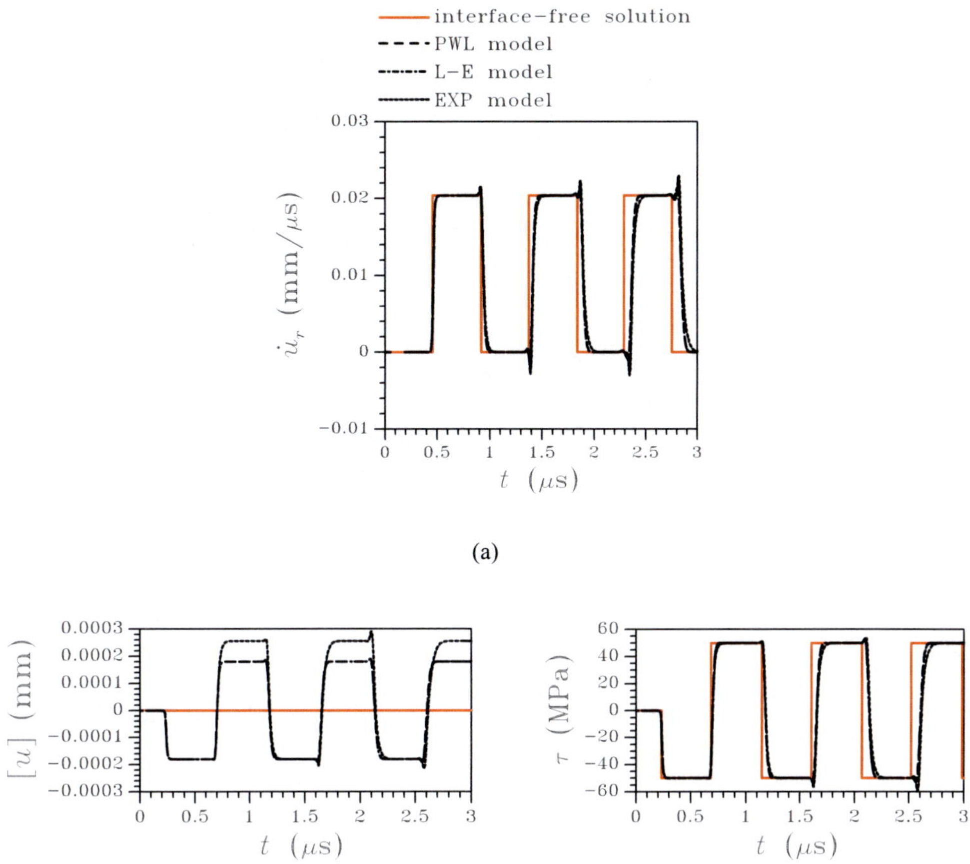

(a)

(b) (c)

Figure 6. impact on a two-layer composite, $\bar{\tau} = 50$ MPa. Effects of the interlaminar laws on the time evolution of (a) the free surface velocity $\dot{u}_r$, of (b) the displacement discontinuity $[u]$ and (c) relevant traction τ along the interface.

leads to a larger opening $[u]$ of the interlaminar surface when subject to a tensile stress. For any constitutive model, the signature of the interface is shown in Figure 6(a) by the delay in the sudden changes of $\dot{u}_r$ with respect to the reference, interface-free solution. This delay, which grows in time, is caused by the compliance of the interface, that is additional to the bulk one.

In case of failure ($\bar{\tau} = 100$ MPa), the free surface velocity $\dot{u}_r$ is affected by the interface model only when the waves, traveling across the interlaminar surface while softening is taking place, reach the rear surface; this occurs in the present case around 1μs after the impact, see Figure 7(a). After failure, the rear lamina detaches from the front one and freely flies off, as testified by the diverging $[u]$ history in Figure 7(b). Part of the shock waves then get confined inside the back lamina: this explains the subsequent doubling of the drops in the $\dot{u}_r$ evolution. It is worth noting that the time elapsed between softening

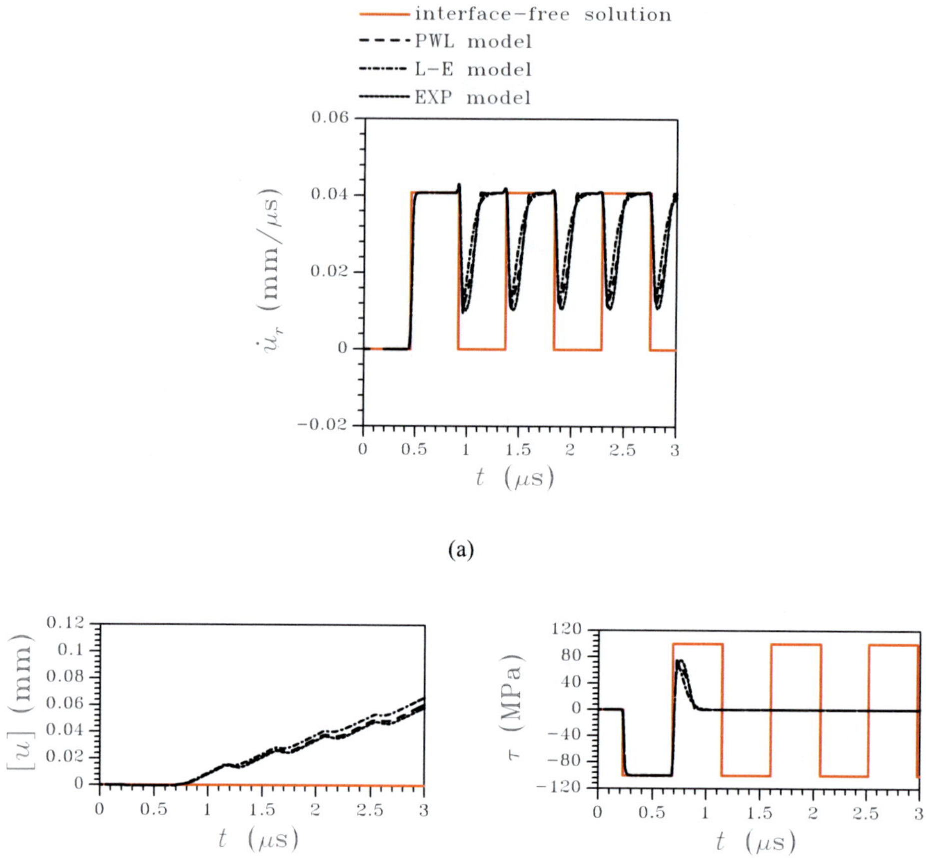

(a)

(b) (c)

Figure 7. impact on a two-layer composite, $\bar{\tau} = 100$ MPa. Effects of the interlaminar laws on the time evolution of (a) the free surface velocity $\dot{u}_r$, of (b) the displacement discontinuity $[u]$ and (c) relevant traction τ along the interface.

inception ($t \cong 0.72$ μs) and whole failure of the interface ($t \cong 0.9$ μs), see Figure 7(c), is very short; only the information on this failure event, constituting the so-called pull-back signal (PBS), conveyed by the shock waves to the free laminate surface, can be used by the filter to calibrate the interface law.

The effects on the PBS of the shape of the tensile envelope and of the strength τ_M and toughness G values need to be assessed. In the absence of any dissipation mechanisms, in this test the free surface velocity $\dot{u}_r$ would drop to zero at $t = 0.917$ μs (see Figure 6(a)); in case of delamination, the minimum attained velocity after the arrival of the unloading tensile wave and the shape of the PBS do furnish information on the interface response. To understand the roles of interface law, strength and toughness, the results of a parametric analysis are shown in Figure 8: for any interface model, τ_M and G are varied by 20% at most with respect to the target values (52). The piecewise linear and linear-exponential

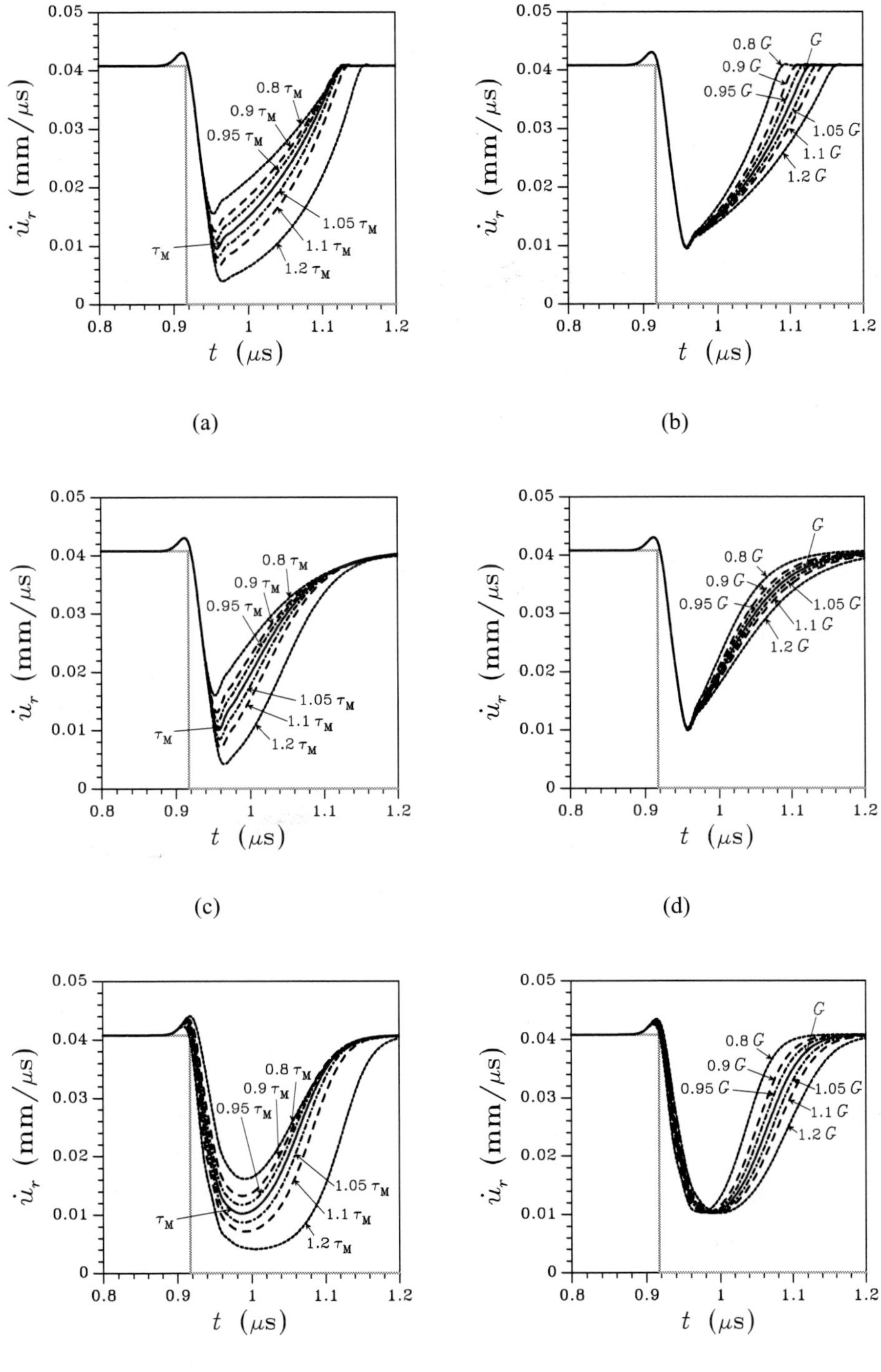

Figure 8. impact on a two-layer composite, $\bar{\tau} = 100$ MPa. Effects of the interface strength τ_M (left column) and toughness G (right column) on the pull-back signal. (a-b) piecewise linear law (12); (c-d) linear-exponential law (13); (e-f) exponential law (14).

laws, having a common initial elastic phase in tension, lead to a common descending branch in the PBS. At variance, the local tangent stiffness of the exponential law in the hardening phase is affected by τ_M and G: the slope of the descending branch of the PBS is therefore affected by τ_M and G too. Independently of the interface law, τ_M turns out to affect the starting stage of debonding, whereas G affects its tail and the time needed to complete it; this is clearly evidenced by the PBS, since τ_M modifies the minimum attained velocity, whereas G influences only the ascending branch with no effects on the pull-back velocity.

From a model calibration perspective, it is clear that the SPKF has chances to improve parameter estimates only in the time interval $0.9 \leq t \leq 1.2$ μs. It is therefore hard to figure out from the previous plots whether the effects of tensile envelope, τ_M and G can be actually interpreted by the filter to improve model calibration. The performances of the SPKF are hence tested here not only looking at parameter estimates, but also checking its capability to detect whether a laminate is failing and, in case of delamination, where it actually takes place.

Typical results of the filtering procedure are depicted in Figures 9 and 10; in this case the pseudo-experimental data, which consist in the free surface velocity alone, have been supposed very accurate, featuring a standard deviation $\delta = 0.33$ m/s (the measurement error covariance matrix $\boldsymbol{W}$ becomes scalar-valued, with entry $W = \delta^2$). As far as the process covariance matrix $\boldsymbol{V}$ is concerned, in case of pseudo-experimental testing it can be assumed to be vanishing, since the filter employs the same structural model adopted to get the pseudo-experimental data. Components of $\boldsymbol{P}_0$ instead need to be finely tuned to enhance filter convergence [11,15].

Figure 9 shows the obtained estimates of τ_M and G as a function of their initial guess in $\hat{\boldsymbol{x}}_0$ (here respectively denoted by $\tau_{M,0}$ and G_0), for all the interface models. These estimates evolve from $\tau_{M,0}$ and G_0 once the PBS is processed by the filter; after that, they become stationary. The tracked state of the laminate is shown in Figure 10 in terms of predicted interface opening $[u]$ and free surface velocity $\dot{u}_r$. Knowing the target response of the composite to the impact, here denoted by the dashed lines, allows to certify stability and convergence of the SPKF, no matter if displacement is diverging in a part of the system and what kind of interface constitutive law has been adopted. It can be seen that estimates get enhanced as soon as the filter senses the PBS: in fact, the sudden changes in the estimate of $[u]$ show up only while processing the PBS, starting from $t \cong 0.9$ μs.

These outcomes testify that the SPKF is very efficient in tracking the state of the laminate, i.e. in understanding whether the structure is failing or not. Model calibration is instead less accurately accomplished: independently of the interface law, τ_M is quite precisely estimated, provided the initial guess $\tau_{M,0}$ is not too far from the target value, whereas G can be hardly inferred. These conclusions are in agreement with the results of the parametric analysis: while τ_M affects the whole PBS, G affects only its ascending branch. Therefore, filtering out from $\dot{u}_r$ the effects of G alone turns out to be extremely difficult.

In case of a much higher scattering of pseudo-experimental data ($\delta = 3.3$ m/s), see Figure 11, results loose accuracy as for the calibration task. Contrariwise, state tracking maintain accuracy: even though measurements contain poor information, the SPKF is again able to provide the evolution of the free surface velocity in the PBS.

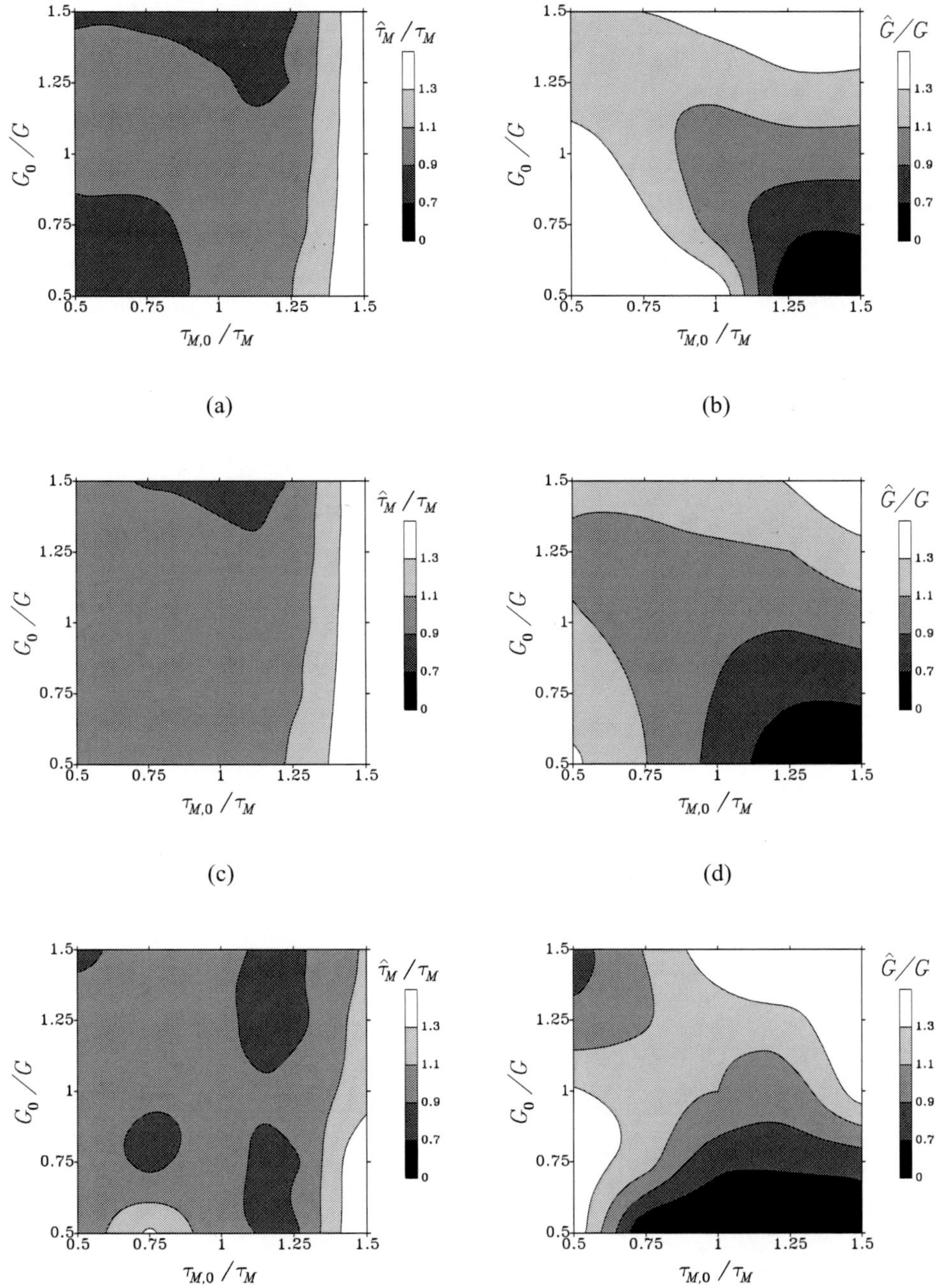

Figure 9. impact on a two-layer composite, $\bar{\tau} = 100$ MPa (W=10^{-1} m^2/s^2). Effects of the initialization values $\tau_{M,0}$ and G_0 on the converged estimates of τ_M (left column) and G (right column). (a-b) piecewise linear law; (c-d) linear-exponential law; (e-f) exponential law.

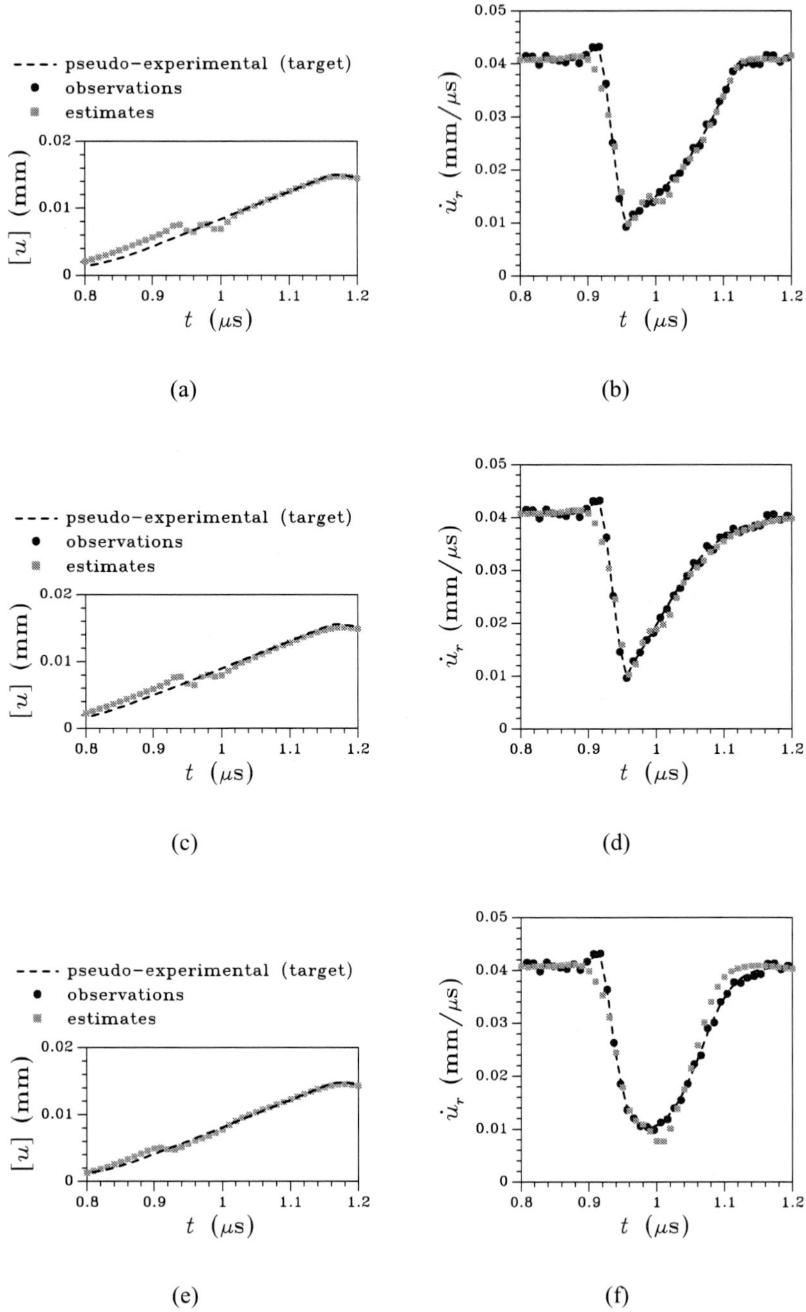

Figure 10. impact on a two-layer composite, $\bar{\tau} = 100$ MPa (W=10^{-1} m^2/s^2). Evolution in time of interface opening $[u]$ (left column) and free surface velocity $\dot{u}_r$ (right column), and comparison among tracked state (orange squares), actual state (dashed lines) and pseudo-experimental data (blue circles). (a-b) piecewise linear law; (c-d) linear-exponential law; (e-f) exponential law.

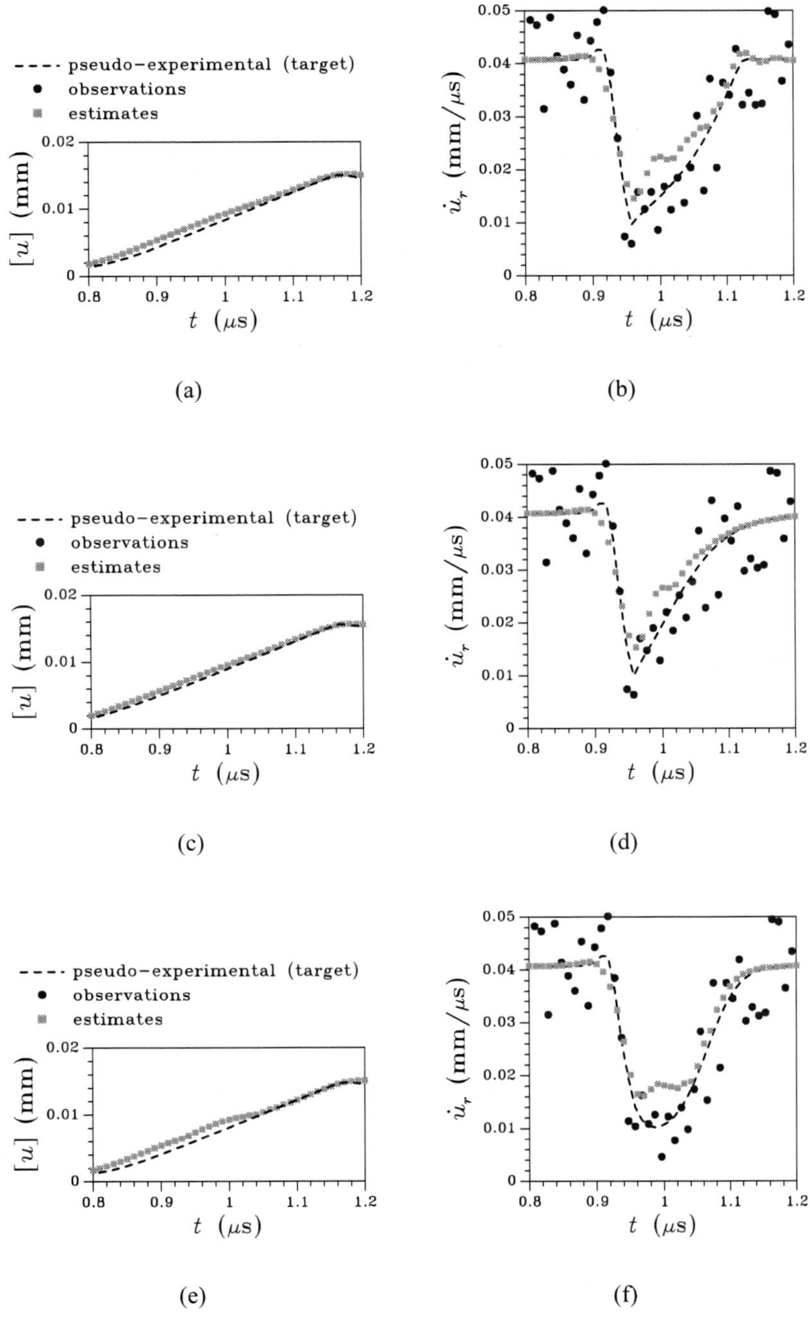

Figure 11. impact on a two-layer composite, $\bar{\tau}$ = 100 MPa (W=10 m^2/s^2). Evolution in time of interface opening $[u]$ (left column) and free surface velocity $\dot{u}_r$ (right column), and comparison among tracked state (orange squares), actual state (dashed lines) and pseudo-experimental data (blue circles). (a-b) piecewise linear law; (c-d) linear-exponential law; (e-f) exponential law.

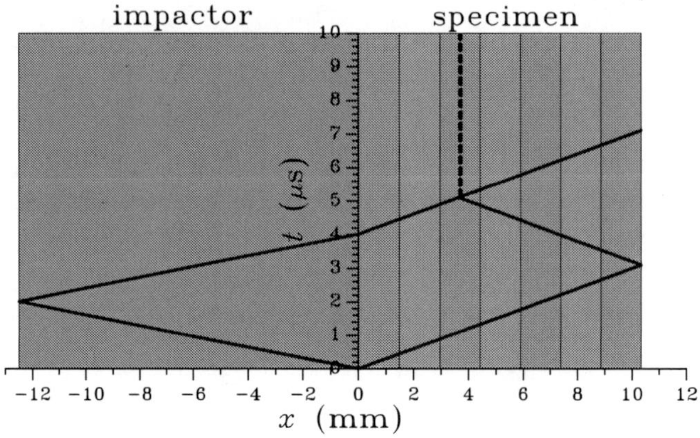

(a)

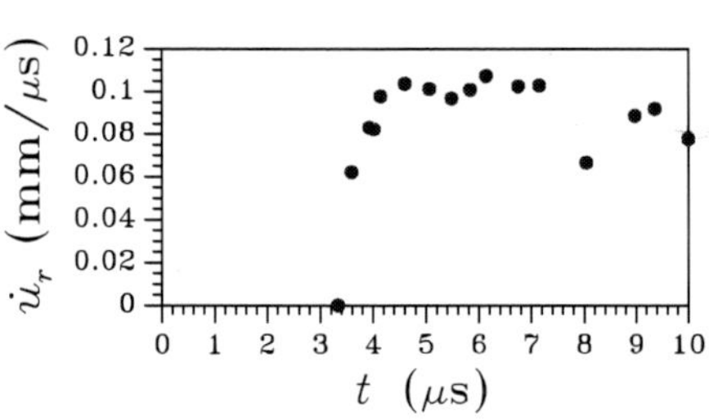

(b)

Figure 12. impact on a 7-layer composite [53]. (a) space-time diagram (the vertical dashed line here represents the possibly debonding surface when a brittle, homogeneous material is subject to the same impact), and (b) experimentally measured free surface velocity.

4.2. Actual Experimental Testings

To finally check the performances of the SPKF when dealing with multi-layered composites, we consider two of the experiments reported in [53] and [14].

In the first experiment (experiment FY06001 in [53]), the specimen is a 7-layer composite plate; each lamina is 1.37 mm in thickness, and is made of a balanced 5-harness satin weave E-glass and LY564 epoxy. The wave speed in the through-the-thickness direction is 3.34 km/s, while the mass density is $\varrho = 1885$ kg/m^3 [53]. This laminate was subject to a plane impact, stricken by an aluminum impactor (12.5 mm thick) flying at velocity $v = 71$ m/s. The relevant space-time diagram is shown in Figure 12, along with the free surface velocity profile measured via a velocity interferometer for any reflector (VISAR).

Account taken of the high accuracy of the experimentally measured $\dot{u}_r$, the identification procedure have furnished the results reported in Figure 13 in terms of time evolution of estimates of τ_M and G as a function of their initialization values within the domain:

$$\mathcal{C}_\vartheta = \{50 \le \tau_{M,0} \le 250 \text{ (MPa)}, 0.5 \le G_0 \le 2.5 \text{ (N/mm)}\} \tag{53}$$

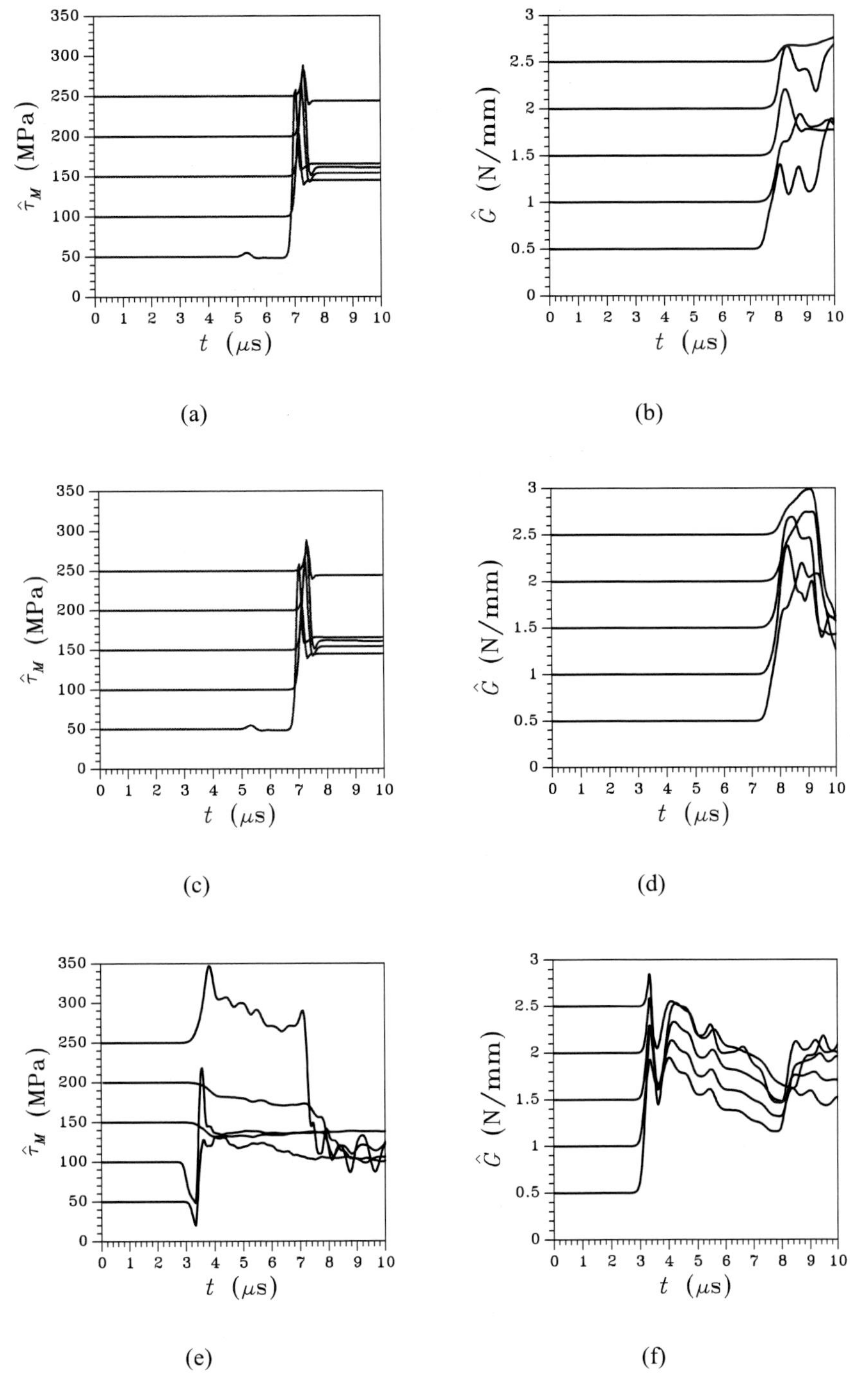

Figure 13. impact on a 7-layer composite [53]. Evolution in time of the estimated values of τ_M (left column) and G (right column). (a-b) piecewise linear law; (c-d) linear-exponential law; (e-f) exponential law.

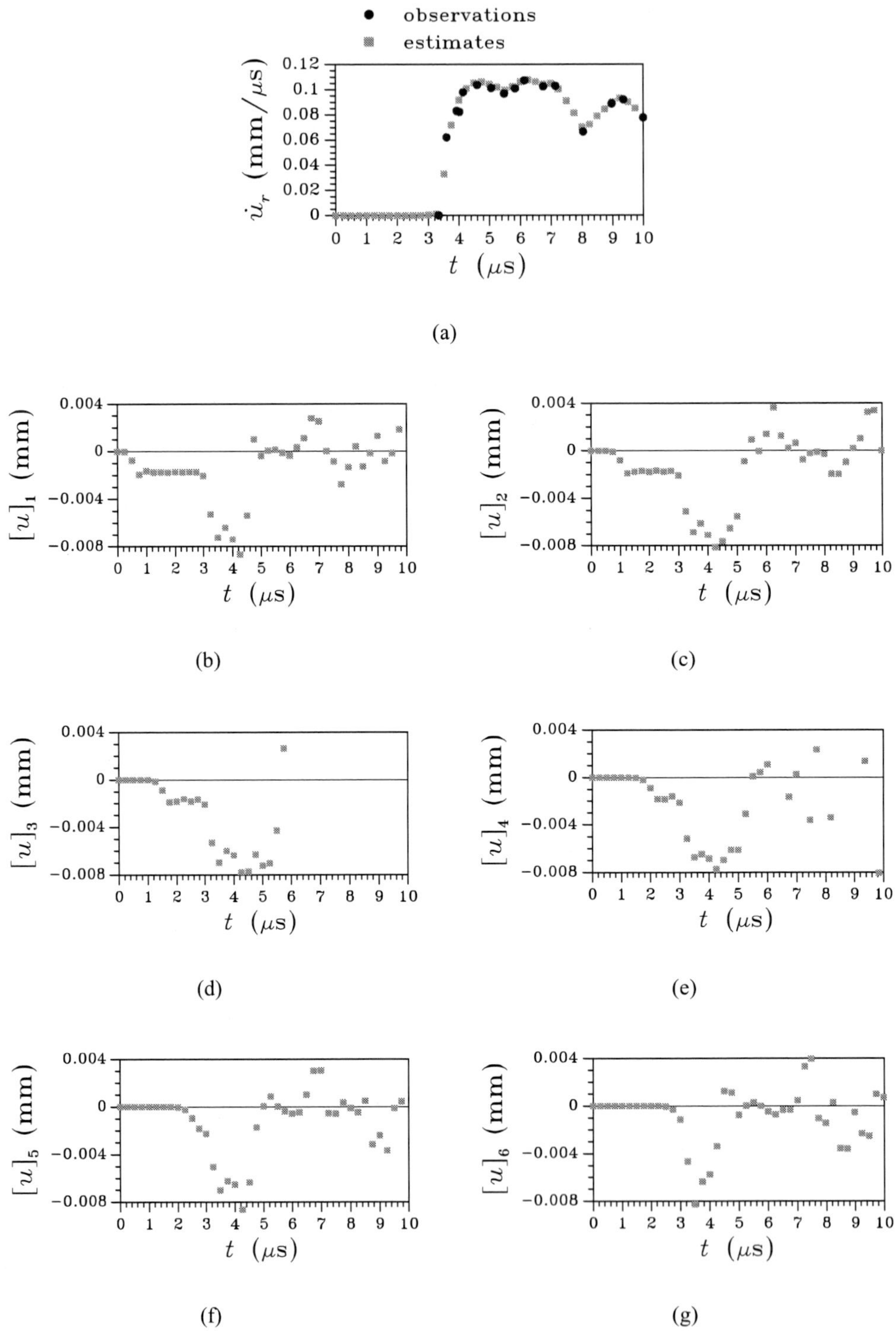

Figure 14. impact on a 7-layer composite [53]. Evolution in time of (a) free surface velocity $\dot{u}_r$ and (b-g) estimated interface openings $[u]_1 - [u]_6$.

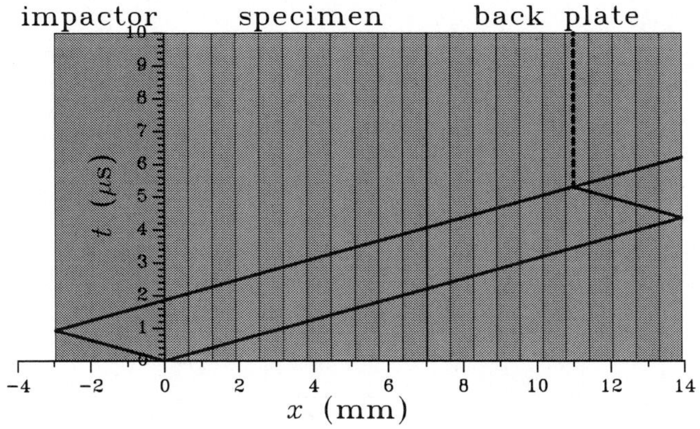

(a)

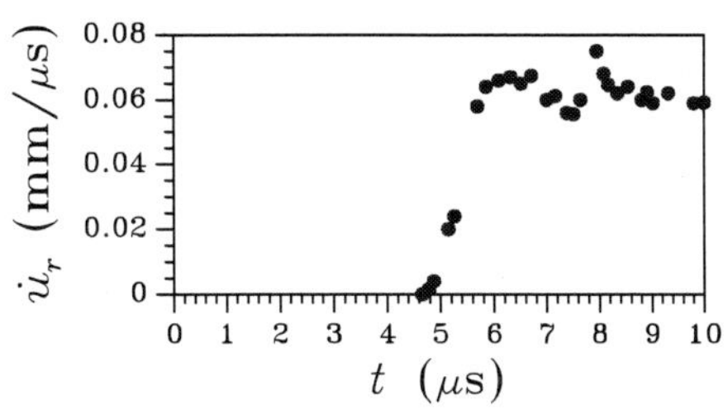

(b)

Figure 15. impact on a 11+11-layer composite [14]. (a) space-time diagram (the vertical dashed line here represents the possibly debonding surface when a brittle, homogeneous material is subject to the same impact), and (b) experimentally measured free surface velocity.

Converged estimates of τ_M are in good agreement with the spall strength of 119.5 MPa reported in [53]; on the other hand, final estimates of G are well representative for this kind of composites. Figure 14 reports the estimated state of the specimen: the capability to track the measured free surface velocity and to foresee delamination along the third interlaminar surface away from the impact plane, is evidenced. This latter result, allowing also for wave dispersion caused by interlaminar surfaces and inner inhomogeneities of the composite, well agrees with the state-space diagram of Figure 12(a).

In the second experiment (experiment 1 in [14]), a GRP specimen, 7.02 mm thick, is backed by another GRP plate, 6.91 mm thick; both laminates are made of 11 plies. The wave speed in the through-the-thickness direction now amounts to 3.19 km/s, and the mass density to $\varrho = 1867$ kg/m^3 [14]. The specimen is stricken by a 5-layer GRP flyer, 2.96 mm in thickness, flying at velocity $v = 85$ m/s. The corresponding space-time diagram, and the free surface velocity profile measured via a VISAR are reported in Figure 15. Because of the test set-up, the release waves interact causing delamination inside the back plate.

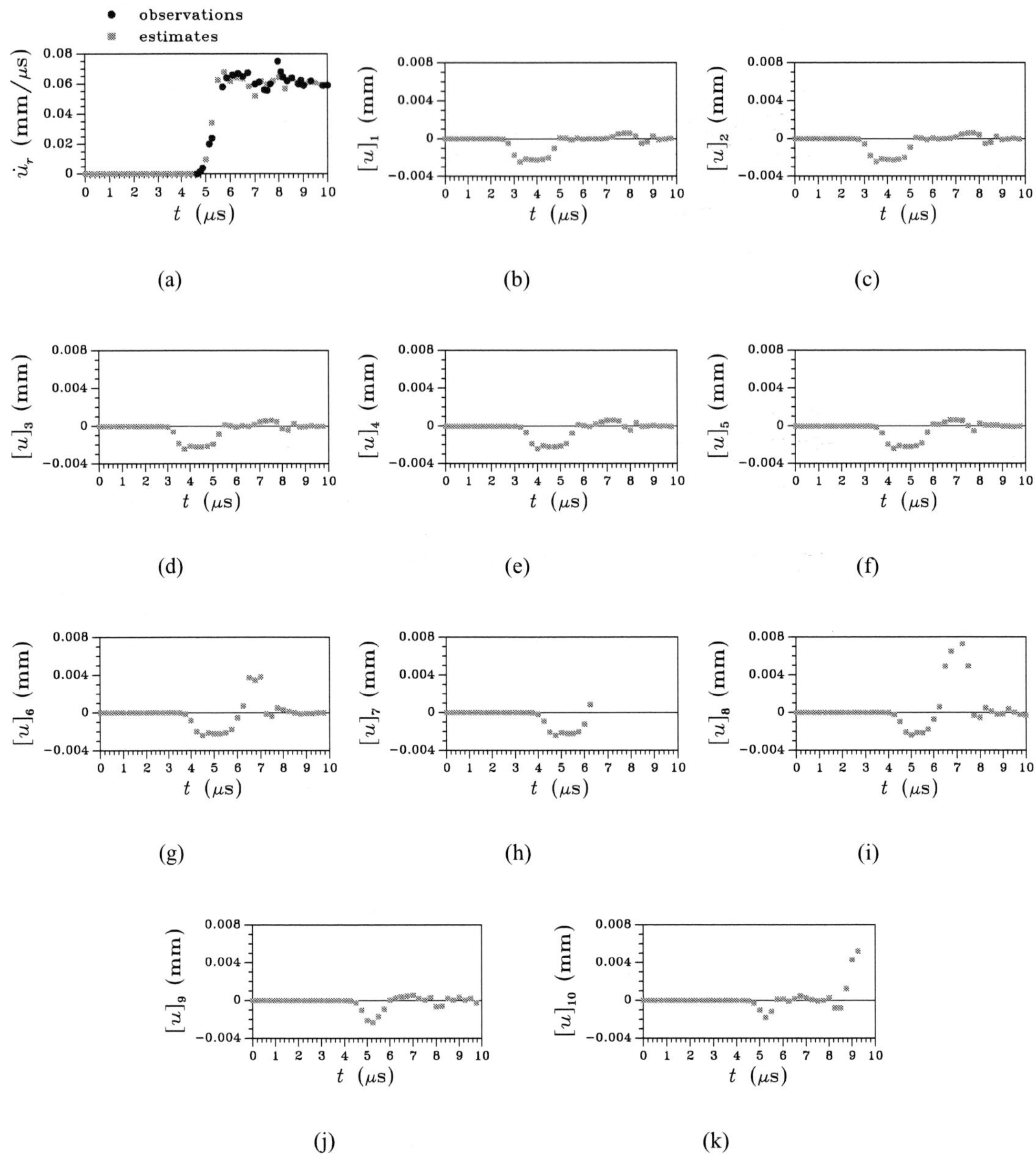

Figure 16. Impact on a 11+11-layer composite [14]. Evolution in time of (a) free surface velocity $\dot{u}_r$ and (b-k) estimated interface openings $[u]_1 - [u]_{10}$ in the back plate.

Results of the filtering process are reported in Figure 16 is terms of tracked free surface velocity and estimated displacement jumps along all the interfaces inside the back plate (sequence starts at the specimen-back plate contact surface). These estimations turn out once again to be independent of the interface law and of the initialization values of τ_M and

G inside the domain:

$$\mathcal{C}_{\vartheta} = \{25 \leq \tau_{M,0} \leq 100 \text{ (MPa)}, 0.1 \leq G_0 \leq 0.6 \text{ (N/mm)}\} \tag{54}$$

While the free surface velocity is accurately tracked, delamination is foreseen to take place along the 7-th interlaminar surfaces, in agreement with the results of [11, 14]. As far as model calibration is concerned, outcomes turn out to be qualitatively in agreement with those already reported for the previous tests.

5. Conclusion

In this Chapter we have addressed some issues related to constitutive modeling and parameter identification in finite element simulations of layered composites subject to impacts. Assuming the impact energy to be high enough to cause damage spreading inside the interlaminar resin-enriched phases, but not high enough to result in penetration of the impactor accompanied by intralaminar damage, a numerical scheme for structural-level analyses has been revised. Within this scheme the laminae are assumed to behave elastically, whereas dissipation mechanisms are lumped onto zero-thickness interlaminar surfaces. Along these interlaminar surfaces strength reduction, eventually leading to delamination, is governed by softening interface constitutive laws linking tractions to displacement jumps.

Interface laws are known to be difficult to calibrate, since direct testing on a single interlaminar phase can not be devised. Here we have offered a sigma-point Kalman filtering approach to estimate uncertain model parameters of the aforementioned interface laws. This technique outperforms most of the customarily adopted ones, since it efficiently deals with nonlinearities, which are a result of interlaminar strength degradation in the case under study.

The performances of the filtering procedure have been assessed through pseudo-experimental testings on a two-layer composite, and through real testings on multi-layer glass fiber reinforced plastic composites. It has been shown that the state of the composite, including delamination inception and growth, is always tracked with a noteworthy level of accuracy. Due to the fast failure events, model calibration is instead less accurately performed and sometimes requires initialization values of uncertain model parameters to be appropriately chosen to avoid biased estimates.

References

[1] S. Abrate. *Impact on composite structures*. Cambridge University Press, 1998.

[2] H.D. Espinosa, S. Dwivedi, P.D. Zavattieri, and G. Yuan. A numerical investigation of penetration in multilayered material/structure systems. *International Journal of Solids and Structures*, **35**:2975–3001, 1998.

[3] O. Allix and P. Ladevéze. Interlaminar interface modelling for the prediction of delamination. *Composite Structures*, **22**:235–242, 1992.

[4] A. Corigliano. Formulation, identification and use of interface elements in the numerical analysis of composite delamination. *International Journal of Solids and Structures*, **30**:2779–2811, 1993.

[5] A. Corigliano and O. Allix. Some aspects of interlaminar degradation in composites. *Computer Methods in Applied Mechanics and Engineering*, **185**:203–224, 2000.

[6] A. Corigliano and S. Mariani. Simulation of damage in composites by means of interface models: parameter identification. *Composites Science and Technology*, **61**:2299–2315, 2001.

[7] G. Alfano and M.A. Crisfield. Finite element interface models for the delamination analysis of laminated composites: mechanical and computational issues. *International Journal for Numerical Methods in Engineering*, **50**:1701–1736, 2001.

[8] A. Corigliano. Damage and fracture mechanics techniques for composite structures. In I. Milne, R.O. Ritchie, and B. Karihaloo, editors, *Comprehensive Structural Integrity*, volume 3, chapter 9, pages 459–539. Elsevier Science, 2003.

[9] A. Corigliano and S. Mariani. Parameter identification of a time-dependent elastic-damage interface model for the simulation of debonding in composites. *Composites Science and Technology*, **61**:191–203, 2001.

[10] A. Corigliano and S. Mariani. Identification of a constitutive model for the simulation of time-dependent interlaminar debonding processes in composites. *Computer Methods in Applied Mechanics and Engineering*, **191**:1861–1894, 2002.

[11] S. Mariani and A. Corigliano. Impact induced composite delamination: state and parameter identification via joint and dual extended Kalman filters. *Computer Methods in Applied Mechanics and Engineering*, **194**:5242–5272, 2005.

[12] S. Bittanti, G. Maier, and A. Nappi. Inverse problems in structural elastoplasticity: a Kalman filter approach. In A. Sawczuk and G. Bianchi, editors, *Plasticity today. Modelling, Methods and Applications*, pages 311–329. Elsevier Applied Science Publishers, 1984.

[13] A. Corigliano, S. Mariani, and B. Orsatti. Identification of Gurson-Tvergaard material model parameters via Kalman filtering technique - I. Theory. *International Journal of Fracture*, **104**:349–373, 2000.

[14] H.D. Espinosa, S. Dwivedi, and H.-C. Lu. Modeling impact induced delamination of woven fiber reinforced composites with contact/cohesive laws. *Computer Methods in Applied Mechanics and Engineering*, **183**:259–290, 2000.

[15] A. Corigliano and S. Mariani. Parameter identification in explicit structural dynamics: performance of the extended Kalman filter. *Computer Methods in Applied Mechanics and Engineering*, **193**:3807–3835, 2004.

[16] S.J. Julier, J. Uhlmann, and H.F. Durrant-Whyte. A new method for the nonlinear transformation of means and covariances in filters and estimators. *IEEE Transactions on Automatic Control*, **45**:477–482, 2000.

[17] T. Lefebvre, H. Bruyninckx, and J. De Schutter. Kalman filters for nonlinear systems: a comparison of performance. *International Journal of Control*, **77**:639–653, 2004.

[18] K. Ito and K. Xiong. Gaussian filters for nonlinear filtering problems. *IEEE Transactions on Automatic Control*, **45**:910–927, 2000.

[19] S. Mariani and A. Ghisi. Unscented Kalman filtering for nonlinear structural dynamics. *Nonlinear Dynamics*, **49**:131–150, 2007.

[20] A. Corigliano, S. Mariani, and A. Pandolfi. Numerical modeling of rate-dependent debonding processes in composites. *Composite Structures*, **61**:39–50, 2003.

[21] A. Corigliano, S. Mariani, and A. Pandolfi. Numerical analysis of rate-dependent dynamic composite delamination. *Composites Science and Technology*, **66**:766–775, 2006.

[22] R. Lund and J. Fish. Space-time multiscale laminated theory. *International Journal of Multiscale Computational Engineering*, **2**:26–47, 2004.

[23] S. Mariani, A. Pandolfi, and R. Pavani. Coupled space-time multiscale simulations of dynamic delamination tests. *Materials Science*, **23**:509–519, 2005.

[24] R.R. Talreja Y.W. Kwon, D.H. Allen, editor. *Multiscale Modeling and Simulation of Composite Materials and Structures*. Springer, 2007.

[25] G.T. Camacho and M. Ortiz. Computational modelling of impact damage in brittle materials. *International Journal of Solids and Structures*, **33**:2899–2938, 1996.

[26] Y. Arun Roy and R.H. Dodds. Simulation of ductile crack growth in thin alluminum panels using 3-D surface cohesive elements. *International Journal of Fracture*, **110**:21–45, 2001.

[27] S. Mariani and U. Perego. Extended finite element method for quasi-brittle fracture. *International Journal for Numerical Methods in Engineering*, **58**:103–126, 2003.

[28] C. Comi and S. Mariani. Extended finite element simulation of quasi-brittle fracture in functionally graded materials. *Computer Methods in Applied Mechanics and Engineering*, **196**:4013–4026, 2007.

[29] S. Li, M.D. Thouless, A.M. Waas, J.A. Schroeder, and P.D. Zavattieri. Mixed-mode cohesive-zone models for fracture of an adhesively bonded polymer matrix composite. *Engineering Fracture Mechanics*, **73**:64–78, 2006.

[30] J.H. Rose, J. Ferrante, and J.R. Smith. Universal binding energy curves for metals and bimetallic interfaces. *Physical Review Letters*, **47**:675–678, 1981.

[31] J.H. Rose, J.R. Smith, and J. Ferrante. Universal features of bonding in metals. *Physical Review B*, **28**:1835–1845, 1983.

[32] X.-P. Xu and A. Needleman. Numerical simulation of fast crack growth in brittle solids. *Journal of the Mechanics and Physics of Solids*, **42**:1397–1434, 1994.

[33] A. Needleman and A.J. Rosakis. The effect of bond strength and loading rate on the conditions governing the attainment of intersonic crack growth along interfaces. *Journal of the Mechanics and Physics of Solids*, **47**:2411–2449, 1999.

[34] C.M. Landis, T. Pardoen, and J.W. Hutchinson. Crack velocity dependent toughness in rate dependent materials. *Mechanics of Materials*, **32**:663–678, 2000.

[35] O. C. Zienkiewicz and R. L . Taylor. *The finite element method*, volume II. McGraw-Hill, London, 1989.

[36] N. Moës, J. Dolbow, and T. Belytschko. A finite element method for crack growth without remeshing. *International Journal for Numerical Methods in Engineering*, **46**:131–150, 1999.

[37] T. Strouboulis, K. Copps, and I. Babuška. The generalized finite element method. *Computer Methods in Applied Mechanics and Engineering*, **190**:4081–4193, 2001.

[38] G.N. Wells and L.J. Sluys. A new method for modelling cohesive cracks using finite elements. *International Journal for Numerical Methods in Engineering*, **50**:2667–2682, 2001.

[39] C. Comi, S. Mariani, and U. Perego. An extended FE strategy for transition from continuum damage to mode I cohesive crack propagation. *International Journal for Numerical and Analytical Methods in Geomechanics*, **31**:213–238, 2007.

[40] M. Jirasek and T. Zimmerman. Embedded crack model: II. Combination with smeared cracks. *International Journal for Numerical Methods in Engineering*, **50**:1291–1305, 2001.

[41] H.M. Hilber, T.J.R. Hughes, and R.L. Taylor. Improved numerical dissipation for time integration algorithms in structural dynamics. *Earthquake Engineering and Structural Dynamics*, **5**:283–292, 1977.

[42] T.J.R. Hughes. *The finite element method. Linear static and dynamic finite element analysis*. Dover Publications, 2000.

[43] L. Ljung. *System identification. Theory for the user*. Prentice Hall, second edition, 1999.

[44] A. Gelb. *Applied optimal estimation*. The M.I.T. Press, 1974.

[45] E.A. Wan and A.T. Nelson. Dual extended Kalman filter methods. In S. Haykin, editor, *Kalman filtering and neural networks*, pages 123–173. John Wiley & Sons, Inc., 2001.

[46] R. E. Kalman and R. S. Bucy. New results in linear filtering and prediction theory. *ASME Journal of Basic Engineering*, **83D**:95–108, 1961.

[47] L. Ljung. Asymptotic behavior of the extended Kalman filter as a parameter estimator for linear systems. *IEEE Transactions on Automatic Control*, **AC-24**:36–50, 1979.

[48] K. Reif, S. Günther, E. Yaz, and R. Unbehauen. Stochastic stability of the discrete-time extended Kalman filter. *IEEE Transactions on Automatic Control*, **44**:714–728, 1999.

[49] S.J. Julier and J.K. Uhlmann. *A general method for approximating nonlinear transformations of probability distributions.* Technical report, RRG, Department of Engineering Science, University of Oxford, 1996.

[50] S.J. Julier, J. Uhlmann, and H.F. Durrant-Whyte. A new approach for filtering nonlinear systems. In *Proceedings of the American Control Conference*, pages 1628–1632, Seattle, June 1995.

[51] E.A. Wan and R. van der Merwe. The unscented Kalman filter. In S. Haykin, editor, *Kalman filtering and neural networks*, pages 221–280. John Wiley & Sons, Inc., 2001.

[52] S.J. Julier. The scaled unscented transformation. In *Proceedings of the American Control Conference*, volume 6, pages 4555–4559, Anchorage, May 2002.

[53] F. Yuan, L. Tsai, V. Prakash, A.M. Rajendran, and D.P. Dandekar. Spall strength of glass fiber reinforced polymer composites. *International Journal of Solids and Structures*, **44**:7731–7747, 2007.

[54] G. Alfano. On the influence of the shape of the interface law on the application of cohesive-zone models. *Composites Science and Technology*, **66**:723–730, 2006.

In: Strength of Materials
Editors: G. Mendes and B. Lago, pp. 133-155
ISBN: 978-1-60741-500-8

Chapter 4

CURRENT STATE OF THE ART OF THE CERAMIC COMPOSITE MATERIAL BIOLOX®*DELTA*

Meinhard Kuntz[1], Bernard Masson[2] and Thomas Pandorf[1]
[1] CeramTec AG, Plochington, Germany
[2] CeramTec AG, Pechabou, France

Abstract

An extensive overview about the state of the art of the ceramic composite material BIOLOX®delta is given. The unique properties rely on a well defined alumina based fine composite microstructure which is mainly achieved by high temperature solid body reaction of the different ceramic phases during sintering. Zirconia comprises 17 % of the total volume. The tetragonal phase of zirconia is stabilized chemically and mechanically.

The high strength and toughness of the material depend on transformation toughening of the zirconia which is clearly shown by various experimental results. The excellent mechanical properties are reproduced batch by batch with a very low scatter.

The outstanding properties of the material BIOLOX®delta support advantageous properties of the final product, e.g. ceramic hard-hard bearings for hip arthroplasty. The burst load of the components is significantly increased. It is shown that the design of the components is also very important for the reliability and the ultimate properties of the system. Wear properties at severe conditions are significantly improved by using the new composite material BIOLOX®delta in comparison to pure alumina.

Phase transformation of zirconia from the tetragonal to the monoclinic phase due to hydrothermal aging is extensively discussed. Due to the particular distribution and stabilization of the zirconia particles instable aging effects are not possible in this material. After very long time of accelerated aging conditions an increase of monoclinic phase is found – however, it is shown that dynamic and static properties of BIOLOX®delta are not influenced by this effect.

1. Introduction

Since 2001 more than 500.000 artificial hip joints with components of the new high performance ceramic composite BIOLOX®*delta* have been successfully implanted on a global basis. Due to the unique strength and toughness of this material the risk of fracture has been substantially reduced when compared to conventional ceramic materials.

The outstanding properties of BIOLOX®*delta* rely on complex reinforcing mechanisms. Therefore, it is necessary to assess if reinforcement is maintained throughout the life-time of the artificial joint which is anticipated to exceed more than 20 years. Furthermore, it is shown that the challenging production of BIOLOX®*delta* is reproduced at a high quality from batch to batch.

Within the scope of this technical contribution the composite ceramic BIOLOX®*delta* is extensively described and analyzed. The composition and the material properties are presented based on data of regular production lots. It is shown that the advantageous properties of this material are based on the reinforcing mechanisms which are activated due to the unique composition of this material.

The particular effect of monoclinic phase transformation and hydrothermal aging is described in detail based on general mechanisms and specific analysis of phase transformation in BIOLOX®*delta.* Furthermore, experimental data are provided which describe the long term properties of the material, in particular with respect to hydrothermal phase transformation of zirconia in combination with wear and cyclic load conditions.

2. International Material Standards

BIOLOX®*delta* is a modern ceramic composite material for biomedical applications. The main components of the composite are alumina and zirconia. There are ISO standards available for bioceramics of high purity alumina (ISO 6474 - 1) and high purity zirconia (ISO 13356). However, these standards are not directly applicable for the composition of BIOLOX®*delta.* The ISO organisation is already on the way to prepare a new standard which is applicable for BIOLOX®*delta* and other similar composite materials (ISO 6474 - 2). The new standard will be released presumably in 2010.

Meanwhile, it is helpful to refer and compare the properties of BIOLOX®*delta* to the draft of the new ISO 6474-2 and the other international material standards which are applicable for related high purity bioceramics.

ISO 6474 – 1 Implants for Surgery – Ceramic Materials - Part 1: Ceramic Materials Based on High Purity Alumina

The current version of this standard was released in 1994. The material properties which are defined here reflect typical properties of high quality pure alumina. Strength and toughness which are required according to this standard are significantly lower than those which are available with the composite material BIOLOX®*delta.*

Today, the experts of the ISO working group agree that some details of the current ISO 6474 do not represent the state of the art. Thus, a new version of ISO 6474 - 1 is being preparedwhich is already published as a Draft International Standard (ISO/DIS). Most technical details in this version are already finally implemented. Some details (fracture toughness, microstructure) are still under discussion. Presumably, the final release of the new standard will be in 2009.

Comment: There is a similar ASTM standard F 603 for the same application and material type. The required material properties are comparable.

ISO 6474 – 2. Implants for Surgery – Ceramic Materials - Part 2: Composite Materials Based on a High Purity Alumina Matrix with Zirconia Reinforcement

A standard which is applicable for alumina zirconia composite materials is under preparation. Such composite materials are distinguished in those where the main phase is alumina (ZTA = zirconia toughened alumina) and those where zirconia is the dominating phase (ATZ= alumina toughened zirconia). Both material types are available for biomedical applications.

The basic physical properties (e.g. hardness, thermal conductivity) of a composite are primarily derived from the main phase. It is thus useful to describe alumina based zirconia toughened materials parallel to pure alumina materials. Consequently, the new standard has been proposed as part 2 of the established standard ISO 6474 for pure alumina. This concept has been discussed at the ISO TC 150 meetings and was approved by the international experts.

The new standard ISO 6474 - 2 is accepted as a Working Draft. It covers all material properties (except of Young´s modulus) which are defined in ISO 6474 - 1. Additionally, the specific subjects of hydrothermal aging and radioactivity, which are relevant for zirconia toughened materials is also included.

ISO 6474 – 2 is defined such that a broad range of inorganic compositions are included. A composition of $\geq$ 60 wt. % alumina and $\geq$ 10 wt. % zirconia is required. Other ingredients are allowed. The new standard is not exclusively designated for BIOLOX®*delta*.

So far there is no ASTM standard for alumina zirconia composite biomaterials available.

ISO 13 356. Implants for Surgery –Ceramic Materials Based on Yttria-Stabilized Tetragonal Zirconia (Y-TZP)

This standard was revised and published in 2008 as an International Standard (ISO).

In contrast to pure alumina, zirconia as a ceramic material can not be produced without a significant amount of other substances for phase stabilization. Several elements are known which are applicable. ISO 13356 is only focussed on Yttria as the stabilizing element. A specific range of Y content is predetermined. It should be noted that the typical range of Y in pure zirconia materials can be different to the required amount of Y in alumina zirconia composites. This issue is thoroughly discussed in chapter 4.

Pure zirconia bioceramics can be applied either for biomedical bearings (e.g. hip or knee) or for dental applications. ISO 13 356 does not distinguish between these different applications.

As a specific issue of zirconia a test for accelerated hydrothermal aging is required.

Table 1 gives an overview of the required properties of the 3 standards. The material properties according to the latest revised versions are chosen

Table 1. Required material properties according to the ISO standards 6474 – 1 (pure alumina), ISO 6474 – 2 (alumina – zirconia composite) and ISO 13 356 (zirconia).

ISO Standard		ISO 6474 - 1	ISO 13 356	ISO 6474 – 2
Material	**Unit**	**Pure Alumina**	**Zirconia**	**Alumina Zirconia**
Average Bulk density		≥ 3,94 g/cm^3	≥ 6,00 g/cm^3	≥ 98,7 %
Chemical Composition	wt%	Al_2O_3 ≥ 99,7 MgO ≤ 0,2 Impurities ≤ 0,1	ZrO_2+HfO_2+Y_2O_3 ≥ 99,0 Y_2O_3 4,5 – 6,0 HfO_2 ≤ 5 Al_2O_3 ≤ 0,5 Others ≤ 0,5	Al_2O_3 60 - 90 ZrO_2+HfO_2 10 - 30 Additives ≤ 10 Impurities ≤ 0,2
Grain Size	µm	MV ≤ 2,5 SD ≤ 40 %	MV ≤ 0,4	Al_2O_3 MV ≤ 1,5 ZrO_2 MV ≤ 0,6 SD ≤ 40 %
Strength Weibull Modulus (4 pt bending)	MPa	≥ 500 ≥ 8	≥ 800	≥ 1000 ≥ 10
Young´s modulus	GPa	≥ 380		
Fracture Toughness	MPa $\sqrt{m}$	≥ 2,5		≥ 4,0
Hardness HV1	GPa	≥ 18		≥ 17
Wear Resistance		Info		Info
Cyclic fatigue limit		No failure at 200 MPa	No failure at 320 MPa	No failure at 400 MPa
Amount of monoclinic phase	%		≤ 20	
Accelerated Aging			≤ 25 % monocl. phase strength decrease not more than 20%	Accomplish requirements described above
Radioactivity	Bq / kg		≤ 200	≤ 100

Note: The values of ISO 6474 – 1 & 2 are not finally fixed at the date of this publication.

3. Description of BIOLOX®*delta*

BIOLOX®*delta* is an alumina based composite ceramic. Approximately 80 vol.-% of the matrix consist of fine grained high purity alumina which is very similar to the well known material BIOLOX®*forte* (ISO 6474). As it is the case in any other composite material, the basic physical properties like stiffness, hardness, thermal conductivity etc. are mainly predetermined from the dominating phase. It was the basic idea for the development of the new material to preserve all the desirable properties of BIOLOX®*forte* - as an excellent bioceramic with more than 30 years clinical experience - but to increase its strength and toughness.

These properties are substantially improved by implementation of reinforcing elements. *Figure 1* shows the microstructure of BIOLOX®*delta*.

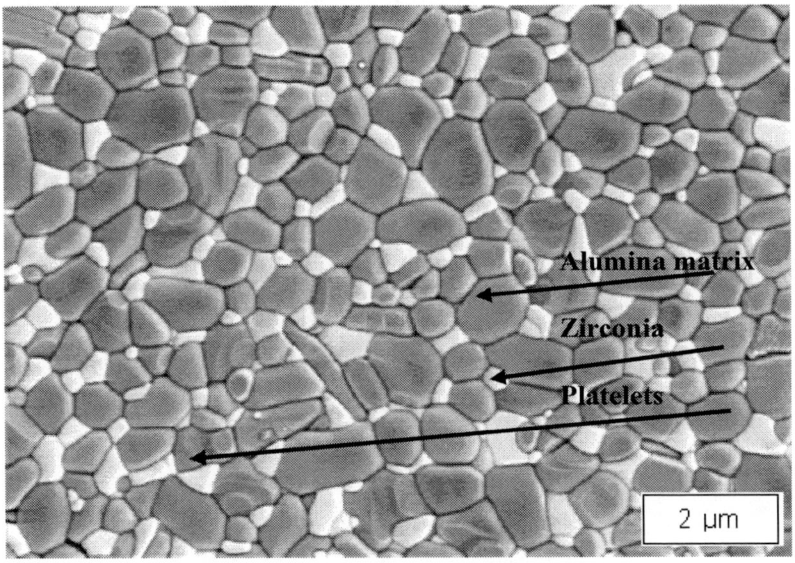

Figure 1. Microstructure of BIOLOX®delta.

Two reinforcing components are integrated into BIOLOX®*delta*. 17 vol.-% of the matrix consist of tetragonal zirconia particles. The average grain size of the zirconia is around 0.27 μm. As a further reinforcing element, approximately 3 vol.-% of the matrix are built by platelet shaped crystals of the ceramic composition strontium aluminate. The platelets stretch to a maximum length of approximately 5 μm with an aspect ratio of 5 – 10. The reinforcing ability of these ingredients is explained below.

Additionally to the reinforcing components, there are also stabilizing elements doped to the material. Chromium is added which is soluble in the alumina matrix and increases the hardness of the composite. The minor amount of chromium is the reason for the pink color of the material, see Figure 2. Furthermore, some yttrium is added to the composite which is solved in the zirconia and supports the stabilization of the tetragonal phase. In Table 2, the composition is given:

Table 2. Raw material specification for BIOLOX®delta

Ingredient	**Formula**	**Weight percent**
Yttrium oxide	Y_2O_3	1,4 – 2,0
Chromium oxide	Cr_2O_3	
Strontium oxide	SrO	
Zirconium oxide	ZrO_2	24,0 – 25,5
Other oxides	TiO_2, MgO, SiO_2, CaO, Fe_2O_3, Na_2O, K_2O	< 0,22
Alumina	Al_2O_3	balance

During thermal treatment of the material the ingredients are transformed to the particular composition with the 3 components. The basic transformation equations are known as follows:

$$2\,YCrO_3 \rightarrow Y_2O_3 + Cr_2O_3 \quad (1)$$

$$SrZrO_3 \rightarrow SrO + ZrO_2 \quad (2)$$

$$Al_2O_3 + Cr_2O_3 \rightarrow Al_2O_3 : Cr \quad (3)$$

$$SrO + 6Al_2O_3 : Cr \rightarrow SrAl_{12-x}Cr_xO_{19} \quad (4)$$

$$ZrO_2 + Y_2O_3 \rightarrow ZrO_2 : Y \quad (5)$$

In Table 3 the volume fractions of the final products according to equations (3) – (5) are given.

Table 3. Components of the final composite BIOLOX®delta

Component of the composite	Formula	Volume percent
Alumina, doped with Chromia	Al_2O_3:Cr	80 %
Zirconia with Y-stabilization	ZrO_2:Y	17%
Strontiumaluminate (minor Cr-content)	$SrAl_{12-x}Cr_xO_{19}$	3 %

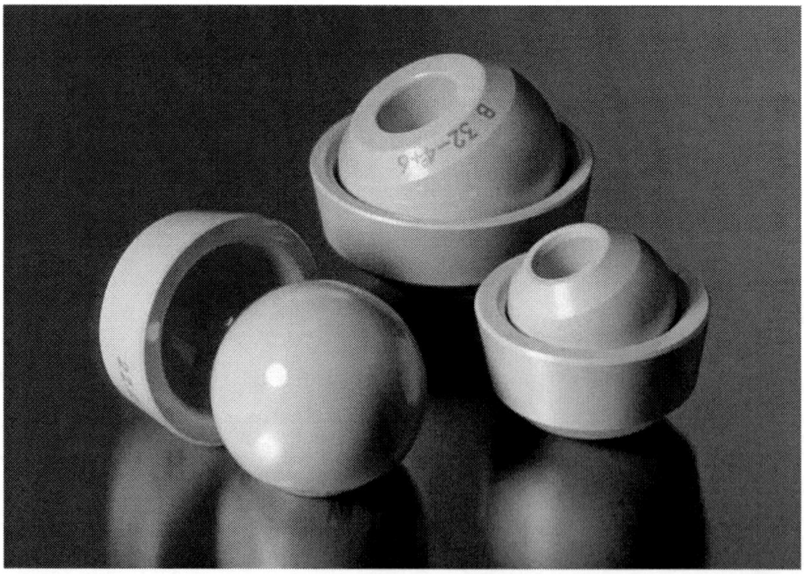

Figure 2. Ball heads and inserts of BIOLOX®delta.

4. Reinforcing Mechanism on BIOLOX® *Delta*

Benefit of Phase Transformation

The reinforcing elements, in particular zirconia, substantially increase fracture toughness and strength of the material. Fracture toughness (K_{IC}) is a measure for the ability of the material to withstand crack extension. Strength (σ_c) is defined as the maximum stress within a structure at failure of the component.

The correlation of strength and toughness is given in the fundamental equation of fracture mechanics:

$$K_{IC} = \sigma_c \sqrt{a_c}\, Y \tag{6}$$

where a_c is the size of a typical critical defect in the material and Y the shape factor. Consequently, when the fracture toughness of the alumina is increased also the strength is directly improved. This basic principle is the concept of the development of BIOLOX®*delta*. The microstructure is designed in order to provide an optimum of resistance against crack extension.

The benefit in crack resistance which is obtained from incorporating zirconia into an alumina matrix (as shown in Figure 3) is well known in the science of high performance ceramics.

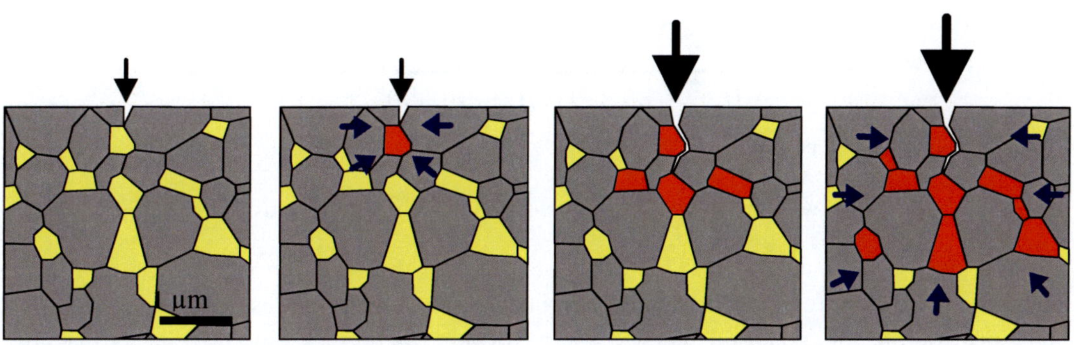

Figure 3. Reinforcing mechanism in BIOLOX®delta at crack initiation and propagation. Yellow particles represent tetragonal zirconia. Color change to red indicates monoclinic phase transformation. Arrows show the region of compressive stresses due to phase transformation.

The figure represents a realistic part of the microstructure. The gray particles refer to the alumina matrix, yellow to tetragonal zirconia. The phase transformation of zirconia is indicated by the change to red color. In the case of severe overloading crack initiation and crack extension will occur. High tensile stresses in the vicinity of the crack tip trigger the tetragonal to monoclinic phase transformation of the zirconia particles. The accompanied volume expansion leads to the formation of compressive stresses which are efficient for blocking the crack extension.

The model as represented in Figure 3 has also been verified experimentally. Pezzotti et.al. [PezTBP] analyzed the monoclinic phase transformation in the vicinity of an artificial crack tip as shown in Figure 4.

As it is demonstrated in Figure 3 this reinforcing mechanism is fully activated within a region of a few micrometers. For the macroscopic performance of the material it is very important that immediately at the beginning of crack initiation also the reinforcing mechanisms are activated. Regarding Figure 3 one should keep in mind that the average distance between the reinforcing zirconia particles is approx. 0,3µm, i.e. similar to the grain size. Thus, the reinforcement is activated immediately when any microcrack is initiated.

The reinforcing ability of zirconia particles is a consequence of the phase transformation, i.e. the spontaneous change from the tetragonal to the monoclinic phase [Han00]. The phase transformation is accompanied by a volume change of 4 % of the zirconia particle. Spontaneous phase transformation is a well known principle in material science. For example, the properties of high performance steels also rely on phase transformation from austenite to martensite.

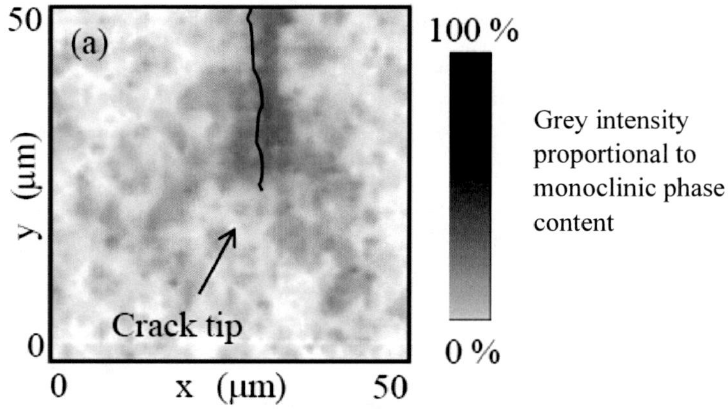

Raman Spectroscopy, G. Pezzotti.

Figure 4. Monoclinic phase transformation in the vicinity of an artificial crack tip.

It should be emphasized that the ability of phase transformation is the precondition for any benefit of the zirconia within the material. The composite is designed such that phase transformation occurs when it is needed, i.e. to prevent microcrack initiation and propagation at a high mechanical stress level.

Experiment: What Happens when Phase Transformation Is Suppressed?

It has been shown experimentally that the ability of zirconia phase transformation in BIOLOX®*delta* is necessary for the excellent mechanical properties. The experiment has been designed such that the experimental material was identically produced to BIOLOX®*delta* but with a significant higher amount of Y_2O_3. Yttria is known for stabilizing the tetragonal phase of zirconia. Consequently, in the case of a too high amount of yttria, the ability of phase transformation is suppressed. This has been shown in the experiment.

The experimental material W3530 has been produced equivalently to the production of BIOLOX®*delta.* In Figure 5 it is shown that the microstructure is identical. In *Table 4* the properties of the two materials are compared.

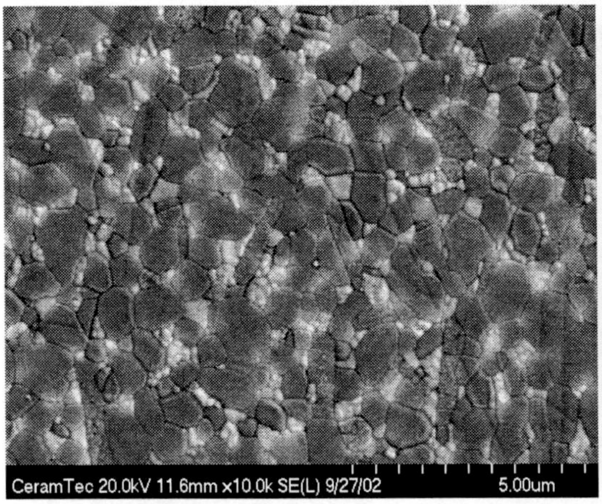

BIOLOX® *delta*

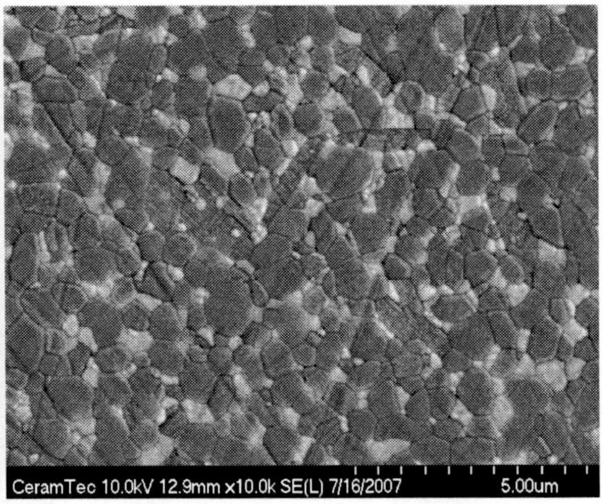

Experimental material W3530

Figure 5. Microstructure of regular BIOLOX®delta and experimental material W 3530

Table 4. Comparison of regular BIOLOX®delta and high stabilized experimental material

Material properties	BIOLOX®*delta*	W3530
Ratio Y_2O_3 / ZrO_2 [mol %]	1,3	3,0
Final density [g/cm^3]	4,37	4,38
Grain size [µm]	0,54	0,55
Strength [MPa]	1392	**777**
Hardness [HV1]	1757	1747
Monoclinic phase content [%]	5	**≤ 1**
Fracture toughness	6,5	**5,1**

The basic properties of regular BIOLOX®*delta* and the experimental material W3530 are identical, i.e. microstructure, density, grain size and hardness. The fundamental difference is the ratio of Y_2O_3/ZrO_2. In BIOLOX®*delta* the amount of yttria is significantly lower. As can be seen from the data, the higher amount of yttria in the experimental material leads - as expected - to a lower content of monoclinic phase, because phase transformation is suppressed. As a consequence the fracture toughness and the strength of the experimental material are much lower than the properties of BIOLOX®*delta*. In particular, the strength of the experimental material W3530 is only 55% of the normal strength of BIOLOX®*delta*. From this result it is immediately clear that the ability of phase transformation is necessary to obtain a high performance composite ceramic. The phase transformation can be easily suppressed by chemical stabilization (using yttria). However, suppressing phase transformation means loosing the excellent mechanical properties of the material.

Stabilization of the Zirconia Tetragonal Phase

In BIOLOX®*delta* the content of yttria has been optimized during the material development. It should be noted that the Y_2O_3/ZrO_2-ratio is lower than in normal pure zirconia materials (3Y-TZP), because the stabilization of the tetragonal phase in BIOLOX®*delta* is also influenced by "mechanical stabilization", i.e. the embedding of zirconia in the stiff alumina matrix.

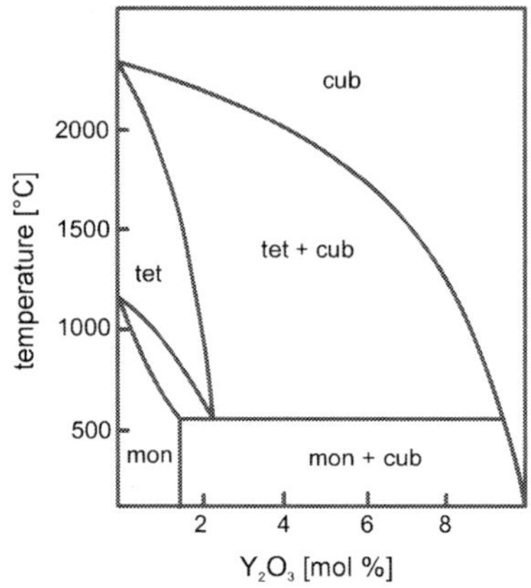

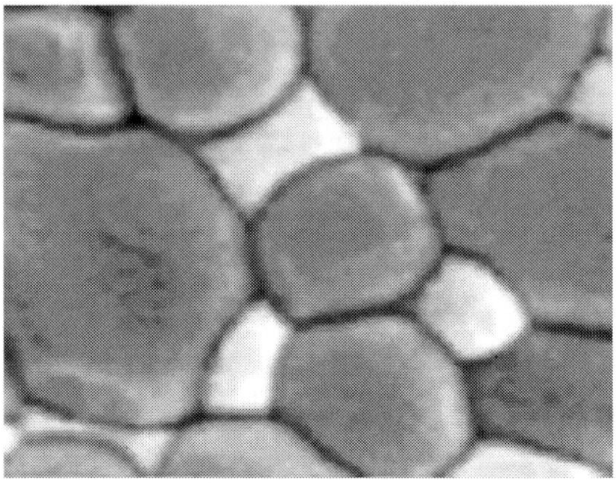

Figure 6. Background of stabilizing effects: Y-doping and embedding of zirconia particles in a matrix.

As can be seen in Figure 6 doping of Y_2O_3 into ZrO_2 shifts the temperature of tetragonal – to - monoclinic phase transformation towards lower temperatures. Thus, doping with Y means "*chemical stabilization*". At Y_2O_3 content lower 10% the stable phase at room temperature is monoclinic. So additionally the embedding of the zirconia particles in a matrix as well as surface stresses in the small particles also act as stabilizing mechanisms. The surrounding material will oppose the transformation and it is the strain energy that is involved in this constraint that allows the tetragonal phase to be retained at room temperature [Gre89]. This effect is referred to as "*mechanical stabilization*"

5. Material Production and Properties

BIOLOX®*delta* is a comparatively complex composite material where 4 different ingredients are mixed during powder preparation and undergo solid phase transformation at high temperature treatment as explained in chapter 2. The ability of reproducing such a material in high quantities with excellent quality batch by batch is the key qualification of CeramTec as manufacturer.

It is the purpose of this section both to summarize the important material properties and to elucidate the reproducibility of the production.

The important production and analytical steps are as follows:

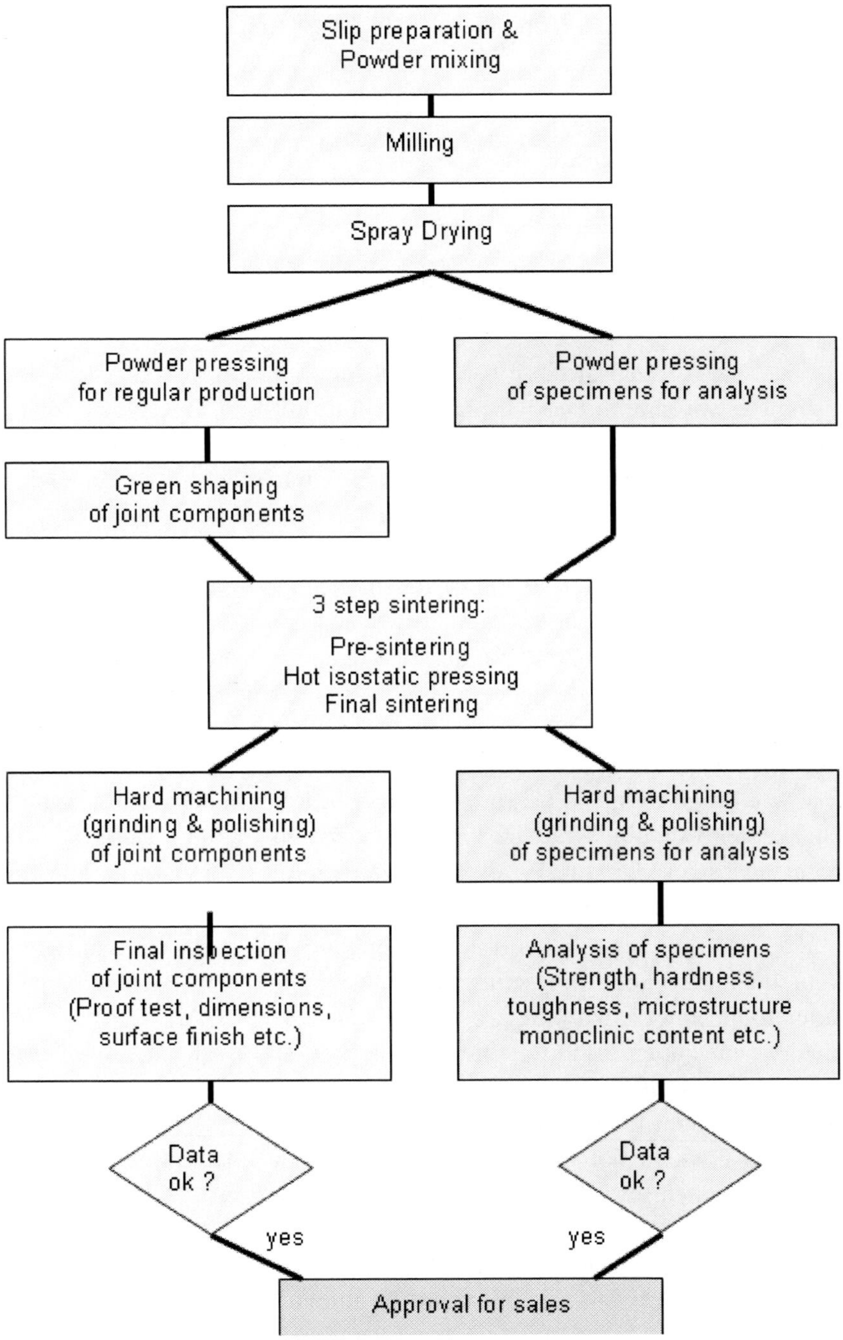

Figure 7. Schematic description of processing and material data generation of a single batch.

As shown in Figure 7 the material data are obtained for every powder batch. In Table 5 the material data as obtained for all powder batches in 2007 are summarized.

Table 5. Material properties of BIOLOX®delta batches in 2007

	Density	4-pt. Bending Strength	Weibull Modulus	Hardness HV 10	Monoclini c content	Grain Size Alumina	Fracture Toughness
	g/cm^3	MPa		GPa	%	µm	MPa m
Average	4,37	1411	14,9	17,2	4,4	0,54	6,4
Std. Dev.	0,007	50	3,1	0,09	1,1	0,027	0,20

The physical background of these parameters should be discussed. There are parameters which are almost invariable due to the physical nature of the property, in particular **density** and **hardness**. In the normal case of regular production there is only very little scatter with these data. However, it is important to analyze these parameters for every batch because any deviation of the expected results would indicate an insufficient production lot.

On the other hand, it is worth highlighting the low **alumina grain size** and the very low scatter of this value. During sintering and final densification of any ceramic the particles build a dense matrix but simultaneously grain growth also occurs. It is the goal of adequate sintering to achieve full density but to suppress grain growth, as a fine microstructure is necessary for good mechanical properties. Obviously, the sintering of BIOLOX®*delta* is very well reproduced batch by batch. The grain size is low in comparison to pure alumina because the dispersed zirconia particles prevent grain growth of the alumina matrix.

The **fracture toughness** is a measure for the reinforcing mechanisms in the material. As described in chapter 3 the high fracture toughness depends on the transformation mechanism of the zirconia particles. Obviously also this value shows very low scatter. The average fracture toughness is 6,4 MPa√m. In contrast, the "overstabilized" material W3530 described in chapter 3 has a significant lower fracture toughness K_{Ic} 5,1 MPa√m. Thus, it can be derived from the evaluation of fracture toughness from batch to batch that the desired transformation toughening is working properly.

As explained in chapter 3 the fracture toughness should be discussed in context with the **monoclinic phase content** which is determined on a polished flat surface of a specimen. The monoclinic phase content [in %] as obtained from the regular X-ray diffraction is *relative to the total zirconia content*, not to the total volume of the material. Thus, in any case the monoclinic phase content of the total material can be determined by simply referring to the zirconia fraction of 17 vol.%. Example: 10% monoclinic phase content is equivalent to 1,7% relative to the total volume of the material.

According to the materials specification the monoclinic phase content after polishing (*intrinsic monoclinic phase content*) is ≤ 10% of ZrO_2 which is regularly determined by X-ray diffraction. The sensitivity of this technique is around 1% monoclinic phase content. As it is evident from the data the average intrinsic monoclinic content of ZrO_2 of 4,4% is above the sensitivity limit. This indicates again that the sound material is in *disposition* of phase transformation. Note that the monoclinic content of the experimental material W3530 is below the sensitivity limit.

In contrast to the other parameters discussed above the **strength** of the material shows significantly higher scatter. It is important to understand that failure in ceramics is always triggered by imperfections of the microstructure. High performance ceramics can only be achieved when the natural defects in the material are very small. Typical relevant imperfections in BIOLOX®*delta* are within a range of 5 – 50µm. Accordingly, the scatter of natural defect size directly matches the scatter of strength which is described by the **Weibull's modulus**. A high modulus indicates low scatter. For the high performance material BIOLOX®*delta* a Weibull's modulus of ≥ 7 is tolerated in the specification. As can be seen from the data the normal scatter of the strength is much lower (i.e. higher modulus).

6. Correlation of Material and Component Properties

In chapter 5 the extensive efforts of analyzing the material properties of BIOLOX®*delta* batch by batch have been discussed. These material properties are determined according to ISO 6474 which is applicable for pure alumina and currently being extended for alumina-composite materials such as BIOLOX®*delta* (ISO 6474-2). This type of data is very familiar for evaluation of the performance of ceramics. In this chapter it is intended to discuss shortly how these material data correlate to component properties, e.g. the strength of ball heads and inserts.

In general, the performance of any system depends on the intrinsic material properties, the design and manufacturing quality of the components and the system, the external load and the particular environment, and finally the quality of mounting and installation. The use of high performance materials inevitably promotes the performance of a system - however, the other factors may be even more decisive for the success of a system. These complex correlations must be necessarily evaluated by design analysis, modeling, simulations, risk analysis and many other tools. In order to eliminate any influences of design features most of the material testing has been performed using 4-point bending bars.

In Figure 9 the setup of the regular 4-point bending test as recommended in ISO 6474-2 and the burst test according to ISO 7206 are shown. The bending test reveals the intrinsic strength of the material whereas the burst test is designed in order to simulate the in-vivo load of ceramic ball heads.

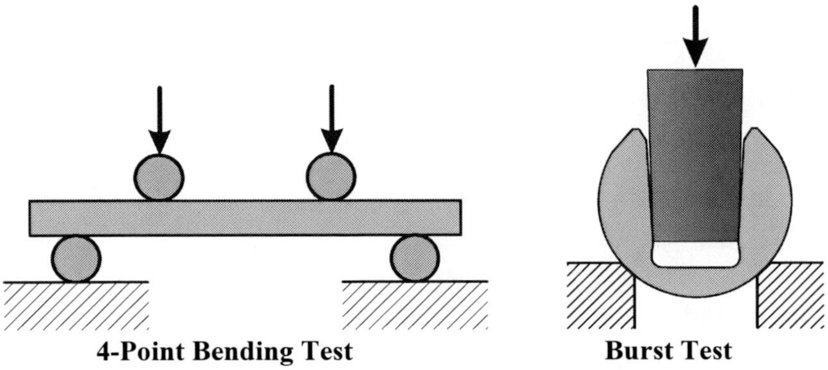

Figure 8. Schematic set-up of 4 point bending test and ball head burst test.

The strength measured with the bending test is – to a first approach - an intrinsic material parameter[1], whereas the burst load as obtained from the burst test depends on the materials strength and the design of the ball head and the taper. This is directly shown by comparison of strength and burst load of different ball heads made of BIOLOX®*forte* and BIOLOX®*delta,* see Table 6.

Table 6. Comparison of strength and burst load of BIOLOX®forte and BIOLOX®delta

Parameter	Test / Design	Unit	BIOLOX®*forte*	BIOLOX®*delta*	*ratio delta / forte*
Strength	4-point bending	MPa	620	1400	*2,3*
Burst load	28-12/14 L	kN	54	85	*1,6*
Burst load	36-12/14 M	kN	110	131	*1,2*

All burst loads are far above the required value of 46kN. The maximum in vivo load at worst case conditions is approximately 10 kN.

The data of the burst tests given in Table 6 are obtained from ball heads with identical geometry, Ti test taper and the same test setup. Thus, the advantage of BIOLOX®*delta* ball heads in the burst load only comes from the higher strength of the material in comparison to the pure alumina BIOLOX®*forte*. The strength of BIOLOX®*delta* is more than twice the strength of BIOLOX®*forte*, whereas the ratio of the burst strength values is lower. This is explained by the ductile deformation of the Ti taper during the burst test which steadily increases the contact area of the conical bore of the ceramic ball head and the metal taper. However, the burst load of identical ball heads is always higher when a high strength material is used.

It is also seen that the burst strength strongly depends on the ball head size. A larger ball head shows a higher burst load. It is concluded that the benefit of using a high performance material is higher when applied to a challenging design, e.g. a ball head with lower wall thickness. Nevertheless, the use of the high strength material always increases the safety margin of the component.

7. Wear Performance of BIOLOX® *Delta*

At normal wear conditions (e.g. standard wear simulator) the wear of a hard-hard couple of BIOLOX®*delta* is identical to the excellent performance of the well proven pure alumina BIOLOX®*forte.* There is only a minor difference in hardness of these both materials which does not compromise the normal wear behavior.

However, a significant advantage of BIOLOX®*delta* is identified in the case of *worst case* wear simulation as shown in *Figure 9.* In this experiment *microseparation* of ceramic ball head and insert during each load cycle has been simulated which leads to highly localized

[1] Due to the statistical nature the strength also slightly depends on the specimen size and the stress distribution. Size effects can be mathematically balanced. In order to obtain results which can be directly compared to each other the standardized set-up of the bending test should be used.

forces at the contact area. This experimental setup was supposed to simulate e.g. low tension of the soft tissue after surgery as it is discussed frequently by orthopedic experts. It was concluded from various retrievals that in some cases a well defined stripe-shaped area shows a more intense worn surface than the normal surface of the ball head. This phenomenon is known as “stripe wear”.

In the experiment, heavy wear conditions were simulated due to the highly localized contact area. It was found after 5 mio microseparation load cycles that the wear volume of BIOLOX®*delta* couples (both ceramic ball head and insert were made of BIOLOX®*delta)* was 7 times lower than that with the coupling made of pure alumina BIOLOX®*forte* [Cla05]. The wear rate of the mixed couplings (either ball head or insert made of BIOLOX®*delta,* the remaining made of pure alumina BIOLOX®*forte)* ranged between those values. It is important to mention that due to the small differential hardness between BIOLOX®*forte and* BIOLOX®*delta* can be combined in a ceramic-ceramic coupling without running the risk of excessive wear or other adverse effects.

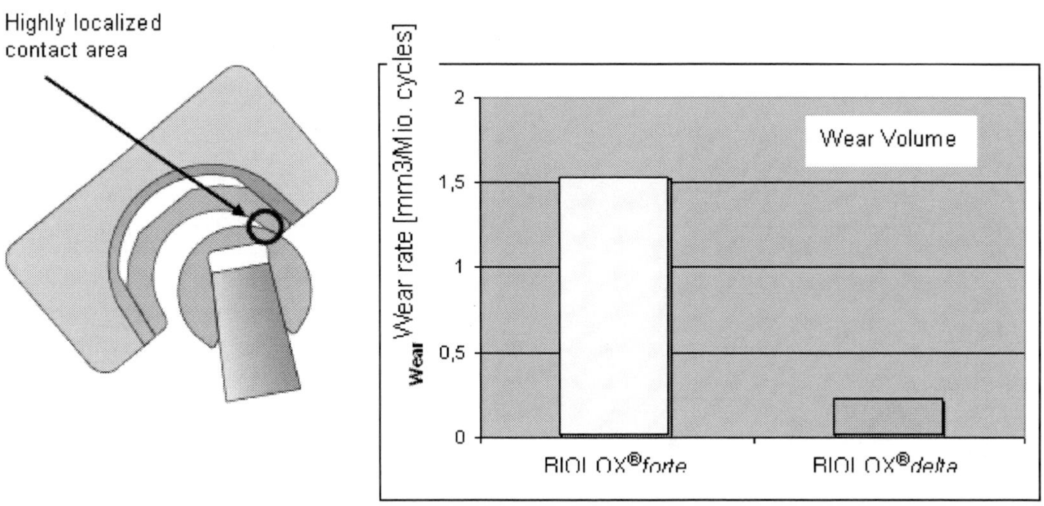

Figure 9. Wear performance of BIOLOX®delta and BIOLOX®forte at simulated micro separation.

Obviously BIOLOX®*delta* shows an excellent “*stripe wear tolerance*”. The analysis as given in [Cla05] shows that in the stripe wear region the monoclinic phase content is strongly increased. This indicates the mechanism behind the excellent stripe wear tolerance of BIOLOX®*delta.* As a first approach it is assumed that at these special test conditions a very high localized stress acts in the contact area which may be able to introduce damage in the surface. In this case the reinforcing mechanism as described in Figure 3 is activated in BIOLOX®*delta* which supports maintaining the surface quality under these extreme conditions. This assumption is supported by the Raman analysis of the worn surface. In the stripe wear region a high monoclinic phase content was found which indicates that phase transformation for reinforcement took place.

8. Discussion of Hydrothermal Aging

Mechanism of Hydrothermal Aging

Aging is a relevant issue for all zirconia containing ceramics. The transformation from the tetragonal to the monoclinic phase can be triggered in a hydrothermal environment. "*Hydrothermal*" means that this particular aging effect only takes place in aqueous environment at elevated temperatures. A critical temperature range for hydrothermal aging is around 120 – 200°C. However, a very slow aging effect is potentially possible even at human body environment. The kinetics and the threshold of aging activation strongly depend on the grain size, volume fraction, amount and type of stabilizing elements (e.g. Y, Ce, Mg, Ca) and mechanical stabilization of the zirconia [Che99].

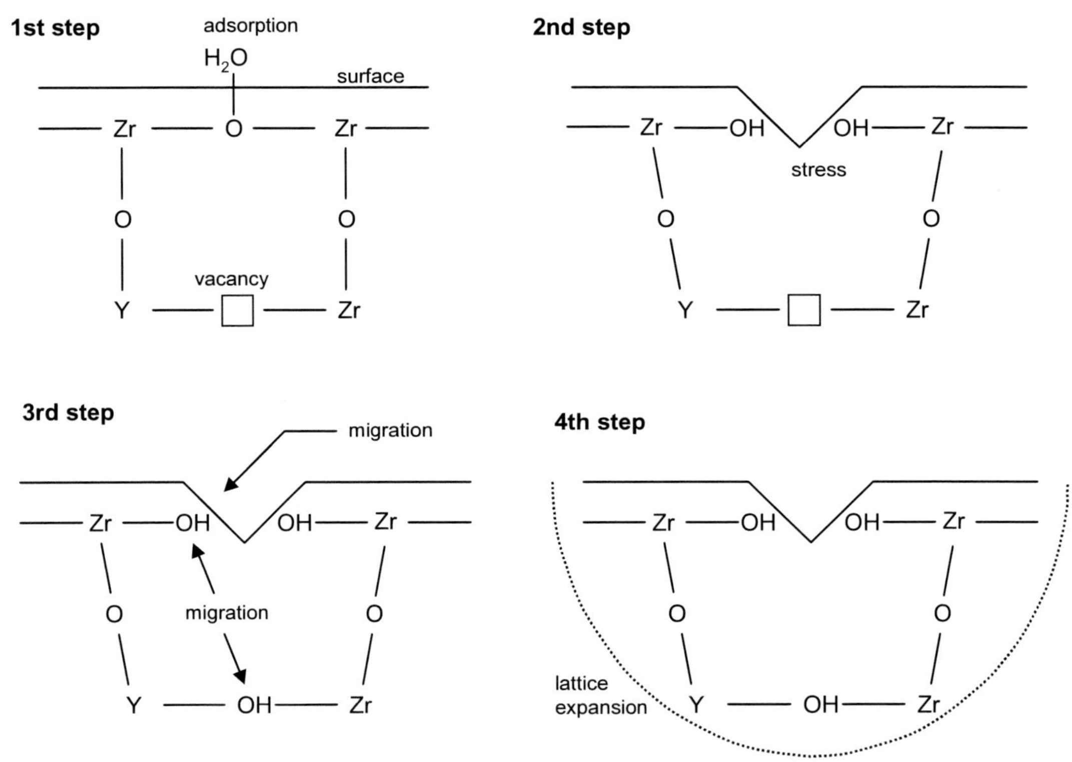

Figure 10. Model of hydrothermal aging.

Albeit the hydrothermal aging is extensively studied in the literature, the exact mechanism behind this effect is still not perfectly described. Most of the available models refer to interaction of the hydroxide ions and oxygen vacancies in the zirconia lattice. The oxygen vacancies inevitably are introduced in the lattice at solution of Y_2O_3 in ZrO_2 due to different valence of the cations. Figure 10 represents a model of the interaction of hydroxide ions with Y-doped zirconia with 4 steps:

1. Adsorption of water molecules at the surface
2. Dissolution of the oxygen bond

3. Migration of hydroxide ions to oxygen vacancy
4. Reorganisation of the atomic lattice

The critical step within this model is the dissolution of the oxygen bond which is very strong in zirconia. Thus, a thermal activation at elevated temperatures is required.

Hydrothermal Aging in BIOLOX®delta

Phase transformation caused by hydrothermal aging is an undesired effect. However, as it has been shown in the previous chapters, monoclinic phase transformation is necessary for the reinforcement and the high strength and toughness of the material BIOLOX®*delta*. Thus, a certain amount of phase transformation in hydrothermal environment is not a matter of concern. It depends on the specific composition of the material if there is a critical level of phase transformation where the material can be damaged. In the years 2001 – 2002 some batches of the pure zirconia material Prozyr® of the company Desmarquest showed catastrophic failure in-vivo due to monoclinic phase transformation at the surface [Che06]. It has to be emphasized that such a damage is impossible in BIOLOX®*delta* due to the fact that only 17vol% consist of zirconia. Consequently, the ultimate upper limit of monoclinic zirconia is only a total of 17% whereas 83% of the material remain not affected by phase transformation and hydrothermal aging.

As mentioned above, hydrothermal aging is accelerated at elevated temperatures. For pure zirconia in biomedical applications a standardized accelerated aging test is recommended in ISO 13 356. The aging conditions are autoclaving at 134°C and 2 bar water steam for 5h. In the new standard ISO 6474-2 being prepared for alumina zirconia composites an aging time of 10h is recommended. As a rough estimate it is proposed that 1h autoclaving is equivalent to 2 – 4 years in human body environment. However, this transfer of aging kinetics depends on individual properties of a material.

The aging behavior of BIOLOX®*delta* has been extensively analyzed using accelerated hydrothermal aging by autoclaving. Moreover, the aging has been combined with severe mechanical static and dynamic testing in order to understand if any material degradation occurs during aging.

In fact, the monoclinic phase content in BIOLOX®*delta* is increased after long term aging. In *Figure 11* quantitative analysis of monoclinic phase transformation using Raman spectroscopy are shown. The aging conditions (121°C, 1 bar) for this analysis were slightly different from the recommendation in the ISO standard. Equal to Figure 4 the gray intensity indicates monoclinic phase distribution. As can be seen the monoclinic content is increased after extreme long exposure time. After 300 h the monoclinic phase content reaches a relative value of 66,7%, i.e. 11,3% of the total volume of the composite are monoclinic zirconia.

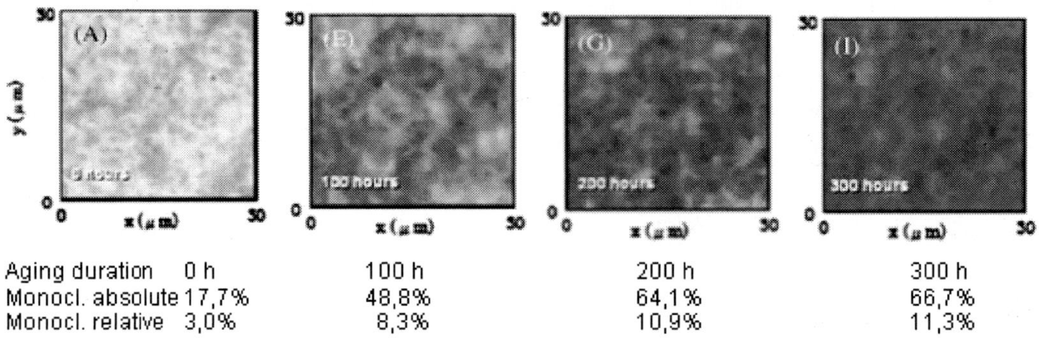

Figure 11. Increase of monoclinic phase content after very long term hydrothermal aging in an autoclave. Autoclaving conditions 121°C, 1 bar. Pictures adopted from [Pez08]. Quantitative monoclinic phase content obtained by direct communication to the author.

The autoclaving time used in this study is extremely long, much longer than recommended in the ISO standard. The monoclinic phase content seems to reach a saturation since the increase from 200h to 300 h is only marginal.

Aging Kinetics of BIOLOX®delta

As described above the aging effect of zirconia containing materials depends on the *individual composition.* It is of particular interest to estimate the aging effect of a bioceramic under in-vivo-conditions, i.e. 37°C in serum. As it was the case in the Prozyr® disaster a high amount of phase transformation occurred after a relatively short time in-vivo. On the other hand, it is well known that even pure zirconia is usually much more stable against hydrothermal aging when produced appropriately.

For a given material, the long term aging (in years) at low temperatures can be predicted based on the assumption that nucleation and growth of phase transformation are thermally activated. The thermal activation can be assessed by a systematic variation of temperature and duration time in hydrothermal environment and measurement of monoclinic phase transformation before and after the test. The most common method for such a prediction is based on the model of Mehl-Avrami-Johnson (MAJ-theory), which was first applied to zirconia based materials by Prof. Chevalier (Lyon, France) [Che99]. For example, it was found that for a conventional zirconia material with 3 mol% Yttria content (3Y-TZP) one hour in autoclaving conditions (134°C, 2 bar water steam) is equivalent to 3-4 years in-vivo. For other zirconia materials, a correlation of only 2 years was found. It is important to understand that this relation is not a universal law for zirconia containing materials but instead the individual outcome of the particular ceramic composition.

The aging kinetics of BIOLOX®*delta* was analyzed in close cooperation with Prof. Chevalier. The complete study will be accomplished and published in 2009. In the following, the preliminary results which are available at the current state are summarized.

The accelerated aging tests were performed in water steam at 142°C, 134°C, 121°C, and 105°C. Additionally one test was performed at 90°C in water. It is assumed that steam and liquid hydrothermal environment are comparable by means of the effect on hydrothermal aging. More tests have also been launched at 70°C and 50°C in water. However, due to the long duration time which is necessary to detect any effect at these temperatures results are expected within the next 2 years. The monoclinic phase content was determined using X-ray diffraction. In Figure 12 the results of the tests are shown on a logarithmic time scale.

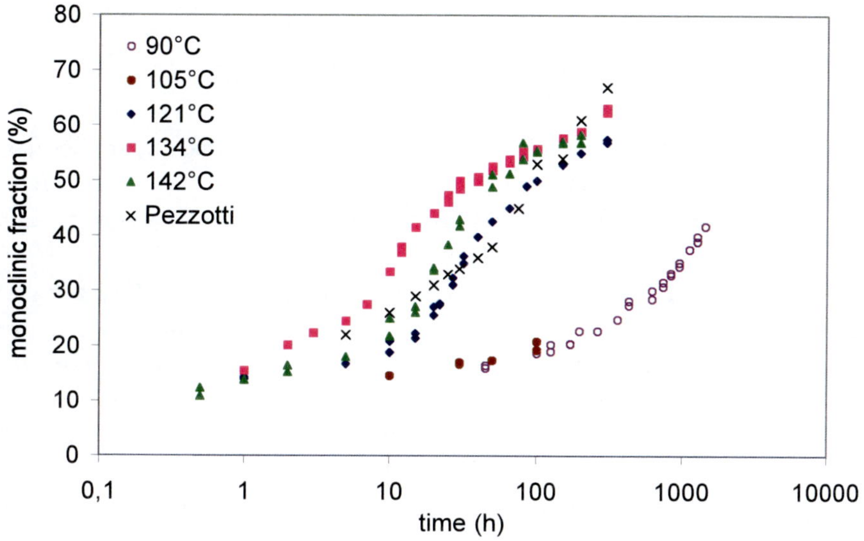

Figure 12.:Monoclinic phase transformation of BIOLOX®delta in hydrothermal environment. For comparison, also data from [Pez08] is included.

For evaluation of these data, the following modified MAJ equation was applied

$$V_m = V_m^0 + (V_{\max} - V_m^0)\left[1 - \exp\left(-(b\,t)^n\right)\right] \qquad (7)$$

where V_m is the monoclinic phase content, V_m^0 is the initial monoclinic phase content prior to the test, $V_{\max}$ is an apparent upper bound of the monoclinic phase, t is the time and n the time exponent. b is the factor which describes the temperature dependence of the aging effect. It is derived from fitting the data according to the following Arrhenius type equation:

$$b = b_0 \exp\left(-\frac{Q}{R\,T}\right) \qquad (8)$$

where b_0 is a material constant, Q is the activation energy, R the universal gas constant and T the temperature.

Evaluation of the data given in Figure 12 reveals an activation energy Q of 108 kJ/mol. This is a particular high value in comparison to a similar material (alumina zirconia composite) which was discussed in [Pez08], where an activation energy of 78kJ/mol is referred. The reason for this discrepancy is, as discussed above, the low amount of oxygen vacancies in BIOLOX®*delta* due to the unique chemical composition. High activation energy means *strong influence of temperature on aging*. Consequently at low temperatures the aging rate of in BIOLOX®*delta* is significantly lower.

From the modified MAJ equations the hydrothermal aging effect under in-vivo conditions can be estimated. The following equivalence to the accelerated hydrothermal aging was found:

1 h at **134°C** is equivalent to **3,9 years** in-vivo

According to the proposal of the new ISO standard 6474-2 hydrothermal aging is simulated under autoclaving conditions at 134°C for 10 hours. These conditions are thus equivalent to 39 years in-vivo for BIOLOX®*delta* which seems to be a realistic upper bound of the expected live time of an artificial joint.

Several accelerated aging tests according to these conditions (134°C, 2 bar water steam, 10 h) have been performed with BIOLOX®*delta*. The initial monoclinic phase content prior to aging depends on the surface finishing. On a ground surface the initial monoclinic phase content is higher than on a polished surface due to the more intensive interaction of the ceramic surface to the diamond tool. There is no effect on strength of the higher initial monoclinic phase content - the test specimens for the regular evaluation of strength are ground, not polished. Table 7 gives the data of monoclinic phase increase after 10 h autoclaving

Table 7. Analysis of zirconia phase transformation at accelerated aging

Surface finishing	Initial monoclinic phase content	Monoclinic phase content after 10 h autoclaving
5 %	8 %	
21 %	33 %	

There is obviously a large difference of initial monoclinic phase content between polishing and grinding. However, the *relative* increase of monoclinic phase transformation does not depend on the surface condition. As discussed in chapter 5 the monoclinic phase content is given relative to the total amount of zirconia. When the effect of phase transformation shall be assessed, it is useful to refer the monoclinic content to the total volume of the material, i.e. taking into account the zirconia volume fraction of 17 vol %. Thus, after grinding and autoclaving the absolute monoclinic phase volume content is 5,6% (33% × 0,17).

The accelerated aging according to Figure 11 was performed at 121°C 1 bar, i.e. a lower temperature and steam pressure than recommended in the ISO standards. It is also possible to evaluate the equivalence of this treatment to in-vivo conditions.

1 h at **121°C** is equivalent to **1,3 years** in-vivo.

It should be noted that there is a strong discrepancy of this equivalence factor to the data which was used and discussed in [Pez08], because in this publication the activation energy of a different material was used. It is evident that the aging kinetics must be carefully analyzed and discussed for the individual material.

Effect of Hydrothermal Aging on Strength of BIOLOX®delta

It is necessary to examine if the phase transformation reveals any damage or loss in strength of the material. For this purpose experiments were designed in order to combine accelerated aging and mechanical loading at a very high stress level. The specimens were prepared according to the 4-point bending configuration as it is shown in Figure 9 (left). These bend bars were exposed to accelerated aging and then to loading-unloading cycles at high stresses. After aging and cyclic loading the residual strength was determined.

Two stress levels (300 MPa and 600 MPa) were chosen for the cyclic loading tests. The lower stress level was applied for 20 Mio cycles, the higher stress level for 5 Mio cycles. All tests were performed in Ringer's solution. The accelerated aging was simulated by 5 h and 100 h treatment in standard autoclaving conditions (134°C, 2 bar water steam). It should be noted that again an aging time much longer than required from the standard was used. For each step of the experiments (as received, after aging, after aging and cycling) the monoclinic phase content was measured using X-ray diffraction.

As the most amazing result the yield of specimens surviving all the tests was 100 % in all cases. Even most severe conditions (i.e. 100 h autoclaving, 600 MPa cyclic load) did not reveal any premature failure. It should be recalled that this stress level represents 4 times the highest load level at worst case conditions in-vivo. We can thus conclude that the reliability of BIOLOX®*delta* exceeds by far the necessary requirements for reliable surgical components.

Table 8. Residual strength and monoclinic phase content after diverse treatments. Monoclinic phase is given relative to the total volume of the material

Autoclaving duration		**no cyclic load**	**300MPa, $20*10^6$ cycles**	**600MPa, $5*10^6$ cycles**
0 h	Strength [MPa]	1346	1433	1284
	Monoclinic phase content [%]	18	*33*	*43*
5 h	Strength [MPa]	1332	1248	1361
	Monoclinic phase content [%]	*22*	*35*	*42*
100 h	Strength [MPa]	1234	1308	1300
	Monoclinic phase content [%]	*30*	*33*	*47*

Table 8 shows the results of the post – test analysis including residual strength and monoclinic phase content. There is –as expected - a marginal natural scatter in residual strength. However, statistical analysis using Student's t-test did not reveal any significant deviation of all strength results.

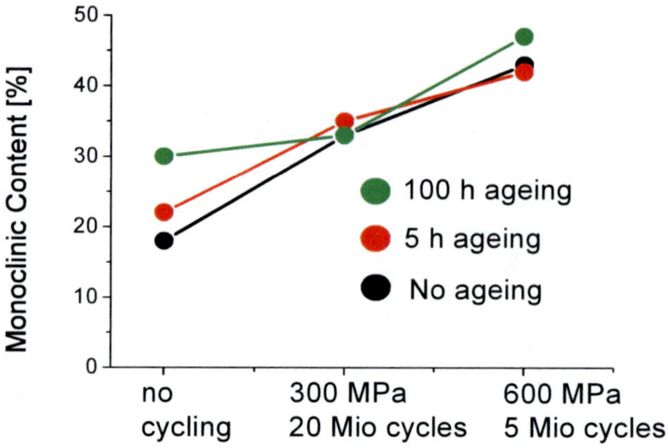

Figure 13. Increase of monoclinic phase content at cyclic loading.

In contrast, there is a clear tendency of an increase in monoclinic phase content both, after autoclaving and after cyclic loading, which is illustrated in Figure 13. For example, the test series without autoclaving shows an increase of monoclinic phase content from 18 % in the initial state to 47 % after 5 Mio cycles at 600 MPa. It must be concluded that the cyclic mechanical loading at a high stress level (600 MPa) which represents already the strength of pure alumina activated the reinforcing ability of the material. As discussed under Figure 3, a high mechanical stress triggers localized phase transformation which prevents any further crack propagation. Obviously the increased amount of monoclinic phase content does not deteriorate the strength of the material. This important conclusion is independent from the source of the phase transformation. In other words, when the phase transformation is activated either by accelerated aging, cyclic fatigue or a combination of both, the residual strength remains at the initial level.

References

[Che99] Chevalier J., Cales B., and Drouin J.M., Low temperature aging of Y-TZP ceramics, *J.Am.Cer.Soc.* **82** (1999) 2150-54

[Che06] Chevalier J., What future for zirconia as a biomaterial, *Biomaterials* **27** (2006) 535-543

[Cla05] Clarke I.C., Pezzotti G., Green D.D., Shirasu H., and Donaldson T., Severe Simulation Test for run-in ear of all-alumina compared to alumina composite THR, *Proceedings 10th BIOLOX Symposium*, 11-20 (2005)

[DeA02] De Aza, A.H., Chevalier J., Fantozzi G., Schehl M., and Torrecillas R., Crack growth resistance of alumina, zirconia and zirconia toughened alumina ceramics for joint prostheses, *Biomaterials* **23**, 937-945 (2002)

[Gre89] Green D.J., Hannink R.H.J., and Swain M.V., Tranformation Toughening of Ceramics, CRC Press Inc., Boca Raton, Florida (1989) ISBN 0-8493-6594-5

[Gre04] Gremillard L., Chevalier J., Epicier T., Deville S., and Fantozzi G., Modeling the ageing kinetics of zirconia ceramics, *J.Eur.Cer.Soc.*, **24**, 3483-3489 (2004)

[Han00] Hannink R.H.J., Kelly P.M., and Muddle B.C., Transformation Toughening in Zirconia-Containing Ceramics, *J.Am.Cer.Soc.* **83** [3] 461-87 (2000)

[Ohm99] Ohmichi N., Kamioka K., Ueda K., Matsui K. and Ohgai M., Phase Transformation of Zirconia Ceramics by Annealing in Hot Water, *J. Cer. Soc. Jap.,* **107** [2] 128-133 (1999)

[Pez06] Pezzotti G. Environmental Phase Stability of Next Generation Ceramic composite for Hip Prostheses, *Key Engineering Materials* Vols. 309-311, 1223-1226 (2006)

[Pez08] Pezzotti G. Yamada K., Sakakura S., and Pitto R.P., Raman Spectroscopic Analysis of Advanced Ceramic Composite for Hip Prosthesis, *J.Am.Ceram.Soc.* **91** [4] 1199-1206 (2008)

[PezTBP] Pezzotti G. Yamada K., Porporati A., Kuntz M., and Yamamoto K., *Raman Spectroscopic Analysis of Advanced Ceramic Composite for Hip Prosthesis: Part II Fracture Behavior* (to be published)

In: Strength of Materials
Editors: G. Mendes and B. Lago, pp. 157-182

ISBN: 978-1-60741-500-8

Chapter 5

PARTICLE MODELING AND ITS CURRENT SUCCESS IN THE SIMULATIONS OF DYNAMICS FRAGMENTATION OF SOLIDS

G. Wang[1*], A. Al-Ostaz[1], A.H.D. Cheng[1] and P. Radziszewski[2]
[1] Department of Civil Engineering, University of Mississippi, Oxford, MS, 38677-1848, USA
[2] Mechanical Engineering Department, McGill University, Montreal, QC, H3A 2T5, Canada

Abstract

Particle modeling (PM) is an innovative particulate dynamics based modeling approach. It has been demonstrated as a robust tool for simulating fracture problems of solids with dynamic fragmentation under extreme loading conditions. These loading conditions can include situations of collapse, impact, blasting or high strain rate tension/compression, as well as thermally-induced breakage problems.

Initially, PM was developed for the purpose of mimicking the microscopic material process at macroscopic level. This method can be conceptually illustrated by fully dynamic particles (or "quasi-particles") placed at the nodes of a lattice network without explicitly considering their geometric size. The potential can be specified for particle-particle interactions via axial springs. Theoretically, PM is an upscale of the molecular dynamics (MD) model applicable to various length scale problems. This is possible if a proper equivalent macroscopic potential is found, and, in case of lattice spacing decreasing to a few Angstroms, a MD model at zero Kelvin with, say, Leonard-Jones potential is recovered. In its current form, PM has been developed as a tool applicable to real engineering problems.

The advantages of PM over the existing discrete element based methods can be summarized as follows:

(1) Sample in theory. Four conservative/equivalent rules (mass, potential energy, Young's modulus and tensile/compression strength) are applied to preserve the equivalent material properties.

(2) Easy for implementation. Since the physical size of each particle is ignored other than its equivalent mass, the algorithm of coding a PM computation is fairly easy.

* E-mail address: gewang@olemiss.edu. Tel: + 662 915 5369, Fax: + 662 915 5523. Corresponding author: Ge Wang

Current research work has exhibited that PM is able to correctly predict dynamic fragmentation of materials with a good accuracy. In modeling an epoxy plate with randomly distributed holes in tension, the PM result of the final crack pattern compared favorably with the associate experiment; for the simulations of impact study of two polymeric materials (nylon, 6-6 and vinyl ester) subject to a rigid falling indenter, the modeling results of resistant force, energy, deflection and drop speed of indenter vs. time quantitatively agree fairly well with the according empirical observations.

1. Introduction

The prompt development of modern computer technology and computational methods enables scientists, more easily than ever before, to numerically examine the nature of materials, to verify or modify the existing theories, and even to discover new phenomena. Dynamic fracture modeling also greatly benefits from this advancement.

The dynamic fracture process itself is an exceedingly complex, multi-scale physical phenomenon. Material failure exhibits non-linearly, localized presence and a dependence on dynamic loading and loading conditions, etc. Hence, an ideal solver of dynamic fracture mechanics is required to be able to handle all the above-mentioned factors, especially to deal with the discontinuity of material that dominantly occurs in a dynamic fracture process which becomes an overwhelming difficulty to numerical approaches.

At the microscopic level, fracturing is a process that material becomes separated due to the successive failures of atomic bonds. Since the intrinsic strength properties at atomic structure level are available, molecular dynamics (MD) analysis has been used in modern scientific research at nano-scale. However, although MD simulation has benefited from the rapid development of modern computer technology and is becoming increasingly popular, the present state of computational power is still far from being able to support simulation at the macroscopic level. For example, we currently still cannot simulate a $1 \times 1 \times 1$ cm^3 cubic copper body at atomic level because the body consists of 10^{24} copper atoms, a number so large that no computer in the world can handle it. The second difficulty is its inability to reach the laboratory time scales. For instance, the laboratory fracture experiments generally last in microseconds (10^{-6} second), while the MD model time steps are typically in the nano (10^{-9}) or pico (10^{-12}) second range. As such, MD is limited narrowly to solving nano- to micrometer scale problems. Hence modeling fracture problems directly at macroscopic scale is a prevalent pursuit.

An all-round search to the existing optional modeling techniques reveals that the modern numerical methods for dynamic fracture simulations at macroscopic level can be generally classified into two categories of approaches. One is continuum mechanics based and the other is discrete element based. Examining the state-of-the-art of the research, we conclude that the continuum mechanics based approaches, such as the finite element method (FEM), have difficulty in solving dynamic fracture problems, particularly in dealing with the simulation of the collapse of structure and its fragmentation under extreme loading conditions. For this reason, another domain of alternative modeling approach —the discrete element based models, has become more and more popular.

Discrete element models share a common concept of "discrete material", which can still be further classified into two sub-categories. The first one is classified as a discrete element method, in which the physical size and shape of each element is considered in the computation, such as the applied element method (AEM) [Meguro & Tagel-Din, 2000], and particle finite element method (PFEM) [Oñate, et al., 2004] - another particle-related model. In practice, this type of models is complex and difficult to be implemented. For instance, AEM is laborious in keeping track of the instantaneous contact positions and the evolving geometry of all elements for a proper updating of the dynamic bonds; PFEM is expensive in remeshing and redefining boundaries at each time step.

The second branch is named particle (or particulate) dynamics method, in which the physical size and shape of each discrete cell is not explicitly considered in the computation. Consequently, each cell is treated as a particle, with the equivalent mass lumped at its center. Smoothed particle hydrodynamics (SPH) [Monaghan, 2005], and particle modeling (PM) are currently the mainly two popular particulate based models. SPH adopts a kernel probability density function to define for each particle a reaction domain at each time step, in terms of a particle number density; it is therefore highly expensive in computation. In contrast, PM is rather simple in theory and easy for implementation; hence, it is attracting increasing interest, despite a good number of SPH codes.

Particle modeling (PM) method was originally proposed by Greenspan [1981, 1997] for the purpose of mimicing the microscopic material process at macroscopic level. PM is fully dynamic with particles (or "quasi-particles") placed at the nodes of a periodic (equilateral triangular) lattice. Conventionally, only the nearest neighboring particles are accounted for in the interaction via axial spring connections. Similar to MD, potential can be specified for particle-particle interactions, and, in the case of lattice spacing decreasing to a few Angstroms, we recover a molecular dynamics (MD) model at zero Kelvin with, say, a Leonard-Jones potential. This implies that PM is an upscaled MD and can be applied to various length scale problems if a proper equivalent macroscopic potential is used. Obviously, by making use of the similar processes as atomic interactions, it is possible to solve fracture problems at macro-scale level. However, as no direct link with the material properties was attempted by Greenspan, Greenspan's PM method remains a conceptual model without a demonstrated success in fracture modeling.

Wang and Ostoja-Starzewski [2005a] developed a newly modified PM with a conventional linkage pattern as shown in Figure 1. In this new PM, equivalent material properties are formulated using four physical conditions to determine continuum-level Young's modulus and tensile strength, while maintaining the conservation of mass and energy of the particle system, and satisfying the interaction laws among all the particles. The theoretical foundations of the current PM can be outlined in the following four aspects:

(a) Derivation is based on equivalence/equality of mass, energy, elastic moduli and tensile strengths between both atomic and quasi-particle systems.
(b) Derivation is carried out in the setting of a 3-D *face-centered cubic* (*f.c.c*) lattice network (Figure 2) for both atomic and quasi-particle structures.
(c) Either Lennard-Jones, polynomial or quadratic potentials can be employed for nonlinear or linear axial linkage.
(d) In principle, it can be applied for various length scale problems of solids.

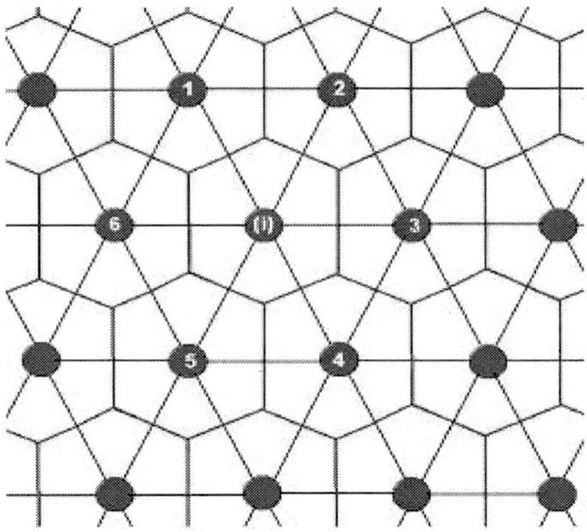

Wang and Ostoja-Starzewski, 2006.

Figure 1. Lattice structure in PM.

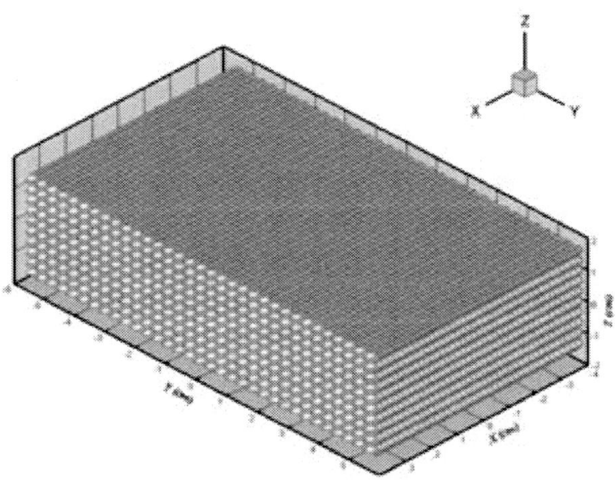

Wang and Ostoja-Starzewski, 2005a.

Figure 2. Meshing system for a 3D material body.

In this paper, current PM modeling development is to be addressed. In the coming sections, first we will introduce the theory of PM; then report some significant success and also defects experienced through the applications. Meanwhile, the potential solutions to the modeling deficiencies are proposed; and finally, in the conclusion part, an outlook of future-going PM development is addressed.

2. Methodology of PM

In particle modeling (PM), the nonlinear interaction force is considered between nearest-neighbor (quasi-) particles and assumed to be of the same form as in MD:

$$F = -\frac{G}{r^p} + \frac{H}{r^q} \tag{1}$$

Here *G*, *H*, *p* and *q* are positive constants, and $q > p \geq 1$ to obtain the repulsive effect that is necessarily (much) stronger than the attractive one, *r* being the distance between two particles.

Ashby & Jones [1980] presented a simple method to evaluate continuum-type Young's modulus *E* and tensile stress $\sigma(r)$ of the material from $F(r)$, namely

$$E = \frac{S_0}{r_0} \tag{2}$$

and

$$\sigma(r) = NF(r) \tag{3}$$

where $S_0 = (dF/dr)_{r=r_0}$, and r_0 is the equilibrium spacing between contiguous particles. *N* is the number of bonds/unit area, equal to $1/r_0^2$. Tensile strength, σ_{TS}, results when $dF(r_d)/dr = 0$, that yields,

$$\sigma_{TS} = NF(r_d) \tag{4}$$

Just as in MD, the non-linear dynamical equation of motion for each particle P_i of the PM system is given by

$$m_i \frac{d^2\vec{r}_i}{dt^2} = \sum \left[\left(-\frac{G_i}{r_{ij}^p} + \frac{H_i}{r_{ij}^q} \right) \frac{\vec{r}_{ji}}{r_{ij}} \right] \quad i \neq j, \tag{5}$$

where m_i and $\vec{r}_{ji}$ are mass of P_i and the vector from P_j to P_i. Note that if an equilateral triangular lattice structure is adopted in 2-D, the resulting Poisson ratio equals 1/4 (or 1/3) when a 3-D (respectively, plane) elasticity formulation is adopted [Ostoja-Starzewski, 2002, 2007].

The derivation of four parameters in Equation (1) from MD structures is conducted on a cubic body with volume $V (= A \times B \times C)$, in Figure 2. A face-centered cubic (*f.c.c*) lattice for both atomic and quasi-particle structures is chosen. If *p*, *q* and r_0 are given, then, by conditions of mass and energy conservation, *G* and *H* can be derived. Consequently, Young's modulus is evaluated by Equation (2) and tensile strength by Equation (4). To represent an expected material property, we would have to do many sets of testing until a unique (p, q) is found to match both Young's modulus and tensile strength of the material. The complete derivation process is described below.

First, for the atomic structure (MD model), we have:

interaction potential energy (*ergs*): $$\phi_a = \left(\frac{G_a r^{1-p_a}}{1-p_a} + \frac{H_a r^{1-q_a}}{1-q_a} \right) \times 10^{-8} \qquad (6)$$

Young's modulus (*GPa*) is obtained from Equation (2) and tensile strength (*MPa*) from Equation (4).

Total number of atoms in $A \times B \times C$ cubic material body:

$$N^* = \left(\frac{A \times 10^8}{r_a} + 1 \right) \times \left(\frac{B \times 10^8}{r_a \sin 60^0} + 1 \right) \times \left(\frac{C \times 10^8}{r_a \sqrt{6}/3} + 1 \right) \qquad (7)$$

In Eqs. (6) and (7), r_a is equilibrium position of the simulated material in atomic structure, and p_a, q_a are the exponential parameters in atomic structure. Note that, for a Lennard-Jones interaction case, $p_a = 7$ and $q_a = 13$.

Next, for the quasi-particle structure (PM model), we have interaction force (*dynes*) as in Equation (1).

Interaction potential energy (*ergs*):

$$\phi = \frac{G\, r^{1-p}}{1-p} + \frac{H\, r^{1-q}}{1-q}, \text{ for } p > 1; \ \phi = G \ln r + \frac{H\, r^{1-q}}{1-q}, \text{ for } p = 1 \qquad (8)$$

total number of quasi-particles in PM system: $$N = i_{\max} \times j_{\max} \times k_{\max} \qquad (9)$$

We now postulate the equivalence of MD and PM models. From the mass conservation, we calculate the mass of each quasi-particle *m* based on atomic mass m_a:

$$m = N^* \times m_a / N \qquad (10)$$

From the energy conservation, we have:

$$(N \times \phi)_{r=r_0} = (N^* \times \phi_a)_{r=r_a} \qquad (11)$$

under the requirement: $$F(r_0) = 0 \qquad (12)$$

From equations (11), (12), we now derive Young's modulus *E*:

for $p = 1$:

$$G = Hr_o^{1-q}, \quad H = \frac{\left(N^* \times \phi_a\right)_{r=r_a}(1-q)}{N(1-q)r_0^{1-q}\ln r_0 - r_0^{1-q}}, \quad E = -Gr_0^{-3} + qHr_0^{-q-2} \tag{13}$$

for $p > 1$:

$$G = Hr_o^{1-q}, \ H = \frac{\left(N^* \times \phi_a\right)_{r=r_a}(1-p)(1-q)}{N(p-q)} r_0^{q-1}, \ E = -pGr_0^{-p-2} + qHr_0^{-q-2} \tag{14}$$

Similarly, tensile strength can be obtained under $dF(r_d)/dr = 0$. Evidently, the four parameters (p,q), r_0 and V affect E and σ_{TS}.

We have established the equations for G, H, p and q, and carried out a parametric study to find the differing effects on p, q, V and r_0 [*Wang and* Ostoja-Starzewski, 2005a]. Herein, we summarize the obtained rules as follows:

(i) The larger the values of (p,q) are adopted, the larger is E generated. This is typically associated with the material becoming more brittle than ductile, albeit there is a range of toughness to choose from. Also, with E going up, there is a fragmentation into a larger number of pieces.
(ii) In the case of $p = 1$, the larger r_0 spacing is adopted, the higher is Young's modulus of the PM material. On the contrary, in the special case of $p \neq 1$, there is an opposite trend. In any case, this increase or decrease does not change very much.
(iii) In the case of $p \neq 1$, while keeping the volume fixed, an increase of r_0 produces a decrease of Young's modulus. The situation is again opposite in the case of $p = 1$.
(iv) A uniform augmentation of volume V by dilation in all three coordinate directions (*xyz*), at any (p,q) combination, results in Young's modulus increasing first strongly and then leveling off.

For brittle materials, a general format of linear dynamical equation is often employed [Wang, et al., 2008a],

$$F = \begin{cases} -S_0(r - r_0) & \text{for } r_c \le r \le r_t \\ 0 & \text{otherwise} \end{cases} \tag{15}$$

with r being the distance between two particles, the stiffness $S_0 = E \bullet r_0$ by Equation (2), E the Young modulus and $r_{\max}$ the failure displacement of material, r_0 the equilibrium spacing between the contiguous particles.

In Equation (15), r_c and r_t are the fracture positions applied for compression and tension, respectively, which need to be empirically determined.

The leapfrog method, with second-order accuracy, is employed in all PM simulations. The leapfrog formulas relating position, velocity and acceleration for particles P_i ($i = 1,2,...,N$) [Greenspan, 1997] are

$$\vec{V}_{i,1/2} = \vec{V}_{i,0} + \frac{(\Delta t)}{2}\vec{a}_{i,0} \quad \text{(starter formula)} \tag{16}$$

$$\vec{V}_{i,k+1/2} = \vec{V}_{i,k-1/2} + (\Delta t)\vec{a}_{i,k}, \quad k = 1,2,3,... \tag{17}$$

$$\vec{r}_{i,k+1} = \vec{r}_{i,k} + (\Delta t)\vec{V}_{i,k+1/2} \quad k = 0,1,2,... \tag{18}$$

where $\vec{V}_{i,k}$, $\vec{a}_{i,k}$ and $\vec{r}_{i,k}$ are the velocity, acceleration and position vectors of particle i at time $t_k = k\Delta t$, Δt is the time step. $\vec{V}_{i,k+1/2}$ stands for the velocity of particle i at time $t_k = (k+1/2)\Delta t$, and so on. Notably, the leapfrog method is of second-order accuracy: $O((\Delta t)^2)$.

The safe time step is after the derivation result by Hockney & Eastwood [1999]:

$$\Omega\Delta t << 2, \qquad \Omega = (\frac{1}{m}\left|\frac{dF}{dr}\right|_{\max})^{1/2} \tag{19}$$

To readily describe the breakage effect on material, we define a concept of fracture density [Wang, et al., 2008b]. By this definition, the local fracture density of particle i, $f_{i_{den.}}$, is equal to the ratio of its current number of broken bonds, N_{b_i} to its original number of bonds, N_{o_i}, i.e.,

$$f_{i_{den.}} = \frac{N_{b_i}}{N_{o_i}} \tag{20}$$

It is clearly seen that a big $f_{i_{den.}}$ value indicates a severe failure locally occurring at *i*.

3. PM Applications: Success and Deficiencies

3.1. Success of PM Applications

Since the establishment of the newly modified PM [Wang and Ostoja-Starzewski, 2005a] up to now, particle modeling (PM) technique has been constantly improved and has found good success in a number of applications including impact induced dynamic crack propagation and fragmentation [Wang et al., 2008a-f; 2007a-d; 2006; 2005a-d; Ostoja-Starzewski and Wang, 2006, 2005, 2004].

3.1.1. Validation Work

The first successful application of PM has been achieved for simulation of dynamic fragmentation in an (elastic-brittle) epoxy plate (8.25 *cm* x 33.02 *cm*), containing non-uniformly distributed circular holes in tension [Ostoja-Starzewski & Wang, 2006]. As demonstrated in Figure 3(a, b), PM forecasting of the crack pattern agrees well with the associated empirical observation [Al-Ostaz & Jasiuk, 1997].

(a) (b)

Ostoja-Starzewski and Wang, 2006.

Figure 3. Experimental and modeling results of epoxy in tension, (a) experiment [Al-Ostaz and Jasiuk, 1997], (b) PM simulation.

Other two validations of the PM are conducted by simulating the impact of a rigid indenter on polymeric materials (nylon-6, 6 and vinyl ester), the associated experiments conducted at the University of Mississippi. Figure 4(a, b) shows the comparison of the experimental result of fracture pattern of nylon-6,6 due to the impact of a rigid indenter while Figure 5(a, b) shows the comparison of the similar impact study of vinyl ester.

From Figure 4(a, b), it found that (1) PM modeling crack pattern agrees with the associated observation, and (2) it is seen that measured load peak happens around $t \approx 1.70$ ms, and the measured deflection at the load peak is 3.03 mm, with the total impact energy equal to 1.2 J. The corresponding PM simulated result shows that the load peak happens around $t \approx 1.66$ ms, and the deflection at load peak is 3.0 mm, and the total impact energy calculated is 1.2 J. Although the simulated load profile is not exactly the same as the experiment, we observe similar characteristics, including the fluctuating profile with roughly the same period. The simulated peak load is also reasonable close to the experimental value. Hence we conclude that the PM simulation compares favorably with the experimental measurements.

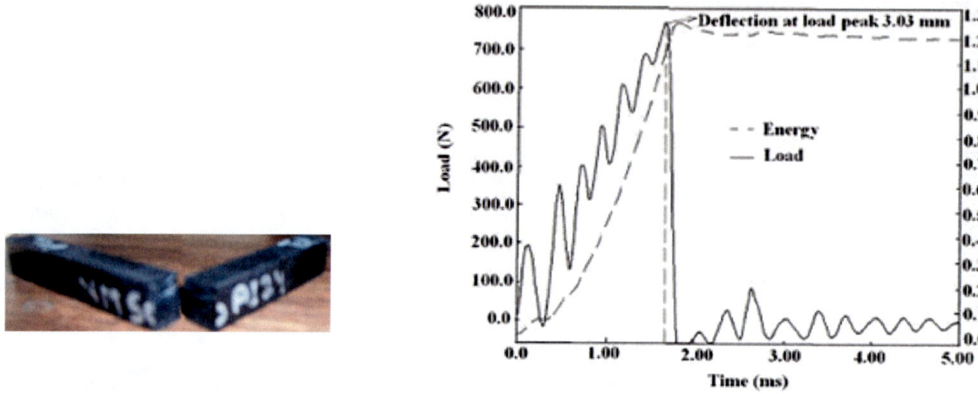

(a) Experimental results

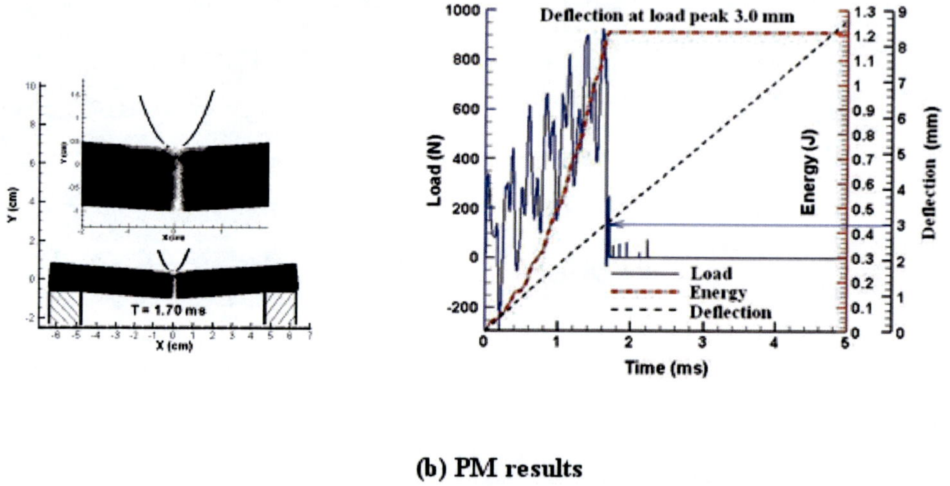

(b) PM results

Wang, et al., 2008a.

Figure 4. The study of the failure of nylon-6, 6 due to the impact of a rigid indenter (a) experimental results; (b) PM results. Maximum drop velocity of indenter is 1.87 m/s.

From Figure 5(a, b), it is seen that measured load peak happens around $t \approx 1.38$ ms, and the measured deflection at the load peak is 2.60 mm, with the total impact energy equal to 0.7 J. The corresponding PM simulated result shows that the load peak happens around $t \approx 1.42$ ms, and the deflection at load peak is 2.66 mm, and the total impact energy calculated is 0.76 J. Similarly to Figure 4(a, b), although the simulated load profile is not exactly the same as the experiment, we observe similar characteristics, including the fluctuating profile with roughly the same period. The simulated peak load is also reasonable close to the experimental value. Hence we conclude that the PM simulation compares favorably with the experimental measurements.

After the confidence with the model is gained, a number of typical extreme loading problems can be explored with PM technique. Next, some selected PM applications will be presented.

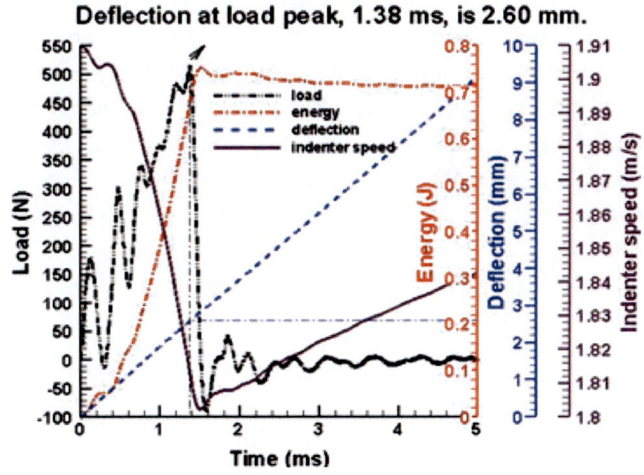

(a) Experimental results

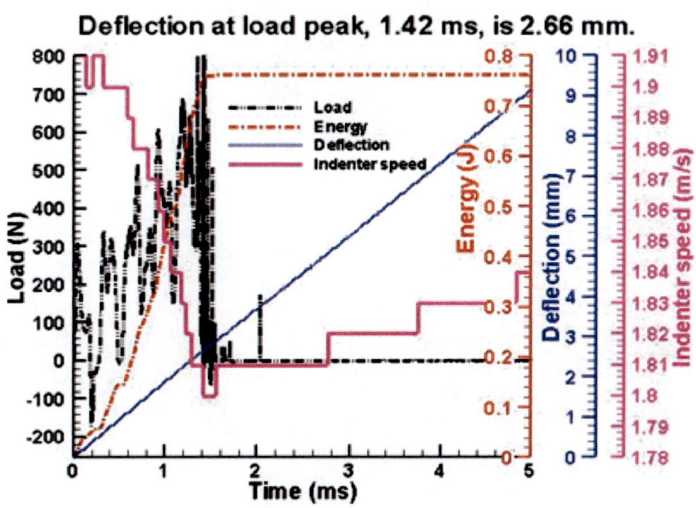

(b) PM results

Experiment conducted at the University of Mississippi.

Figure 5. Comparison of experimental and modeling results of vinyl ester due to the impact of a rigid indenter: load, energy, deflection and indenter speed curves versus time. Maximum drop velocity of indenter is -1.91 *m/s*.

3.1.2. Miscellaneous Applications

(A) High Speed Collision and High Strain Rate Tension/Compression of Material Blocks

Simulation of high speed collision of material blocks is quite a challenge to continuum based models as dynamic discontinuity of material geometry is impossibly predefined and stress singularity must be elaborately considered in the vicinity of cracks. However, the basic treatment of PM, in which fracture is created when a bond (spring) is broken by translational force, provides PM a unique power to be able to quite easily overcome these problems. In PM, for each particle, the fracture density of bond illustrated in Equation (20) can plot exactly its connection with the remaining assembly of material. Thus, the discontinuity of material can be easily traced. Else, since the consideration of PM is not based on the stress intensity factor, energy release rate, or the plastic process zone near or at the crack tip, therefore, in reality PM does not create a stress singularity. It is worth to point out that, although the near fracture tip behavior cannot be modeled as the classical continuum fracture mechanics, following Equation (1) PM can model the nonlinear constitutive behavior near the crack, as well as everywhere else.

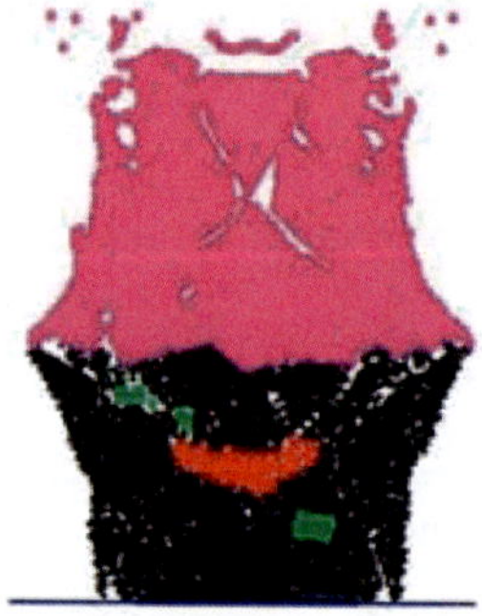

(a)

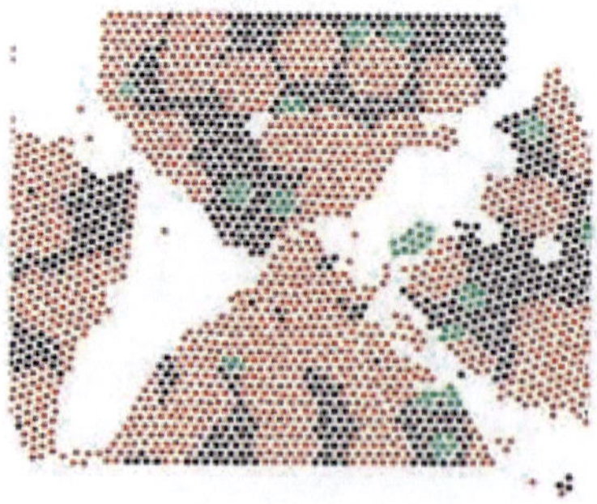

(b)

Figure 6. Continued on next page.

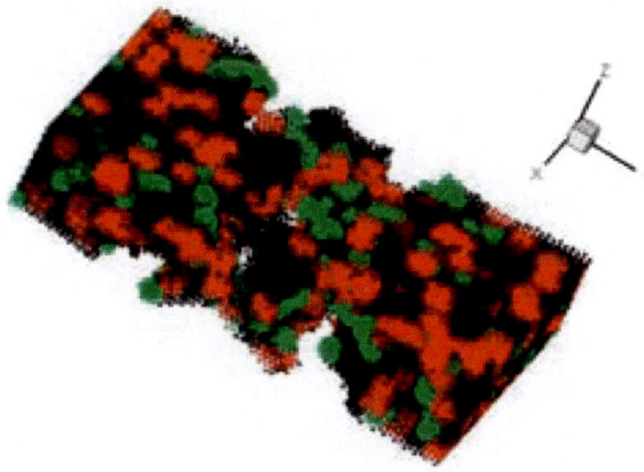

(c)

Figure 6. PM simulations: (a) high speed collision of two materials blocks, (b) high strain rate compression, and (c) high strain rate tension.

Figure 6(a-c) displays PM simulations of high speed collision, tension and compression of material blocks with heterogeneity. Different material properties plotted in different colors in Figure 6 are assigned with different (*p*, *q*) into Equation (1).

(B) Blasting Simulations

Simulation of blasting problems is another important application of PM. Figure 7(a, b) displays the PM blasting simulations of a wall with and without a retrofitting treatment. From Figure 7(a) it is found that for the case without the retrofitting enforcement attached to infrastructure, debris flow of the material will fly into the area behind the wall after the structure is fractured, and consequently can cause severe damage to the creatures wherein. In contrast, for the case with the retrofitting consideration, due to the resistance from the

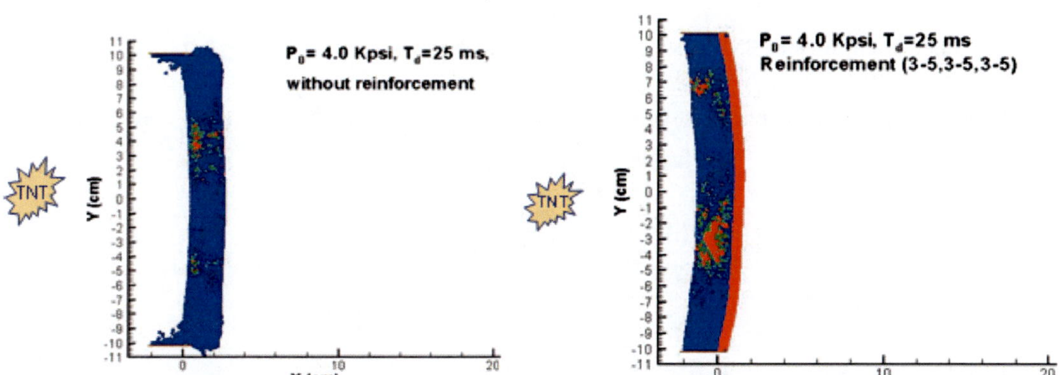

(a) homogenous material **(b) retrofitting material structure**

Figure 7. Examples of PM simulations of blasting (a) homogeneous material, (b) retrofitting material structure.

enforcement of the layered material, there are no material fragments intruding into the area behind the wall, as shown in Figure 7(b); hence, people and facilities in the area will be well protected.

(C) Crack Formation and Propagation in Different Materials

Crack formation and propagation can result in a sudden catastrophic failure of material. It is obvious that many lives may survive if the failure propagation speed can be postponed to have more time for the victims to escape or be rescued from the disaster once it happens. Thus, investigations of crack formation and propagation within different materials are necessary. PM technique can be easily applied for the study of crack formation and propagation in different materials, say ductile or brittle. Equation (1) with different (*p*, *q*) can result in different materials. Table 1 illustrates the physical outcomes by using (*p*, *q*) = (3, 5), (5, 10) and (7, 14), under equilibrium lattice spacing $r_0 = 0.2\,cm$. Figure 8 displays the according interaction force profile vs. the three above-employed (*p*, *q*) values.

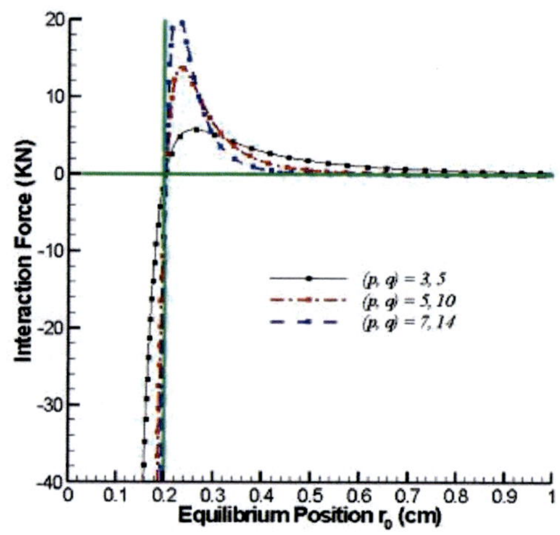

Figure 8. Interaction force of PM under $r_0 = 0.2\,cm$, with respect to (*p*, *q*) = (*3*, *5*), (*5*, *10*) and (*7*, *14*).

Table 1. Physical outcomes with (*p*, *q*) = (3, 5), (5, 10) and (7, 14) under equilibrium lattice spacing $r_0 = 0.2$ *cm*

(*p*, *q*)	(3, 5)	(5, 10)	(7, 14)
G	2.473×10^7	1.781×10^6	1.102×10^5
H	9.892×10^5	5.698×10^2	1.411
$E\,(GPa)$	15.457	69.557	150.706
$\sigma_{TS}\ (MN/m^2)$	86.205	263.570	441.534
Necking position	$1.29\bullet r_0$	$1.15\bullet r_0$	$1.11\bullet r_0$

Table 1 illustrates that with the increase of (p, q), (i) the Young's modulus and tensile strength values of the resultant material also increase. So does the necking position of the interaction force profile shown in Figure 8 as well as in Table 1. This indicates that the material with big (p, q) tends to be brittle and vice versa.

Figure 9(a-c) displays, respectively, the initial crack formation, propagation and final crack pattern under (p, q) = (5, 10); Figure 10(a-c) shows, respectively, the initial crack formation, propagation and final crack pattern under (p, q) = (7, 14).

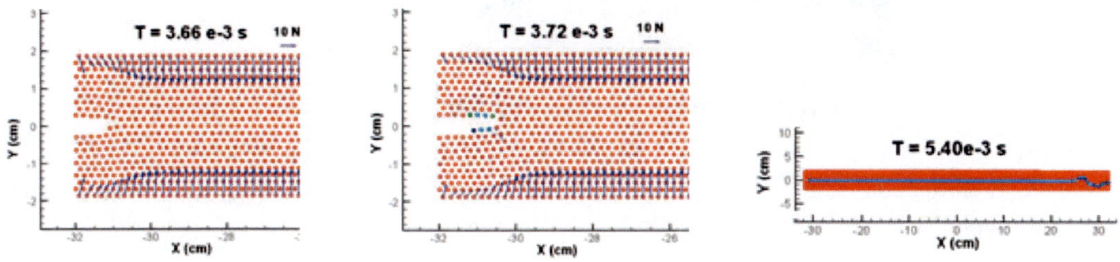

(a) (b) (c)

Figure 9. Time-dependant fracture of 2D plate with initial crack-tip under $r_0 = 0.2$ cm, (p, q) = (5, 10). Streching rate = 40 cm/s. (a) $T = 3.66\,ms$, (b) $T = 3.72\,ms$, and (c) final crack pattern.

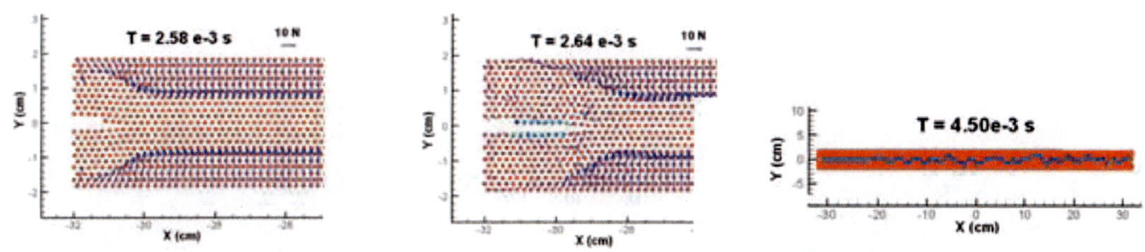

(a) (b) (c)

Figure 10. Time-dependant fracture of 2D plate with initial crack-tip under $r_0 = 0.2$ cm, (p, q) = (7, 14). Streching rate = 40 cm/s. (a) $T = 2.58\,ms$, (b) $T = 2.64\,ms$, and (c) final crack pattern.

Comparing Figures 9 and 10, we obtain some very interesting results:

(a) Crack develops sooner and propagates faster in brittle materials than in ductile materials (compare Figs. 9(a, b) with 10(a, b)).
(b) Crack propagation tends to behave in an unstable manner in brittle materials while in a steady fashion in ductile materials (compare Figs. 9(c) with 10(c)).

The benefit from this research is that the above-obtained results may help fabricate a high-resistance retrofitting layered structure optimally comprised of different materials. This fabricated enforcement structure is then coated to the infrastructure to postpone the failure propagation speed, as shown in Figure 11. Prolonging the failure process with this treatment, more lives can escape from the disaster once failure occurs.

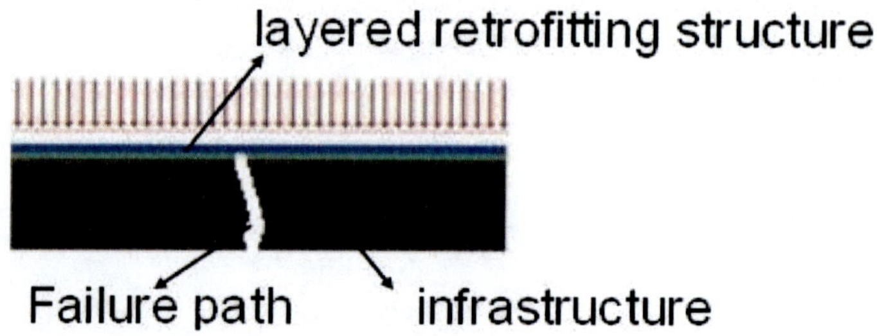

Figure 11. A fibrication of a retrofitting layered structure with different material properties attached to the infrastructure to postpone the failure propagation speed.

(D) Thermally-Induced Breakage of Ores

It has been found that most minerals on the earth or on the other planets in space (e.g. Moon, Mars, etc.) are composed of two types of materials: little or no heat generated, such as calcite, etc., and heat generated, such as pyrite, etc. Therefore, a rapid heating of ore minerals in a microwave-transparent matrix can generate thermal stress of sufficient magnitude to create micro-cracks along grain boundaries, and this type of microcracking might have the potential to improve ore grindability and increase liberation of individual mineral phases.

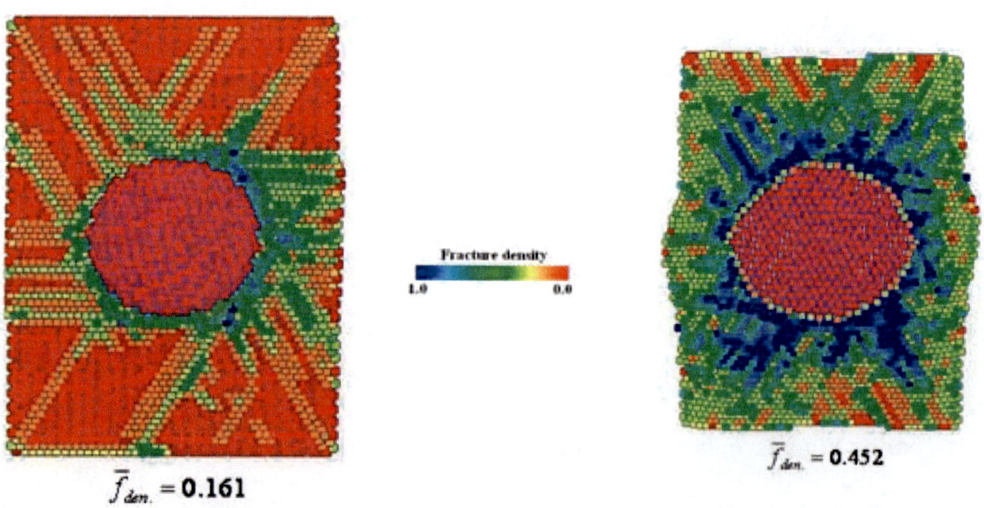

(a) $P_d = 1.0\times10^9$, $T = 1.37\times10^{-2}s$ **(b)** $P_d = 1.0\times10^{11}$, $T = 7.5\times10^{-3}s$

Wang, *et al*, 2008b.

Figure 12. PM simulation of fracturing efficiency results of rock using different microwave power density. (a) $P_d = 1.0\times10^9$, $T = 1.37\times10^{-2}s$; (b) $P_d = 1.0\times10^{11}$, $T = 7.5\times10^{-3}s$.

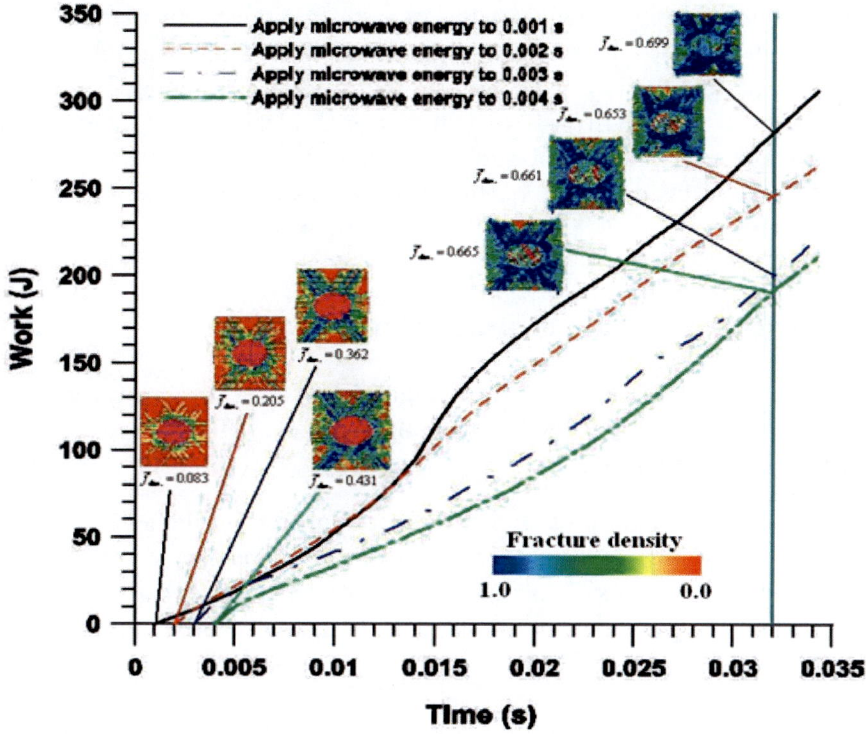

Wang, *et al*, 2008b.

Figure 13. Mechanical energy consumption of breakage by a constant uniaxial compression of $20.0 cm/s$ applied after different time length of microwave radiation on the heterogeneous material (a) $T = 1.0\times10^{-3} s$, (b) $T = 2.0\times10^{-3} s$, (c) $T = 3.0\times10^{-3} s$, (d) $T = 4.0\times10^{-3} s$. Microwave power density $P_d = 1.0\times10^{11} W/m^3$.

If microwave energy can indeed induce microcracking around phase grain boundaries of ores, the reductions in required comminution energy and enhanced liberation of valuable mineral would occur. In practice, experimental approaches cannot easily reach a precise insight into the entire thermal fracture process of ores because most minerals are brittle materials with a complex three dimensional structure. In this situation, numerical investigations are powerful. To meet this demand, a thermal-based PM has been established and been employed to explore this problem [Wang, et al, 2008b].

Figure 12 (a, b) shows the global averaged fracture density results with respect to the two different microwave input powers. It shows that heating first causes thermal expansion of pyrite (interior circular part). Consequently the calcite part (exterior rectangular part) is fractured by the pyrite expansion from the inside. In detail, we see that breakage originates from the region of interface between the two material phases and propagates into the calcite phase. It demonstrates (i) that heating pyrite in a microwave-transparent matrix, calcite, indeed creates micro-cracks along grain boundaries, and this type of microcracking can improve ore grindability and increase liberation of individual mineral phases; and (ii) that the bigger power density of microwave is applied, the sooner the fracture is generated and the higher breakage efficiency is reached.

Figure 13 illustrates the modeling results of a study in which, first we use microwave to heat the pyrite/calcite sample in different durations, and then a mechanical breakage approach is applied. We find that the longer we heat up the material, the more mechanical energy is saved from the breakage cost. This reveals that it is an ideal option to combine microwave breakage technique with mechanical methods to aid the mechanical breakage of ores and mineral assemblages. This new technique not only can help enhance mining efficiency on the earth, but also can be applied to aid human space exploration activities.

3.2. Deficiencies of PM and Potential Solutions

Since PM employs the simple, nearest-neighbor and axial linkage mechanism, and the according computations are conducted on a regular triangular lattice network, three following technical deficiencies are unavoidable:

1) A bias in the fracture propagation direction inherent to the geometry of the lattice network employed, say, along the 60° direction for the equilateral triangular, nearest neighborhood network. This deficiency of the regular lattice model has been demonstrated by Jirasek, *et. al.* [1995a, b] and Schlangen [1995]. A mesh bias example is illustrated in Figure 14.
2) A fixed Poisson ratio. Lattice model (LM) theory has addressed that the equivalent Poisson's ratio on a 2-D equilateral triangular lattice is fixed to the value of 1/3 whereas it is ¼ for a 3-D structure [Ostoja-Starzewski, 2002].
3) Isotropic solver. An equilateral triangular lattice network is limited to a valid description of isotropic materials [Ostoja-Starzewski, 2007].

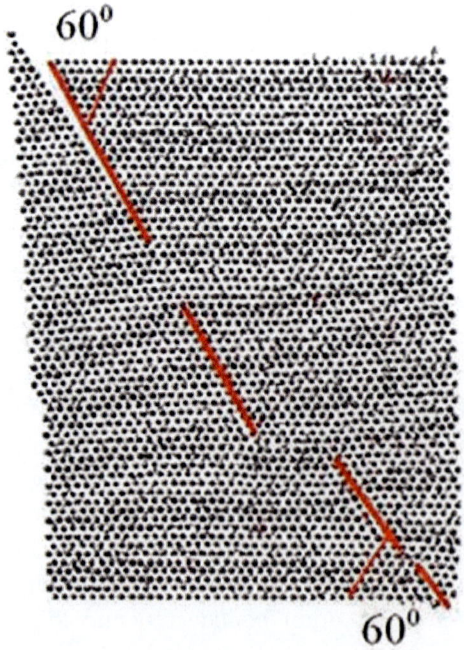

Wang, 2005b.

Figure 14. Example of mesh effect by lattice model: compression simulation (red line indicates crack path).

The above three situations can be remedied by the use of the more advanced lattice models. For example, Ostoja-Starzewski [2002, 2007] has manipulated several types of spring systems, including central (α), angular (β) and the mixed (α - β) interactions, coupled with different lattice networks (triangular, rectangular, etc.) as well as multiple neighborhoods (nearest, second neighboring particle interaction, etc.), that allows the modeling of various Poisson ratio materials, elimination of mesh bias problem, as well as the potential for applying to anisotropic materials.

Ostoja-Starzewski [2002] addressed combined axial-angular (α - β) scheme that theoretically works for various Poisson ratio and anisotropic materials as in Figure 15. For simplicity, assigning both of six axial and six angular springs equal, an isotropic LM solver is obtained that works for Poisson ratios ranging from (-1, 1/3]. For the cases with Poisson's ratio ranging from 1/3 up to 1, a 'triple honeycomb lattice' can meet this demand [Ostoja-Starzewski, 2002]. As is shown in Figure 16, this isotropic technique considers nearest neighbors but sets up three axial spring constants α_1, α_2 and α_3 in each triangular unit cell, respectively. Synder *et al* [1992] derived the Young's modulus and the Poisson's ratio from this technique as follows.

$$E = \frac{2\sqrt{3}(\alpha_1 + \alpha_2 + \alpha_3)}{3\{1 + 2(\alpha_1 + \alpha_2 + \alpha_3)/9[(1/\alpha_1) + (1/\alpha_2) + (1/\alpha_3)]\}} \tag{21}$$

$$\nu = 1 - \frac{2}{\{1 + 2(\alpha_1 + \alpha_2 + \alpha_3)/9[(1/\alpha_1) + (1/\alpha_2) + (1/\alpha_3)]\}} \tag{22}$$

We have proposed a newly developed hybrid lattice particle modeling (HLPM) technique that combines the strengths of the LM and the PM [Wang, et al., 2008c,d]. In Table 2, the strengths and weaknesses of the traditional LM and PM are summarized and compared. It is clearly seen that HLPM contains the strengths of both LM and PM.

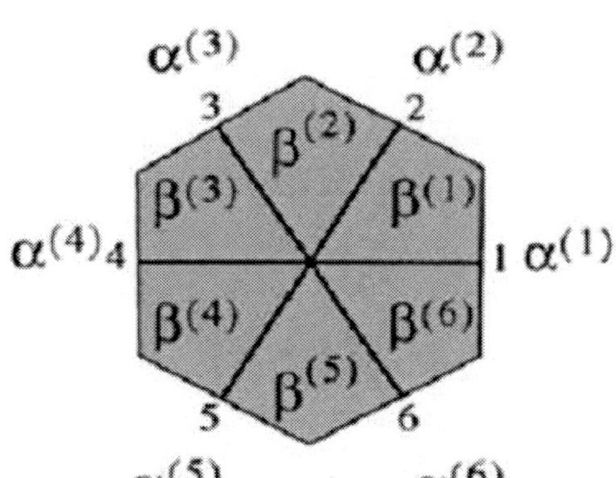

Ostoja-Starzewski, 2002.

Figure 15. Schematic of axial-angular (α - β) scheme.

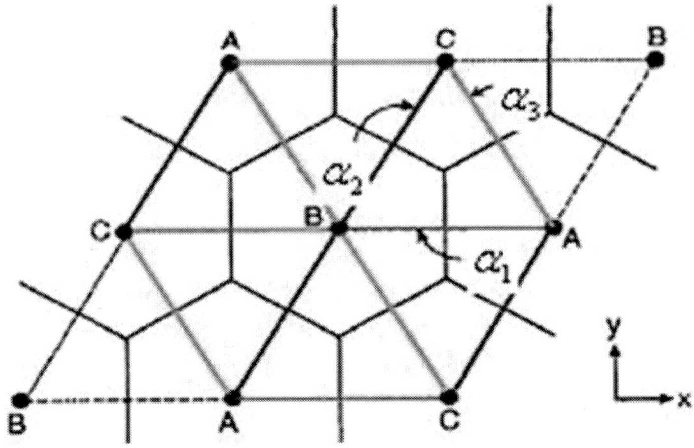

Figure 16. A triangular lattice with hexagonal unit cell for $\alpha_1 - \alpha_2 - \alpha_3$ model, considering nearest neighbor particle interactions; after Snyder *et al* [1992].

Table 2. Comparison of the lattice model (LM), the particle model (PM), and the hybrid lattice particle model (HLPM)

	Lattice Model (LM)	Particle Model (PM)	Hybrid Lattice Particle Model (HLPM)
Particle interaction	spring (axial/angular), beam, etc.	Lennard-Jones potential (axial only)	spring (axial/angular) mimicking the Lennard-Jones potential
Interaction neighborhood	not limited to nearest neighbor	nearest neighbor only	not limited to nearest neighbor
Mesh system	Eulerian	Lagrangian	Lagrangian
Poisson's ratio	flexible	fixed	flexible
Time process	static	dynamic based on Newton's second law	dynamic based on Newton's second law
Force-displacement relation	Displacement (strain) interpreted from force (stress)	Force interpreted from displacement (distance between particles)	Force interpreted from displacement (distance between particles)

The principle of HLPM can be described as follows: the particle-particle interaction is derived from lattice modeling (LM) theory whereas the computational scheme follows particle modeling (PM) technique. The current demonstration of the HLPM is based on a linear elastic model with an ultimate translational strength (i.e., tension/compression). Once the translational strength is exceeded, the spring is broken and a fracture is created. The linear model is created by using a quadratic form instead of the Lennard-Jones or polynomial potential. The nonlinear constitutive laws are not implemented at the present stage because the lack of independent measurement of the nonlinear material properties of the material tested. Once such constants are available, the implementation of nonlinear constitutive law is inherent in the Lennard-Jones or polynomial potential approach of the PM, and is not a difficulty for the HLPM at all.

The above-introduced Figures 4 and 5 are the successful outcomes of HLPM that eliminate mesh bias problem after employing a two-layer neighboring particle interaction scheme [Holnicki-Szulc and Rogula, 1979a] shown in Figure 17.

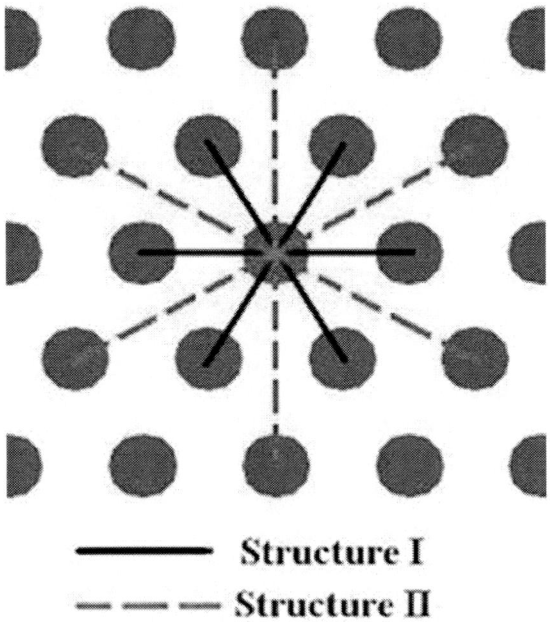

Holnicki-Szulc and Rogula, 1979a.

Figure 17. Schematic of a two-layer neighboring particle interaction scheme.

Inputting the spring constants α_1, α_2 and α_3 of Eqs. (20) and (21) into HLPM, a preliminary modeling result of the analogous case to Figure 4 while with a large Poisson ratio $\nu = 0.443$ is obtained as illustrated in Figure 18.

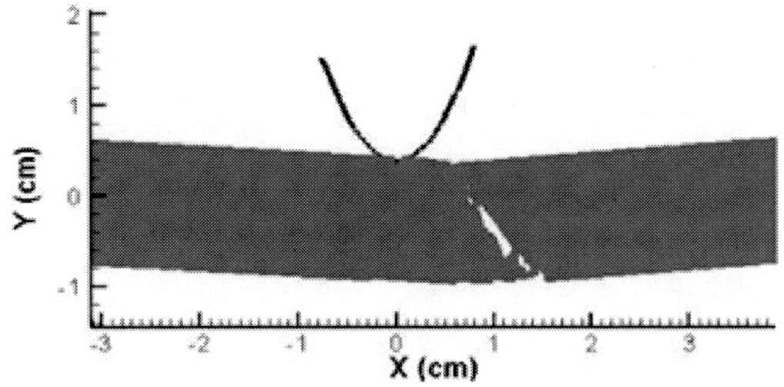

Figure 18. A preliminary HLPM result using $\alpha_1 - \alpha_2 - \alpha_3$ scheme. Poisson ratio $\nu = 0.443$.

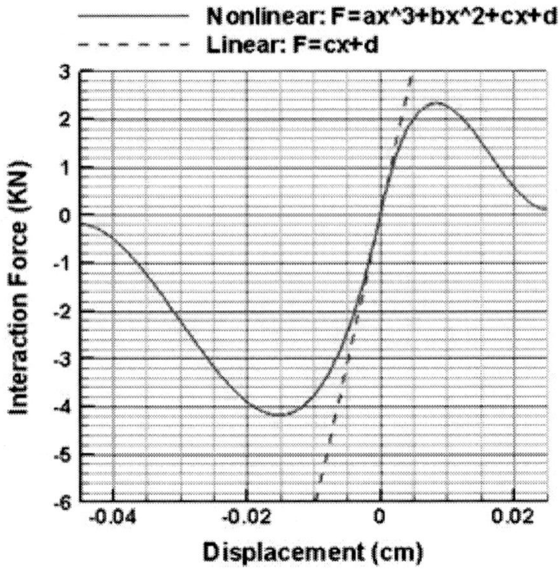

Figure 19. A nonlinear constitutive polynomial for HLPM.

As has been addressed, HLPM can be easily extended to inelastic considerations if a nonlinear constitutive polynomial is alternatively employed, instead of a Lennard-Jones type, as a definition of particle-particle interactions in Figure 19. A preliminary inelastic numerical result of HLPM is given herein. Adopting a 3-order polynomial into HLPM shown in Figure 20(a) to present a material with weak inelasticity, the modeling result of the analogous case to Figure 4 is predicted in Figure 20(b).

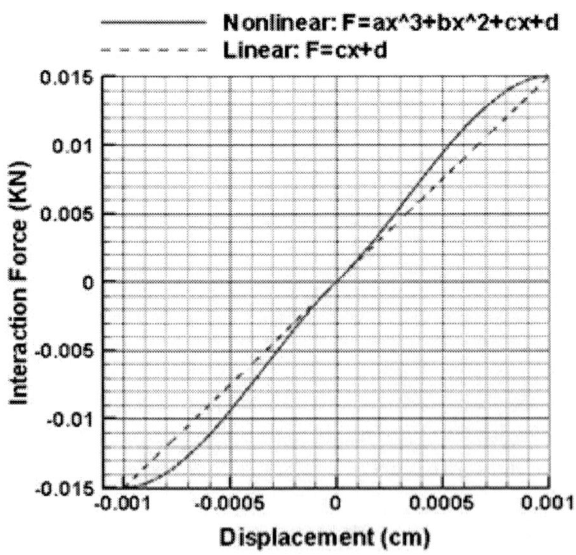

(a)

Figure 20. Continued on next page.

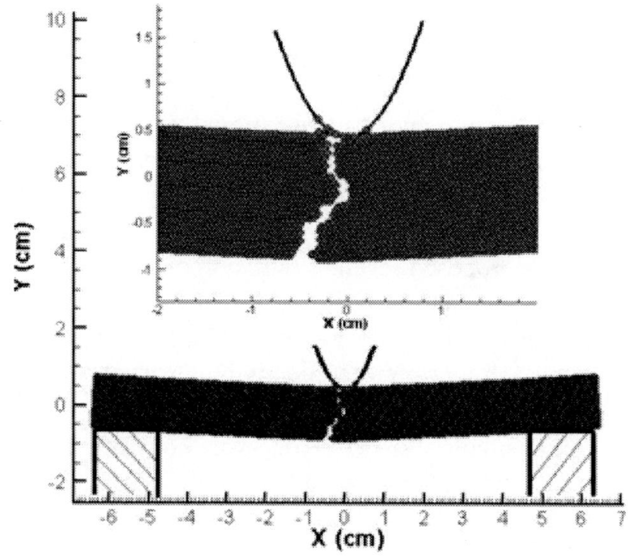

(b)

Figure 20. A weakly nonlinear modeling result of HLPM, (a) constitutive equation, (b) HLPM result.

4. Conclusion

This paper systematically introduce a novel particulate dynamics based modeling approach, particle modeling (PM) technique, and its successful applications in a number of fracture problems of solids with dynamic fragmentation under various extreme loading conditions. These loading conditions can include situations of collapse, impact, blasting or high strain rate tension/compression, as well as thermally-induced breakage problems. Meanwhile, the deficiencies of PM are also addressed and the associated solution, a hybrid lattice particle modeling (HLPM) scheme, is proposed.

PM is a numerical technique similar to the molecular dynamic (MD) simulation; but rather than simulating actual atoms, it is based on lumped mass particles distributed on a grid to allow macro scale modeling. The PM utilizes an equivalent Lennard-Jones or polynomial potential to model the nonlinear constitutive law at the continuum, macroscopic level. The mass has inertia that obeys Newton's second law of motion. It is a Lagrangian model that keeps track of particle location and velocity. The advantages of PM over the existing discrete element based methods can be summarized as follows:

1) Sample in theory. Four conservative/equivalent rules (mass, potential energy, Young's modulus and tensile/compression strength) are applied to preserve the equivalent material properties.
2) Easy for implementation. Since the physical size of each particle is ignored other than its equivalent mass, the algorithm of coding a PM computation is fairly easy.

Despite its success, the PM has a few deficiencies: (i) a bias in the fracture propagation direction inherent to the geometry of the lattice network employed, say, along the 60°

direction for the equilateral triangular, nearest neighborhood network; (ii) a fixed Poisson ratio. In the modeling of solid with PM, the potential type formulation allows only one elastic constant to be modeled, which is typically selected as the bulk modulus, or Young's modulus. The second elastic constant present in an isotropic material, say, Poisson's ratio, becomes the property of the grid system used, such as the triangular or rectangular networks; and (iii) an isotropic solver, as a result of the geometry of the adopted lattice network.

To overcome the three major above-mentioned shortcomings, a hybrid lattice particle modeling (HLPM) approach is proposed for the simulation of dynamic fracture phenomena in homogeneous and heterogeneous materials at macro-scales with a variable Poisson's ratio effect and anisotropic properties. It is concerned with the mathematical derivations of employing elastic interaction formula between contiguous particles in 2-D lattice networks accounting for different linkage mechanisms and different shapes of lattice. For instance, axial (α) and combined axial-angular (α - β) models are considered coupling with different lattices (triangular and rectangular, etc.) as well as multiple neighborhoods for the particle dynamic interactions. The principle of HLPM can be described as follows: the particle-particle interaction is derived from lattice modeling (LM) theory whereas the computational scheme follows particle modeling (PM) technique. The newly proposed HLPM is free from the above-mentioned deficiencies and can be applied to a wide range of impact and dynamic fracture failure problems.

An outlook of future PM development will be outlined as follows:

1) Implementation of HLPM with all the above-mentioned linear schemes from lattice modeling (LM) and other nonlinear constitutive laws.
2) Validation of HLPM with more real tests.
3) Improvement of PM by feedbacks from modeling applications.

Acknowledgement

This work was partially supported by Natural Sciences and Engineering Research Council of Canada (NSERC), COREM (Quebec Mining Technology Research Institute), and also by the funding received under a subcontract from the Department of Homeland Security-sponsored Southeast Region Research Initiative (SERRI) at the Department of Energy's Oak Ridge National Laboratory, USA.

References

[1] Al-Ostaz, A. and Jasiuk, I. (1997), Crack initiation and propagation in materials with randomly distributed holes, *Engineering Fracture Mechanics*, **58**, 395-420.

[2] Ashby, M.F. and Jones, D.R.H. (1980), *Engineering Materials 1: An Introduction to Their Properties and Applications*, Pergamon Press.

[3] Greenspan, D. (1997), *Particle Modeling*, Birkhäuser Publishing.

[4] Greenspan, D. (1981), *Computer-Oriented Mathematical Physics*, University of Texas at Arlington, Pergamon Press.

[5] Hockney, R.W. and Eastwood, J.W. (1999), *Computer Simulation Using Particles*, Institute of Physics Publishing.

[6] Holnicki-Szulc, J. & Rogula, D. (1979a), Non-local, continuum models of large engineering structures, *Arch. Mech.* **31**(6), 793-802.

[7] Jirasek, M., and Bazant, Z. P. (1995a), Macroscopic fracture characteristics of random particle systems, *Int. J. Fract.*, **69**(3), 201-228.

[8] Jirasek, M., and Bazant, Z. P. (1995b), Particle model for quasi-brittle fracture and application to sea ice, *J. Eng. Mech.*, **121**(9), 1016-1025.

[9] Meguro, M. and Tagel-Din, H. (2000), Applied element method for structural analysis: theory and application for linear materials, *Structural Eng./Earthquake Eng., JSCE*, **17**(1), 1-14.

[10] Monaghan, J. (2005), Smoothed particle hydrodynamics, *Rep. Prog. Phys*, **68**(1), 1703-1759.

[11] Oñate, E., Idelsohn, S.R., Pin, F. D, and Aubry., R. (2004), The particle finite element method. An Overview, *International Journal Computational Method*, **1**(2), 267-307.

[12] Ostoja-Starzewski, M. (2007), *Microstructural Randomness and Scaling in Mechanics of Materials*, in Modern Mechanics and Mathematics Series, Chapman & Hall/CRC/Taylor & Francis.

[13] Ostoja-Starzewski, M. and Wang, G. (2006), Particle modeling of random crack patterns in epoxy plates, *Probabilistic Engineering Mechanics*, **21**, 267-275.

[14] Ostoja-Starzewski, M. and Wang, G. (2005), Probability and materials: from nano- to macro-scale, *NSF Workshop*, Baltimore, MD, USA, January 5-7.

[15] Ostoja-Starzewski, M. and Wang, G. (2004), Particle modeling of dynamic fragmentation, *International Symposium on Developments in Plasticity and Fracture: Centenary of M. T. Huber Criterion*, Cracow, Poland, August 12-14.

[16] Ostoja-Starzewski, M., Lattice models in micromechanics, *Appl. Mech. Rev.*, 2002, 55(1), 35-60.

[17] Schlangen, E. (1995), Computational aspects of fracture simulations with lattice models, *Fracture Mechanics of concrete structures (Proc. FraMCoS-2, Zurich)*, F. H. Wittmann, ed., Aedificatio, Freiburg, Germany, 913-928.

[18] Wang, G., Al-Ostaz, A., Cheng, A.H.-D. and Mantena, P.R. (2008a), Particle modeling of a polymeric material (nylon-6, 6) due to the impact of a rigid indenter, *Computational Materials Science* (in press).

[19] Wang, G., Radziszewski, P. and Ouellet, J. (2008b), Particle modeling simulation of thermal effects on ore breakage, *Computational Materials Science* (In press).

[20] Wang, G., Al-Ostaz, A., Cheng, A.H.-D. and Mantena, P.R. (2008c), A hybrid particle-lattice model for dynamic fracture problems, *AAM'08*, Louisiana, June 17-20.

[21] Wang, G., Al-Ostaz, A., Cheng, A.H.-D. and Mantena, P.R. (2008d), Hybrid lattice particle modeling: theoretical considerations for a 2-D elastic spring network for dynamic fracture simulations, *Computational Materials Science*. (revised).

[22] Wang, G., Al-Ostaz, A., Cheng, A.H.-D. and Mantena, P.R. (2008e), Particle modeling of crack propagation in materials at a macroscopic level, *AAM'08*, New Orleans, Louisiana, June 17-20.

[23] Wang, G., Al-Ostaz, A., Cheng, A.H.-D. and Mantena, P.R. (2008f), Particle modeling of dynamic fracture simulations of a 2D polymeric material (nylon-6,6) subject to the impact of a rigid indenter, *EM'08*, Minneapolis, Minnesota, May 18-21.

[24] Wang, G., Radziszewski, P. and Caron, S. (2007a), Particle modeling of material comminution, *Discrete Element Methods (DEM) '07*, Brisbane, Australia, August 27-29, 8pp.

[25] Wang, G., Radziszewski, P. and Ouellet, J. (2007b), Exploring DEM model development for the simulation of thermal effects on ore breakage, *Discrete Element Methods (DEM) '07*, Brisbane, Australia, August 27-29, 8pp.

[26] Wang, G., Radziszewski, P. and Ouellet, J. (2007c), Particle modeling approach for the simulation of thermal effects on ore breakage, *Mid-South Area Engineering and Sciences Conference (MAESC) '07*, the University of Mississippi, Oxfod, May 17-18.

[27] Wang, G., Radziszewski, P. and Caron, S. (2007d), Particle modeling of material comminution process, *Mid-South Area Engineering and Sciences Conference (MAESC) '07*, the University of Mississippi, Oxfod, May 17-18.

[28] Wang, G., Ostoja-Starzewski, M., Radziszewski, P. and Ourriban, M. (2006), Particle modeling of dynamic fragmentation – II: fracture in single- and multi-phase materials, *Computational Materials Science*, **35**, 116-133.

[29] Wang, G. and Ostoja-Starzewski, M. (2005a), Particle modeling of dynamic fragmentation – I: theoretical considerations, *Computational Materials Science*, **33**, 429-442.

[30] Wang, G. (2005b), Particle modeling of dynamic fragmentation, *Ph.D. Dissertation*, McGill University, Canada.

[31] Wang, G., Ostoja-Starzewski, M. and Radziszewski, P. (2005c), Particle modeling of comminution, *20th Canadian Congress of Applied Mechanics*, May 30th to June 2nd, Montreal, Canada, 149-150.

[32] Synder, K.A., Garboczi, E.J., and Day, A.R., The elastic moduli of simple two-dimensional composites: computer simulation and effective medium theory, *J. Mech. Phys. Solids*, 1992, 72, 5948-5955.

In: Strength of Materials
Editors: G. Mendes and B. Lago, pp. 183-215
ISBN: 978-1-60741-500-8

Chapter 6

NON-ORIENTED ELECTRICAL STEELS: MATERIALS FOR SAVING ENERGY AND CONSERVING THE ENVIRONMENT

Taisei Nakayama*
Sumitomo Metal Industries, Ltd.
Wakayama Steel Works
1850 Minato, Wakayama, 640-8555 Japan

Abstract

Electrical steels are the core materials for electrical motors or transformers. Those materials for motors are played an energy conversion roll from electricity to motion. However, energy losses are accompanied with this conversion. To minimize these losses is a key technology to conserve our environment.

Numerous researches on the grain-oriented electrical steels reported. Those researches especially for transformers are focused on the reducing the losses at supplying the electricity from power plants. On the other hand, home or industrial appliances are the power consuming devices, and the most effective point on the energy loss reduction. These home appliances are used small motors using non-oriented electrical steels.

In this review, several researches on the non-oriented electrical steels are discussed and focused on the metallurgical control of the steels to reduce the core loss for generating waste heats and motor building innovation technologies for decreasing the building factor of the core losses.

In the metallurgical part, some additive elements as phosphorus, aluminum and manganese for improving magnetic properties reviewed. Moreover some contaminating elements as vanadium, titanium and zirconium are discussed especially for precipitation studies in the steels have been done. These precipitations are inhibited the grain growth at final annealing or stress relief annealing. These inhibited small grains increase the core losses.

For studying motor building technologies, compression stress effect, shearing stress effect are discussed. Even though the best core materials are used for manufacturing motors, those building deteriorations make worse for the motor efficiency. Therefore, those technologies are also important for reducing the carbon dioxide emission.

* E-mail address: nakayama-tis@sumitomometals.co.jp. Phone:+81-73-451-2400. Fax:+81-73-451-2412.

Introduction

Electrical steels are the core materials for electrical motors or transformers. Those materials for motors are played an energy conversion roll from electricity to motion. However, energy losses are accompanied with this conversion. To minimize these losses is a key technology to conserve our environment. Reducing total carbon dioxide emission is an important issue identified since the 3rd Conference of Parties (COP3) of United Nation Framework Convention on Climate Change (UNFCCC) in Kyoto in 1997. Thus, much attention is now paid to the motor or transformer efficiency.

Numerous researches on the grain-oriented electrical steels have been reported [1-5]. Those researches especially for transformers are focused on the reducing the losses at supplying the electricity from power plants [1,4]. Metallurgical studies on grain oriented electrical steels such as the domain control technologies [2] or secondary recrystallization [3,5] to improve the magnetic properties for reducing core loss.

On the other hand, electrical motors are the power consuming devices. To reduce the carbon dioxide emission, these devices should be performed more effectively. Most of the electrical motors are used for home appliances. They are installed for motion controlling or power devices. These small motors are made of non-oriented electrical steels as core materials. Recent eco-applications need to install such high efficient motors [6]. Same as home appliances, large motors in industrial use should be focused on the motor efficiency. Moreover, hybrid electrical vehicles [7,8], or trains and trams for transportation [9] are used some small size high power high efficient motors.

Thus, the core loss reductions of electrical motors is necessary for reducing carbon dioxide emission, and to adopt more energy effective materials is a key to solve the problem. Much attention is now paid to developing magnetic properties through the control of chemistry [10], grain size [11,12] and texture [13,14], processes [15]. Several new products with low core loss and high induction have been developed [16-18]. These improvements have been achieved with the advances in steelmaking technology, which turn out to be the relationship between structure and magnetic properties.

In this paper, several researches on the non-oriented electrical steels are reviewed. There are two parts of the major topics. The first part refers to the effects of additive and contaminating trump elements in the silicon steels, and the second part is reviewed the deteriorations by the core manufacturing process

In the first part, several effects of some positive elements for improving magnetic properties such as manganese [19] or aluminum [20] or phosphorus [21,22], and effects of trump elements as contaminants for deteriorating magnetic properties such as vanadium [23], titanium [24] or zirconium [25] are reviewed.

The key technology of using positive additives, such as manganese or aluminum is how to prevent to form harmful inclusions and precipitates. These inclusions or precipitates inhibit to the grain growth during final annealing. Small grains and precipitates themselves lead to the increase of the core loss, especially hysteresis loss. The effect of phosphorus is the grain orientation control. It assists magnetically favorable textures during final annealing.

Contaminating elements such as vanadium, titanium and zirconium easily form the precipitations with carbon or nitrogen in the steels. These precipitations inhibit the grain growth during final annealing or stress relief annealing. These inhibited small grains increase the core losses. The contaminating trump element controls are important technologies to reduce the core loss. To clarify the maximum limit of these contaminants in the steels is valuable for the commercial production.

Even though using the superior core material for manufacturing small motors, the core building affect the core loss. In this review two deterioration cases, the magnetically deterioration of interlocking lamination stacking [26,27] and the deterioration by the compression force [28] after the core heat-shrunk into the case are referred.

To solve those deteriorations of core loss, some special steels with excellent stampability and machinability controlled by chemistry are introduced [29,30].

Part 1. Effects of Additive or Contaminating Elements in the Silicon Steels

Non-oriented electrical steels are classified as fully processed type and semi-processed type [31,32]. Among fully processed type, some improved grades such as high magnetic induction type are developed recently. Figure 1 [16,31] shows the magnetic properties of fully processed steels. Magnetic inductions of the improved steels are higher than that of regular one. Therefore, the improved steels make the motor or transformer size smaller than regular steels with the same performances.

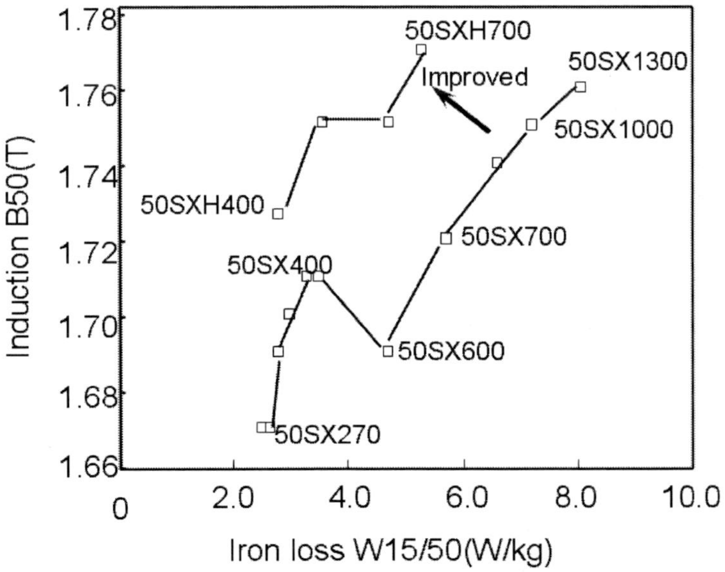

Figure 1. Non-oriented electrical steels manufactured by Sumitomo Metals.

Semi-processed steels are used to be manufactured with final skin-pass process. These skin-passed steels relatively non-sensitive to the contaminant or inclusions. However, innovations of the steel making technologies make the semi-processed electrical steels free from skin-pass process. Figure 2 [31] shows the magnetic properties of semi-processed steels.

Controlling the chemistry in the steels especially, prevention from the contaminants makes the semi-processed steels same process as fully processed steels.

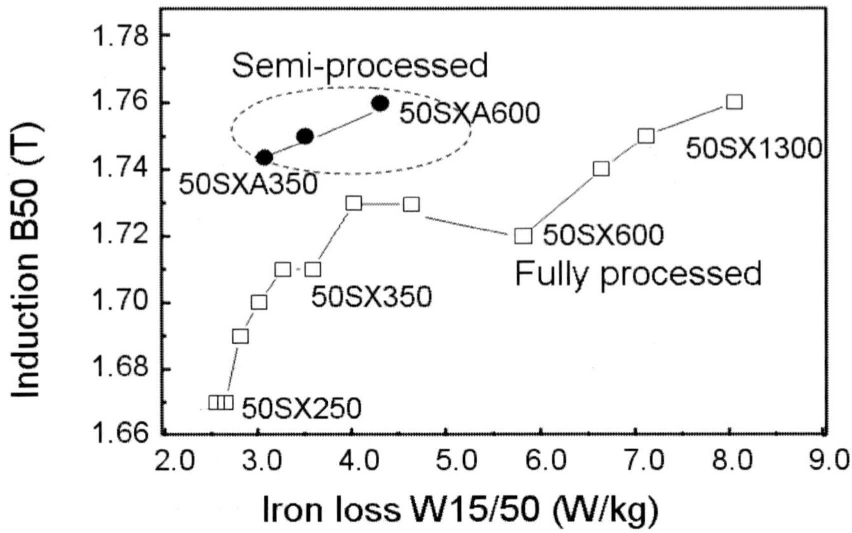

Figure 2. Magnetic properties of semi-processed steels.

1.1. Additive Elements for Improving Magnetic Properties

1.1.1. Effects of Phosphorus [21 ,22]

Non-oriented electrical steels with low silicon (less than 1.0mass %) usually contain phosphorus as additives for controlling mechanical properties. Low silicon steels with ultra-low carbon lead to the low yielding strength if they do not have any other additive elements. Due to prevent the magnetic aging carbon cannot be added more than 0.003 mass % [33]. However core making such as high speed stamping requires the high yielding strength materials. To solve this problem phosphorus is widely used as an element for increasing the yielding strength in low costs.

However higher silicon products more than 1% are usually non-phosphorus bearing steels, because their hardness or tensile strength increases by silicon itself. If the steels contain both silicon and phosphorus, the mechanical properties of the steels are high strength and they are harmful for stamping with die systems.

Marvelous die systems [34] enable to make high speed core stamping with such high strength silicon steels. Recently some high efficient interior permanent magnet synchronous motors whose rotor core shapes are sophisticated are developed [6,35], and these high efficient motors require the special die system with special core materials.

In this case, core materials for high efficient motor with low core loss and high magnetic induction are realized not only by the phosphorus content but also by the process especially for the hot band annealing heat cycles.

In this study, two different level phosphorus bearing steels with 0.002 mass % carbon, 2.0 mass % silicon, 0.3 mass % aluminum and 0.2 mass % manganese are used.

Figure 3 shows the relationship between cold rolled reduction and core loss at 1.0 T and 400Hz (W10/400) and magnetic induction at 5000 A/m (B50) in comparison with phosphorus bearing steel (P=0.1mass %) and regular steel (P=0.01mass %).

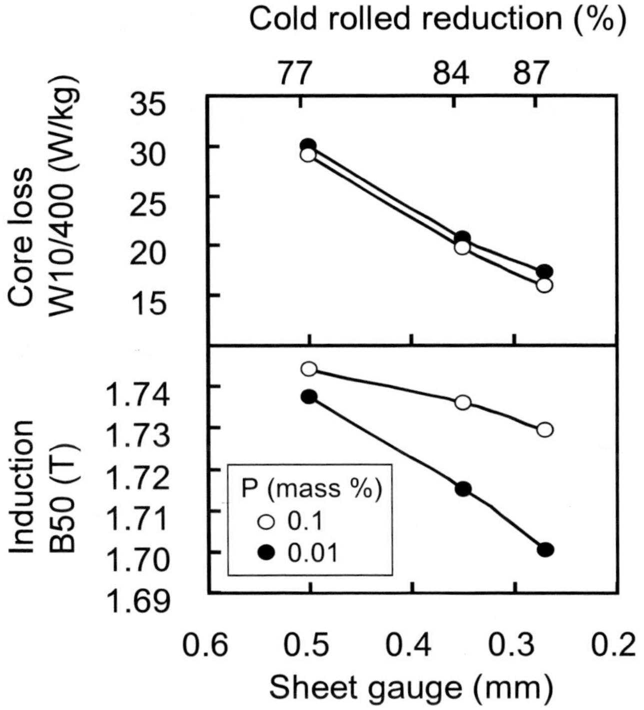

Figure 3. Effect of cold rolling reduction on the magnetic properties of phosphorus bearing steels.

In the case of regular silicon steel (P=0.01mass %) the magnetic induction is decreased as increasing the cold rolled reduction for reducing core loss especially for eddy current loss, while the magnetic induction of phosphorus bearing silicon steel (P=0.1mass %) is less decreasing than that of regular silicon steel.

This less magnetic induction decrease with thin gauge (or high cold rolled reduction) sheet requires a suitable hot band annealing condition for the texture control during the final annealing Figure 4 (A) and (B) show the texture diagrams in the final products in comparing with regular silicon steel and phosphorus bearing steel. The grains in the regular silicon steel are oriented the magnetically unfavorable texture such as {111}<112> in the case of high cold roll reduction (thin gauge) sheet, while that of phosphorus bearing steel are not.

Even though the phosphorus bearing steel without adequate annealing, this unfavorable texture evolution occurs during the final annealing. Figure 4 (B) and (C) show the texture diagrams comparison between with and without hot band annealing. Hot band annealing affects the grain orientation of the final products. In this case, hot band annealing induced the magnetic favorable texture after cold rolling and final annealing.

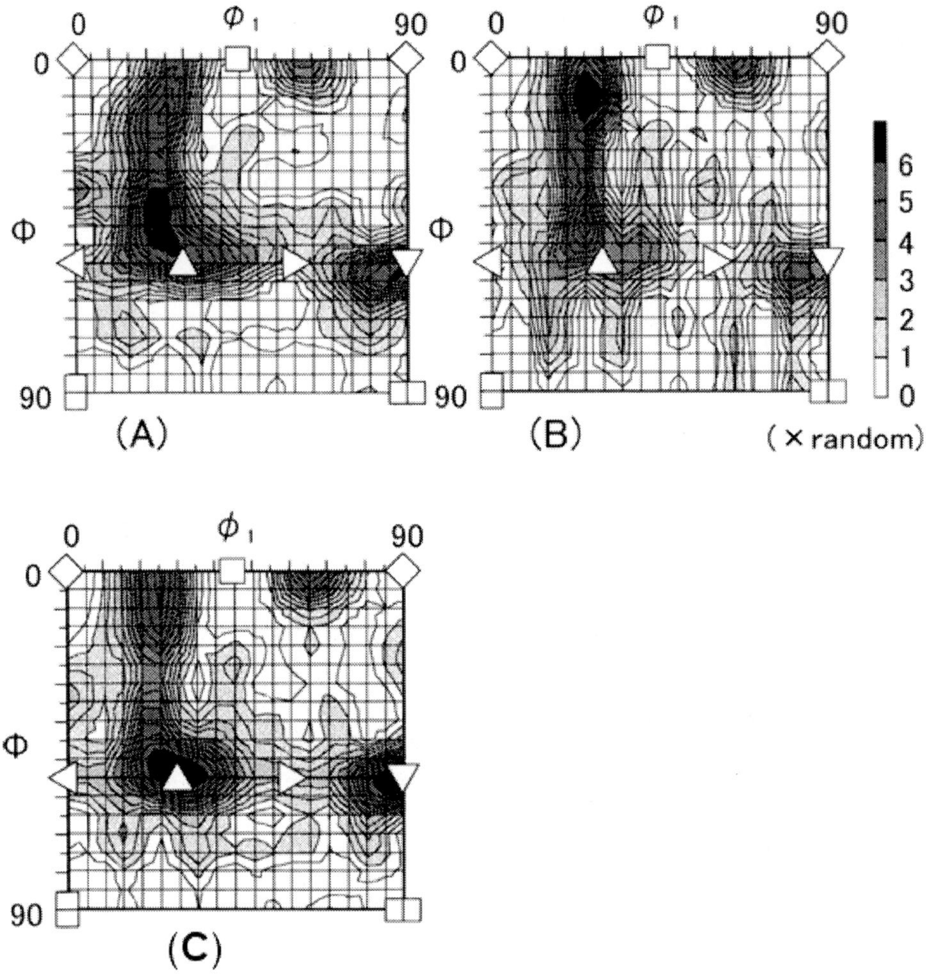

(A)regular steel (0.01 mass% P) with batch type hot band annealing; (B)improved steel (0.1 mass % P) with batch type hot band annealing ; (C)improved steel without hot band annealing.

Figure 4. Texture comparison between phosphorus bearing steel and a regular steel.

1.1.2. Effects of Aluminum [20]

Aluminum plays a roll to improve the magnetic properties as an additive element. It increase the resistivity of steels same as silicon. However, aluminum combines easily with nitrogen, forms aluminum nitrides. Thus, controlling this precipitates is important to improve the magnetic properties.

In steels containing 0.001mass % carbon, 0.3 mass % silicon, 0.3 mass % manganese, 0.08 mass % phosphorus, 0.005 mass % sulfur and 0.0005-0.0042 mass % nitrogen with 0.001-0.30 mass % aluminum, Figure 5 shows the relationship between core loss at 1.5 T and 50 Hz (W15/50), and aluminum and nitrogen content after stress relief annealing. The core loss of the steels with <0.001 mass % nitrogen keeps low in the range of aluminum < 0.01 mass %, and it increases with an increase of the aluminum content toward 0.03 mass %, then it peaks at 0.03 mass %, over there, the core loss decreases with an increase of the aluminum. In the case of the steels with the nitrogen > 0.002 mass %, the core loss increase in the range of aluminum <0.01 mass %, then it peaks at 0.01 mass %, over there, it decrease with an increase of aluminum. Hou *et al.*[36] investigated the effects of aluminum on the magnetic properties of lamination steels and concluded that the aluminum addition to the steels decreased the core loss through increasing the electrical resistivity to reduce the eddy current loss and coarsening grain size to decrease the hysteresis loss. However in that study, any comment of the effect of nitrogen did not mention, despite of studying the effect of aluminum nitride precipitates.

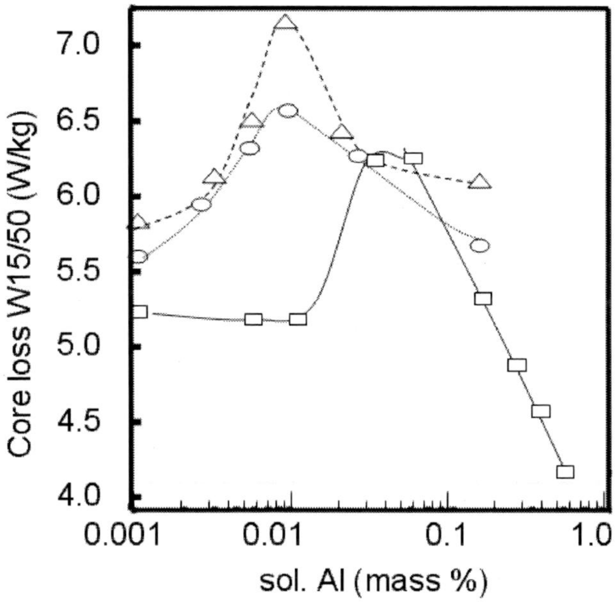

Figure 5. Effect of aluminum and nitrogen contents on core loss W15/50 after stress relief annealing. (□: N=0.0005-0.0009 mass %; ○:N=0.0021-0.0028 mass %; △:N=0.0035-0.0043 mass %;).

On the point of the magnetic induction, Figure 6 supported the Hou's study. As increasing the aluminum content, the magnetic induction at 5000 A/m (B50) decreased in any nitrogen level. Yashiki and Okamoto [37] investigated the effect of hot band annealing and concluded that the grain size of the hot band affect the magnetic induction, however in this study as shown in Figure 6, hot band grain size does not affect the magnetic induction in the range of the nitrogen content in this study (<0.0043 mass %). However magnetic induction B50 is slightly decreasing as an increase of aluminum contents over 0.1 mass % in steels with <0.0043 mass % nitrogen. This decreasing magnetic induction is affected by a decrease of the saturated magnetic induction as increasing aluminum content.

The deterioration of the core loss was caused by the small grains pinned down by the precipitation of small aluminum nitrides (< 0.5 μm). To compare with 0.001 mass %, 0.078 mass % and 0.30 mass % aluminum steels with < 0.001 mass % nitrogen, grain size in these 3 steels are different. Steels with 0.001 mass % and 0.30 mass % aluminum are almost same grain size (50-80 μm), while steel with 0.078 mass % aluminum is small (20-30 μm). To make the reason why steel with 0.078mass % aluminum has small grains clear, precipitates in these steels were investigated by TEM. In the steel with 0.001 mass % aluminum, the major inclusions or precipitates are MnS, and in the steel with 0.078 mass % aluminum, many 'harmful AlN' (< 0.5μm in size) were observed along the grain boundaries, while in the steel with 0.30 mass % aluminum, some 'harmless large AlN' (>1.0μm in size) were observed in the grain. This means the small AlN around 0.5 μm were pinned down the grain boundaries, therefore, the grains were not coarsened at all at the stress relief annealing.

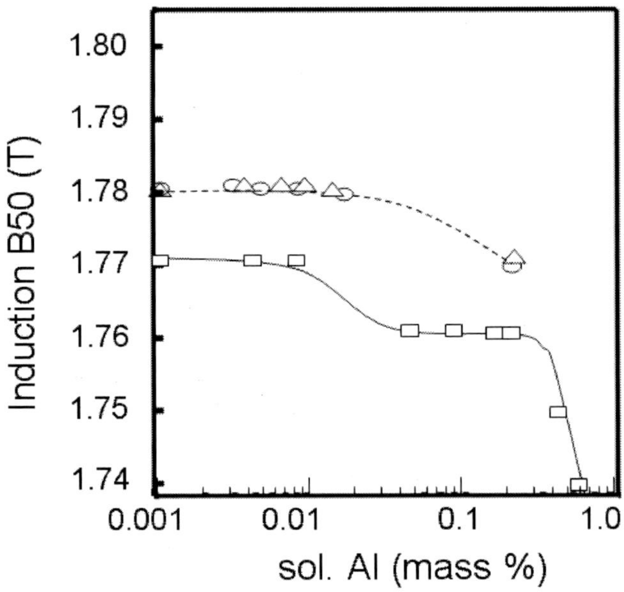

Figure 6. Effect of aluminum and nitrogen contents on induction B50 after stress relief annealing. (□: N=0.0005-0.0009 mass %; ○:N=0.0021-0.0028 mass %; △:N=0.0035-0.0043 mass %;).

As Sawamura and Mori [38] showed the AlN solubility of 1.0 mass % Si steel with 0.01mass % Al and 0.005 mass % nitrogen and concluded that the all Al and N were precipitated as AlN at the temperatures < 900 K. Following that study, aluminum and nitrogen should be precipitated as no 'harmful AlN' on the magnetic properties with < 0.0024 mass % AlN. Darken *et al.* [39] studied the behavior of AlN in austenite steel and introduced the equation of AlN solubility. According to the Darken's equation, all AlN were soluted with < 0.005 mass % nitrogen and < 0.01 mass % aluminum at the slab reheating temperature (1450 K). The magnetic properties and the precipitates studies indicated that the 'harmful AlN' (< 0.5μm in size) in the steels with > 0.0024 mass % AlN during the hot rolling process. These results follow Darken's equation statistically. Moreover, Darken studied nitriding in aluminum killed steel and pointed out the nitrogen diffusivity was low and formed 'subscale layer' as the surface nitrided band, however, this layer was not observed in the hot band. In the case of precipitating AlN during the hot rolling process, the 'harmful AlN' in the slab were soluted, while 'harmless large AlN' (>1.0μm) were not. In steels with Al > 0.1 mass % in Figure 11, this 'harmless large AlN' (>1.0μm) which keep large in size at slab reheating temperatures (1450 K) with following hot rolling process, did not pinned down the grain boundaries, and did not affect on the magnetic properties. These 'harmless large AlN' precipitates were observed with > 0.1 % Al in this study.

In steels with excess nitrogen, $MnSiN_2$ were observed (Figure 7), this is why the core loss of 0.001 mass % aluminum and high nitrogen steel, which have no 'harmful AlN', were higher than steel with < 0.001 mass % nitrogen. These manganese silicon nitrides were the same as Yashiki and Kaneko [40] reported in the electrical steels.

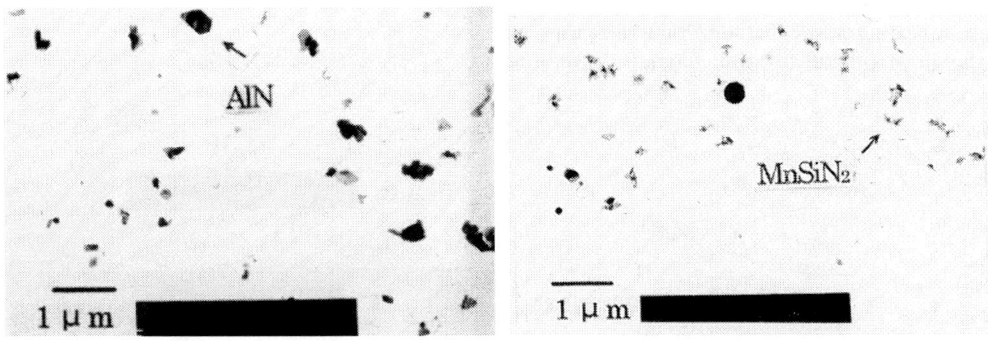

Al: 0.043 mass %
N: 0.0009 mass %

Al: 0.001mass %
N: 0.0025 mass %

Figure 7. Precipitates in 0.3 mass % silicon steels with aluminum.

1.1.3. Effects of Manganese [19]

Same as aluminum, manganese is also an element for increasing a resistivity of the steels. However, its resistivity increase per weight addition is about a half of the silicon's or aluminum's [33]. Moreover, manganese also forms harmful precipitates as manganese sulfides. It affects the magnetic properties. In steels containing 0.001 mass % carbon, 0.3 mass % silicon, 0.3 mass % aluminum, 0.08 mass % phosphorus, and 0.001 mass % nitrogen with 0.12-0.91 mass % manganese and 0.0005-0.038 mass % sulfur, Figure 8 shows the relationship between core loss at 1.5 T and 50 Hz (W15/50), and manganese and sulfur content after stress relief annealing. The core loss of the steels increases with an increase of the sulfur in each steel with different manganese content. The core loss decreases with an increase of manganese. These phenomena are explained that the decrease of core loss with an decrease of sulfur mainly are derived from an decrease of hysteresis loss by an decrease of MnS inclusions and coarsened grains, and the decrease of core loss with an increase of manganese are from a decrease of eddy current loss by increasing resistivity. For instance the resistivity of the steels containing 0.3 mass % Si, 0.12% Mn, and 0.3 mass % Al is $20.9\,\Omega\text{m} \times 10^{-8}$, while that of steels cintaining 0.3 mass % Si, 0.93% Mn, 0.3 mass % Al is $25.9\,\Omega\text{m} \times 10^{-8}$, and the calculated core loss of steels with 0.93 mass % Mn decreased to 94% that of 0.1 mass % Mn, on the basis of eddy current decrease by the increase of resistivity. However, observed core loss of the steels with 0.93 mass % Mn is 90% that of the 0.12 mass % Mn, and the difference between the calculation and observation supposed to be a decrease the hysteresis loss by coarsen grains in the steel with 0.93 mass % Mn [41].

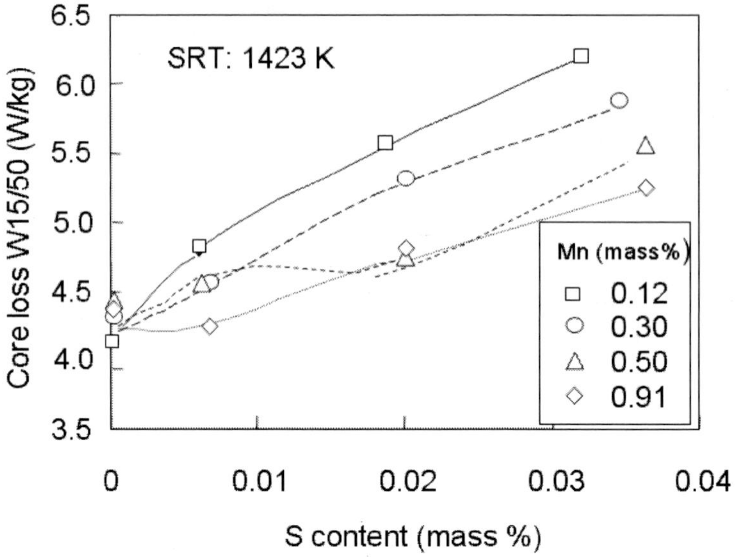

Figure 8. Effect of manganese and sulfur on core loss W15/50 after stress relief annealing.

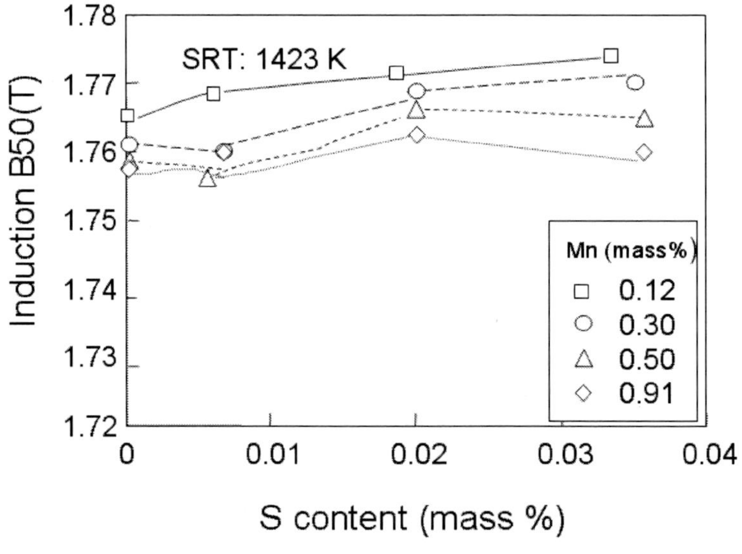

Figure 9. Effect of manganese and sulfur on induction B50 after stress relief annealing.

On the point of the magnetic induction, Figure 9 shows the effects of manganese and sulfur. As increasing the sulfur content in each manganese bearing steel, the magnetic induction B50 is almost no change in each manganese level. Yashiki and Okamoto [37] investigated the effect of hot band annealing and concluded that the grain size of the hot band affect the magnetic induction. However, the grain size of the hot band is slightly decreased as increasing the sulfur content in the steels bearing 0.31% manganese, while magnetic induction at 5000 A/m (B50) is slightly increasing as increasing the sulfur content. This might be another effect, such as textures, on the magnetic induction B50.

In these MnS precipitations, slab reheating temperatures (SRT) strongly affect on the magnetic properties. To clarify the precipitation, TEM observation was carried out in those different SRT with different sulfur-bearing steels (Figure 10). MnS is observed very seldom in steels with S<0.001mass % processed any SRT, however, steels with >0.001 mass % sulfur processed SRT at 1273 K has some 'coarse MnS' (ca. 0.5μm), while those steels processed SRT at 1423, or 1523 K many 'fine MnS' (ca. 0.1μm) were observed. This 'fine MnS' retard the grain growth in recrystalization annealing and deteriorated the core loss.

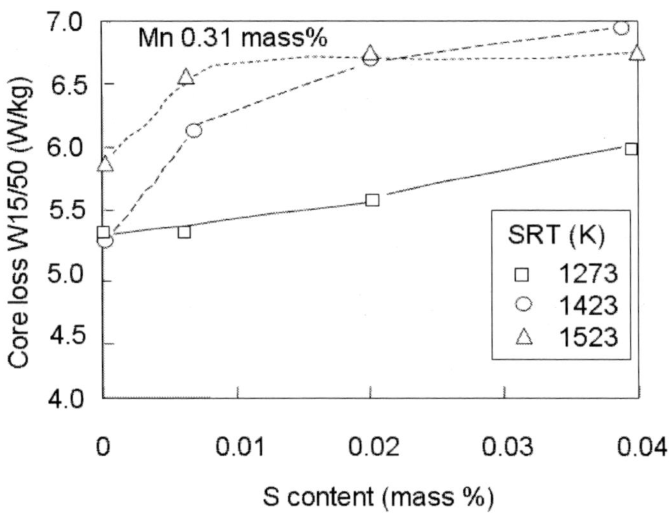

Figure 10. Effect of sulfur content and slab reheating temperatures (SRT) on core loss W15/50 in 0.31 mass % Mn steels. (□:SRT at 1273 K; ○: SRT at 1423 K, □: SRT at 1523 K).

1.2. Trump Elements for Deteriorating Magnetic Properties

1.2.1. Effects of Vanadium [23]

To control the contaminants such as vanadium, titanium or zirconium is important to improve the magnetic properties. In the case of vanadium as contaminants in steels containing 0.001mass % carbon, 0.3 mass % silicon, 0.3 mass % manganese, 0.07 mass % phosphorus, 0.005 mass % sulfur and 0.001 mass % nitrogen with 0.001-0.124 mass % vanadium, vanadium nitride precipitates influenced the final grain size. The grain size decreases with an increase in vanadium content < 0.016 mass %. However, when the vanadium content is > 0.016 mass %, the steel grains again growth towards the lower vanadium content size. It is well known that the grain size affects magnetic properties [42,43]. In steel containing vanadium 0.016 mass %, vanadium carbonitrides were observed along the grain boundary (Figure 11) when it was pinned down. Therefore, the effect of vanadium carbonitrides in steel with 0.016 mass % vanadium on the retardation of grain growth is more pronounced than that of compound precipitates in 0.001 mass % vanadium steel. However, the pinning effect is weak in steels with vanadium contents > 0.016 mass %, whose carbonitrides are larger than those with lower vanadium contents. This behavior is very similar to that with aluminum nitrides. Nakayama and Honjou [19] or Hou *et al.*. [36] studied the effect of aluminum on the magnetic properties of lamination steels and pointed out that the size and distribution of aluminum nitride precipitates in hot-rolled plates influenced the final grain size.

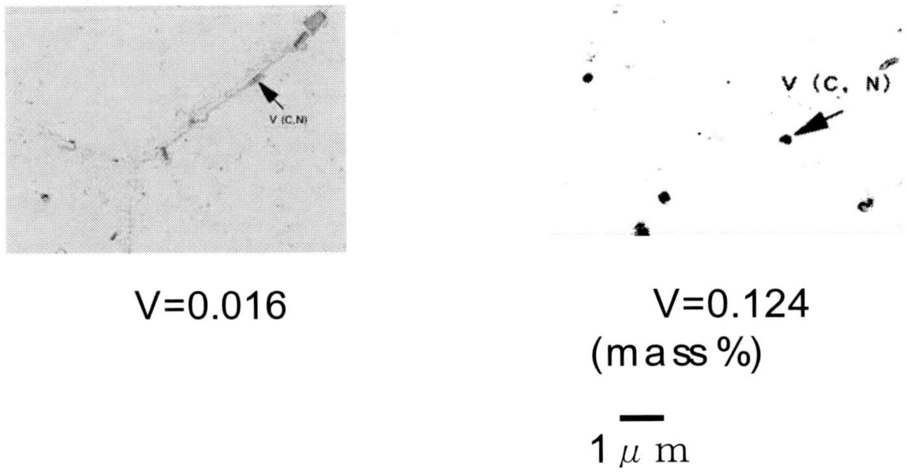

Figure 11. Precipitations of steels with vanadium.

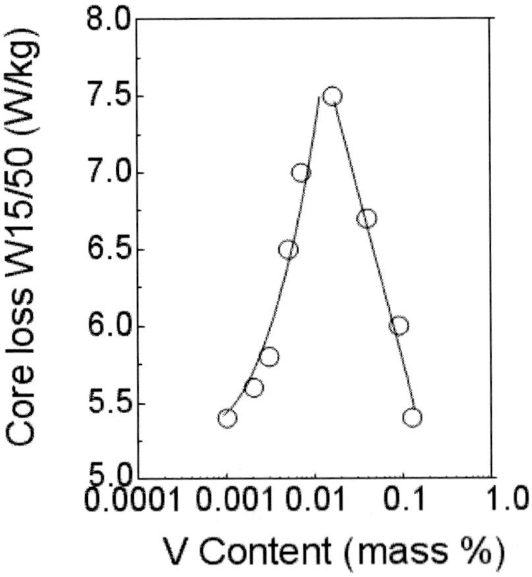

Figure 12. Effect of vanadium addition on core loss W15/50 after stress relief annealing.

Figure 12 shows the relationship between core loss at 1.5 T and 50 Hz (W15/50) and vanadium content after stress-relief annealing. The core loss is low at 0.001 mass % vanadium content, as there is no vanadium carbonitrides region; at 0.124 mass % vanadium content there is a large vanadium carbonitrides region in the precipitates study. By the contrast with the core loss, magnetic induction at 5000 A/m (B50) decreased with an increase in vanadium content < 0.016 mass %, but increased at contents > 0.016 mass % (Figure 13). Magnetic induction or permeability is affected by the grain size and texture of hot bands. Transmission electron microscope (TEM) study (Figure 11) made it clear that the vanadium carbonitrides precipitated during hot rolling pinned down the grain boundary in the steel containing vanadium 0.016 mass %. Therefore, these small hot band grains lead to a low magnetic induction B50.

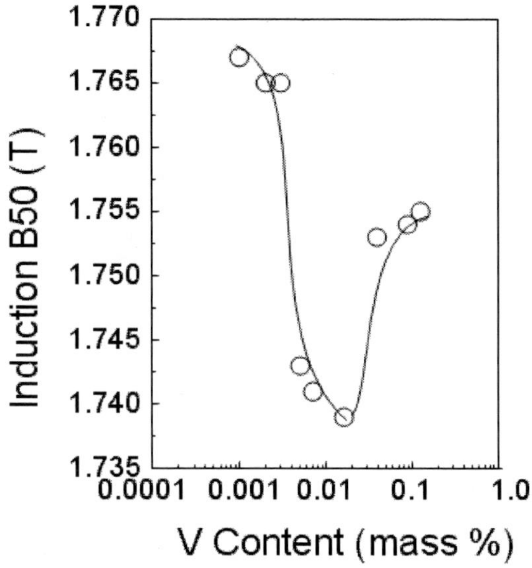

Figure 13. Effect of vanadium addition on induction B50 after stress relief annealing.

1.2.2. Effects of Titanium [24]

Same as vanadium, titanium also forms harmful precipitates to deteriorate the magnetic properties. In the case of titanium as contaminants in steels containing 0.001mass % carbon, 0.3 mass % silicon, 0.3 mass % manganese, 0.1 mass % phosphorus, 0.004 mass % sulfur, 0.3 mass % aluminum and 0.001 mass % nitrogen with 0.001-0.11 mass % titanium, the grain size decreases with the increase of the titanium content. Although in the steel contains less than 0.016 mass % titanium, the shape of grains was equiaxed, while in the range of the titanium content over 0.016mass %, the grains were elongated. Vanderschueren [44] studied the mechanism of recrystallization of IF steel and pointed out that the deformed ferrite grain was very difficult to recrystallize. Moreover, Park *et al.* [45] investigated phosphorus in titanium stabilized IF steel and summarized that phosphorus made a retardation on the recrystallization of cold-rolled steel. Furthermore, Brun et al.[46] and Jeong and Chung [47] investigated this deterioration and concluded that the precipitation of phosphides as (Fe, Ti)P was inevitable in the steel containing much titanium and phosphorus unless coiling temperatures was lower than 773 K on the hot rolling.

Figure 14 shows the TEM replica of the various titanium containing steel sheets after continuous annealing. In steels containing 0.006 mass % titanium, titanium carbonitride, pinning down the grain boundary, were observed along it. The effect of titanium carbonitride on the retardation of grain growth is more pronounced than that of compound inclusions in steel containing 0.002 mass % titanium. However, in steels containing more titanium (> 0.016 mass %) numerous iron-titanium phosphides (Fe,Ti)P precipitates were observed both in grains and along grain boundaries. In the case of non-oriented semi-processed steels, some phosphorus (ca.0.1 mass %) is added to improve the stampability. Therefore, in these steels, the precipitating behavior turns at the point between 0.006 mass % titanium, and major precipitates change from titanium carbonitrides to iron-titanium phosphides.

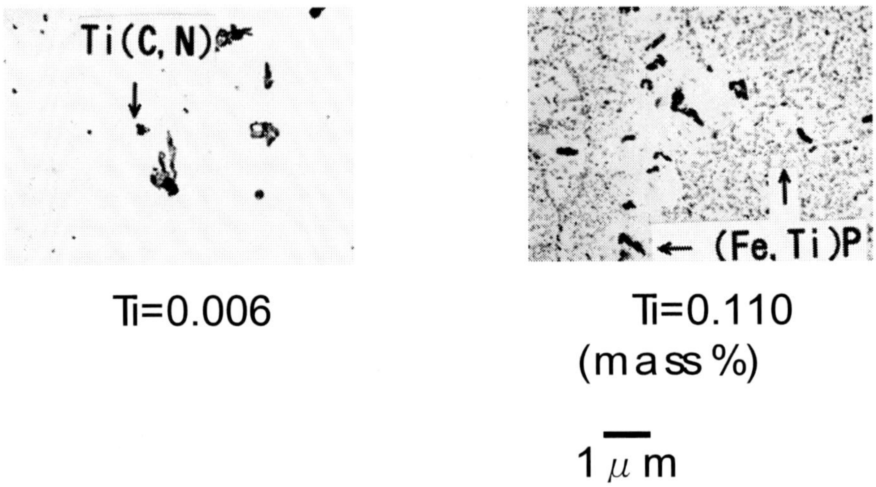

Figure 14. Precipitations of steels with titanium.

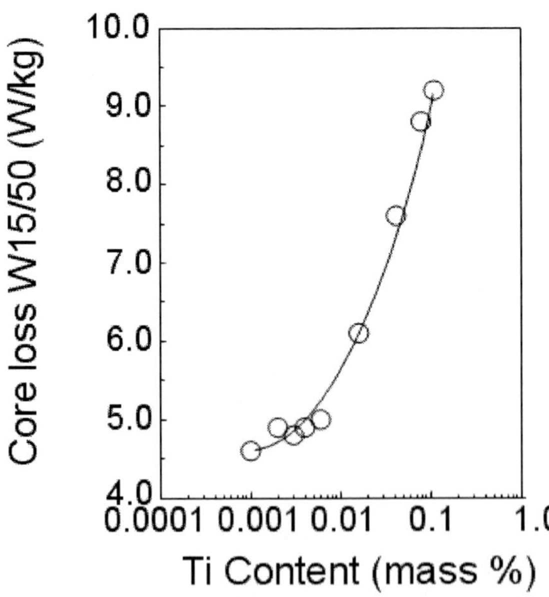

Figure 15. Effect of titanium addition on core loss W15/50 after stress relief annealing.

Figure 15 shows the relationship between core loss at 1.5 T and 50 Hz (W15/50) and titanium contents after stress relief annealing. Same as vanadium carbonitrides, titanium carbonitrides affect the core loss drastically. Matsumura *et al.*[41] reviewed the non-oriented electrical steel sheet and mentioned that the grain size affected the core loss and the lowest core loss was achieved at the size about 150μm. The grain size smaller than 150 μm, core loss was increasing with the decease of the grain size. In these case, the grain size was less than 40μm, so that the core loss is increasing with the increase of titanium content. Furthermore, Matsumura *et al.* [41] pointed out that core loss increase with increase of the number of inclusions. In this case, steels containing lower than 0.016 mass % titanium have less inclusions such as titanium carbonitrides, while steels containing over 0.016 mass % titanium have numerous inclusions as (Fe,Ti)P. This is the reason why the core loss increases dramatically in the steel containing over 0.016 mass % titanium.

By contrast with the core loss, magnetic induction at 5000 A/m (B50) was decreasing with an increase in titanium content < 0.016 mass % and was dramatically dropped at contents > 0.016 mass % (Figure 16). Magnetic induction or permeability is affected by the grain size and texture of hot bands. Steels containing <0.016 mass % titanium were equiaxed grain, while steels containing >0.016 mass % titanium recrystallized partially. Therefore, these unrecrystallized hot bands lead the low magnetic induction B50. Moreover,{222} texture, which is not easy to magnetize, developed in the finished sheets. This is another factor of the deterioration.

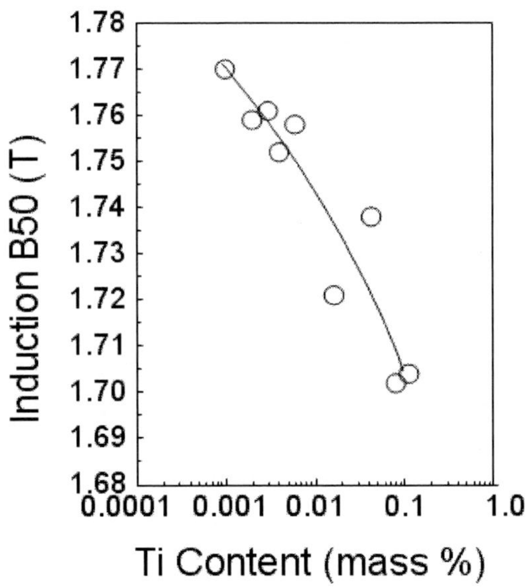

Figure 16. Effect of titanium addition on induction B50 after stress relief annealing.

1.2.3. Effects of Zirconium [25]

Zirconium is also well-known element to form nitrides or carbonitrides. In the case of zirconium as contaminants in steels containing 0.001mass % carbon, 0.3 mass % silicon, 0.3

mass % manganese, 0.08 mass % phosphorus, 0.003-0.005 mass % sulfur, 0.3 mass % aluminum and 0.0011-0.0036 mass % nitrogen with 0.001-0.133 mass % zirconium, the grain size in steels containing >0.04 mass % zirconium are smaller than that in steels containing < 0.01 mass % shown in Figure 17. Numerous precipitates were observed both along the grain boundaries and in grains. These precipitates were analyzed by the TEM replica method with EDAX and proved that they are Zr_3Fe. These precipitates did not inhibit the grain growth compared to those of V(C,N) and Ti(C,N). It is well known that zirconium is one of the easiest elements to combine with nitrogen and form zirconium nitrides, or zirconium carbonitrides, but any zirconium nitrides or carbonitrides were observed in these steels [42]. Therefore, in steels with zirconium < 0.01 mass % , no zirconium containing precipitates but aluminum nitrides and manganese sulfides were observed in this range , while in steels with zirconium > 0.01 mass % , numerous Zr_3Fe were observed and increased the core loss.

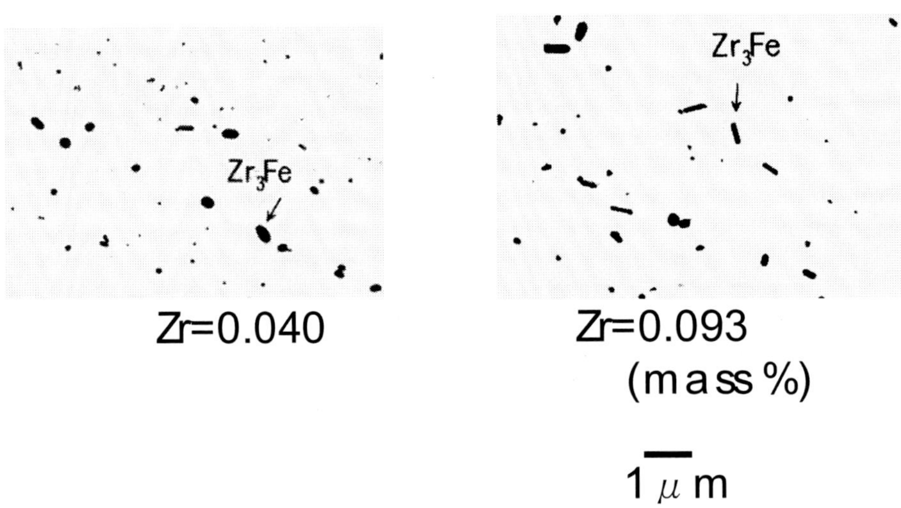

Figure 17. Precipitations of steels with zirconium.

Figure 18 shows the relationship between core loss at 1.5 T and 50 Hz (W15/50) and zirconium content after stress relief annealing. Core loss is almost no change with an increase in steels containing zirconium < 0.01 mass %, while in steels containing zirconium > 0.01 mass %, the core loss is increasing drastically by the numerous precipitates both along the grain boundaries and in grains. The effect of nitrogen content on the core loss is independent of that of zirconium content, however, the iron loss in the higher nitrogen containing steel was more deteriorated than that of lower one, due to the increasing hysterisis loss by aluminum nitrides precipitates. On the other hand, the induction at 5000 A/m (B50) is very little change by zirconium addition (Figure 19), due to no effects of zirconium addition on the grain size of hot bands. As increasing the zirconium content, there is no change on both grain size and texture, therefore, the magnetic induction B50 is no change in steels containing zirconium < 0.133 mass %. Contrarily, as shown in Figure 18, the core loss is deteriorated in steels containing zirconium > 0.01 mass %.

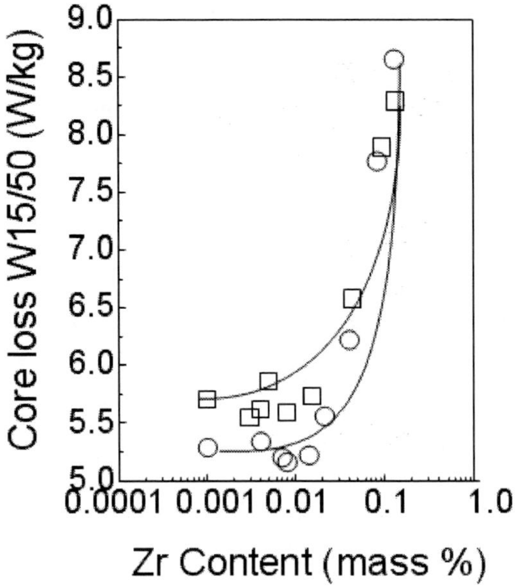

Figure 18. Effect of zirconium addition on core loss W15/50 after stress relief annealing. (○:N=0.0011-0.0020 mass %; □: N=0.0027-0.0036 mass %).

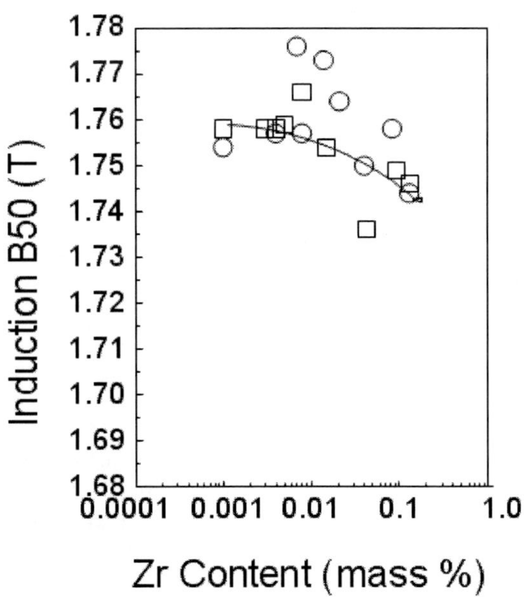

Figure 19. Effect of zirconium addition on induction B50 after stress relief annealing. (○:N=0.0011-0.0020 mass %; □: N=0.0027-0.0036 mass %).

Part 2. Core Manufacturing Technologies

The other problem on using electrical steels is core manufacturing. Even though using the steels with the excellent magnetic properties, poor manufacturing leads the poor results. Figure 20 shows a diagram of making motor cores. The following part is some key technologies on the core manufacturing topics.

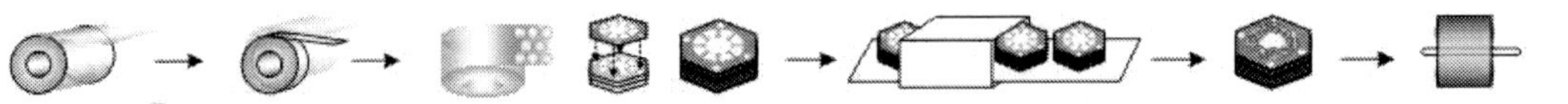

Figure 20. A process flow of motor core manufacturing.

2.1. Magnetic Properties Deterioration by Interlocking Lamination [26]

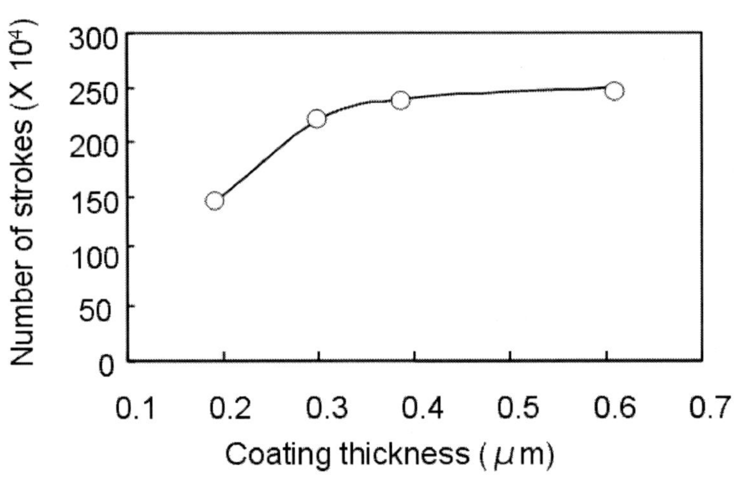

Figure 21. Effect of surface insulation coating thickness on the stamping performance with stamping conditions; evaluated number of strokes : stamped number of strokes when burrs height reach to 50 μm; blanking shape: 17mm X 17mm square core; die materials: SKD11; knife clearance: 5- 7% of the sheet thickness; stamping speed: 350 strokes/min; lubrication oil: kerosene; core material: 50SX1300)

Electrical steels sheets are usually coated with organic, inorganic or organic and inorganic mixture insulation on the surface. Stamping performance or die worn rate of the electrical steels with organic material as the surface insulation is better than that without organic insulations. Japanese Industrial Standard (JIS) C-2552 classified the surface insulation to CS-1(inorganic) and CS-2(organic and inorganic mixture), and CS-2 designated to better performance in high speed continuous stamping. This means that the organic insulation plays a lubrication role between steels and die at stamping. Figure 21 shows a relationship between insulation coating thickness and number of strokes as a result of die worn in continuous stamping. As increasing the coating thickness, stamping performance was improved due to the highly lubrication by thick organic layer.

However this lubrication cause the interlocking performance to reduce the fastened strength. Figure 22 shows the relationship between coating thickness and fastened strength. As increasing the coating thickness, the fastened strength was decreased. In order to check the lubricating role, we investigate the cross-section of sheared surface. Figure 23 shows the Cr distribution map on the cross-section of stamped edges by electron probe micro-analyzer (EPMA). To compare the cross-section after stamped, the electrical steel sheet with thick coating (ca. 0.6μm thick) is more widely rubbed the coating fragments (detected as Cr) over the sheared area than that with thin coating (ca. 0.3μm thick). These coating fragments play a rubricating role and affect the fastened strength.

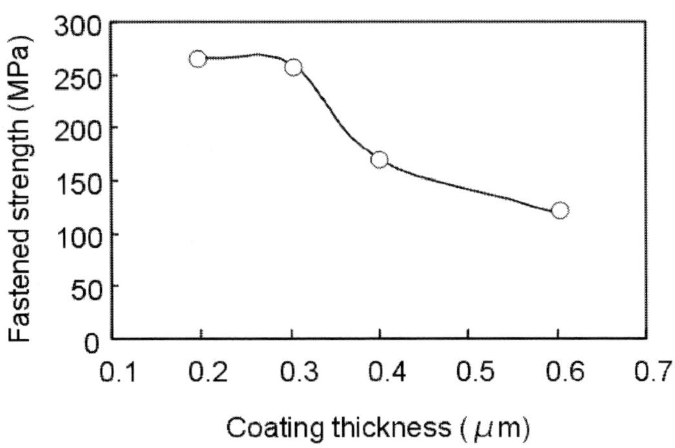

Figure 22. Effect of surface insulation coating thickness on the interlocking fastened strength of the ring core with 4 V-cut bottom rectangular shape protuberances (VR) illustrated in fig.25 (material: 50SX1300).

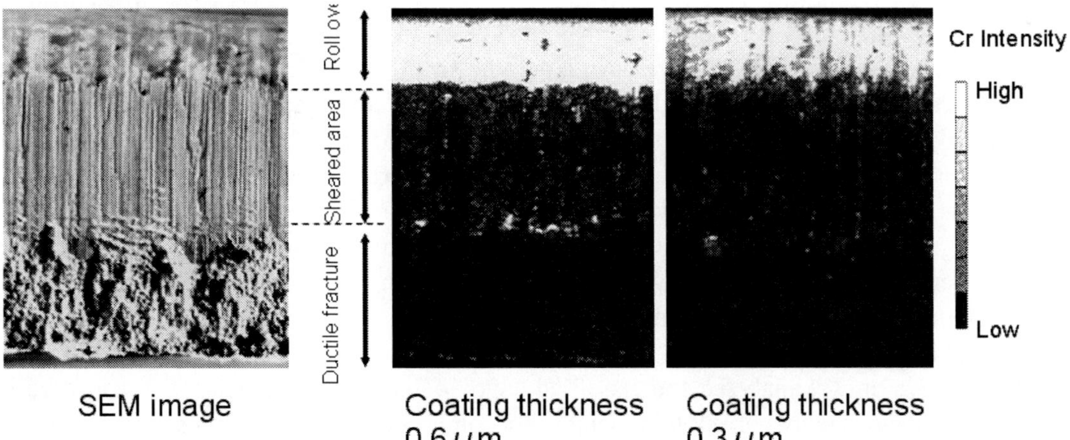

Figure 23. Cr distribution on the cross-section of interlocking protuberances by electron probe micro-analyzer (EPMA).

Since interlocking is a fastening by the friction at the cross-section, thin gauge sheets are disadvantage on the stacking strength. Figure 24 shows the relationship between sheet gauge and stacking strength. Thin gauge sheet such as 0.2 mm thick was so weak fastened that it needs careful handling on core making or coil winding. On this handling-easy point of view, 0.27mm thick sheet is the thinnest gauge on the usual high-speed stamping for building a small size motor. Kabasawa et al. [8] reported to apply the 0.27 mm thick non-oriented electrical steel sheet for hybrid electrical vehicles and mentioned that the 0.27 mm thickness sheet has the optimal balance between stacking strength and low core loss at high frequency.

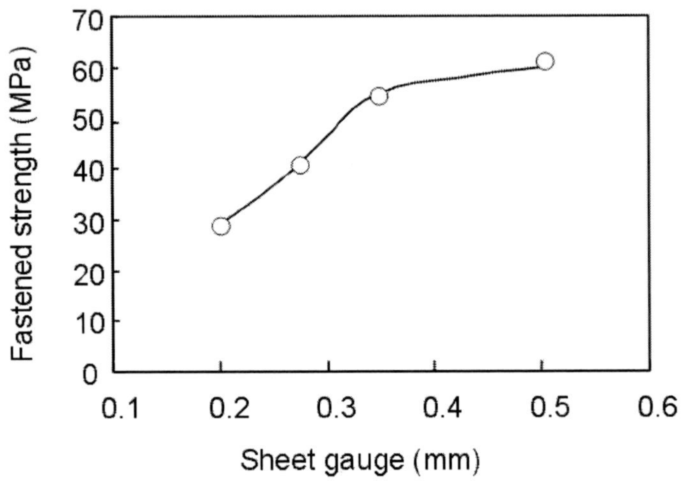

Figure 24. Gauge effect on interlocking fastened strength of the ring core with 4 V-cut bottom circle shape protuberances (VC) illustrated in fig.25 (core material: 2.0 % Si steels).

The deterioration of magnetic properties caused by not only simple obstacle of magnetic flux flow but by flow from lamination to lamination through interlocking protuberances and by compression or expansion stress between two fastened protuberances.

Fastened by flat bottom type interlocking (FC and FR in Figure 25) is weaker than that by V-cut type (VC and VR). However, the magnetic deterioration of flat bottom interlocking is less than that of V-cut one (Figure 26). Although the size of interlocking protuberance affects the deterioration of core loss strongly, obstacle area on the core magnetic circuit is estimated as no flow in protuberances as the model in Figure 27. Magnetic flux is concentrated in the narrowed cross-sectional area beside protuberances. As increasing magnetic flux density, the core loss is increased at the narrowed point. In the case of using the circular protuberances and average magnetization at 1.0T, the narrowed area near protuberance, magnetic flux density is calculated about 1.78 T, while in the case of using the rectangular type, it is about 1.2 T. The calculated core loss deterioration in Figures 28 and 29 is based on this core loss increase at 4 protuberance points.

Fujimura et al.[27] analyzed and classified the core loss and concluded that the hysteresis loss and anonymous eddy current loss increase by compressing or expanding stress. Therefore in this study, to clarify the core loss deterioration by interlocking, all cores were measured after stress relief annealing at 1023 K to eliminate the deterioration by all stress. The calculated core loss is blank core loss without interlocking and deterioration by increasing

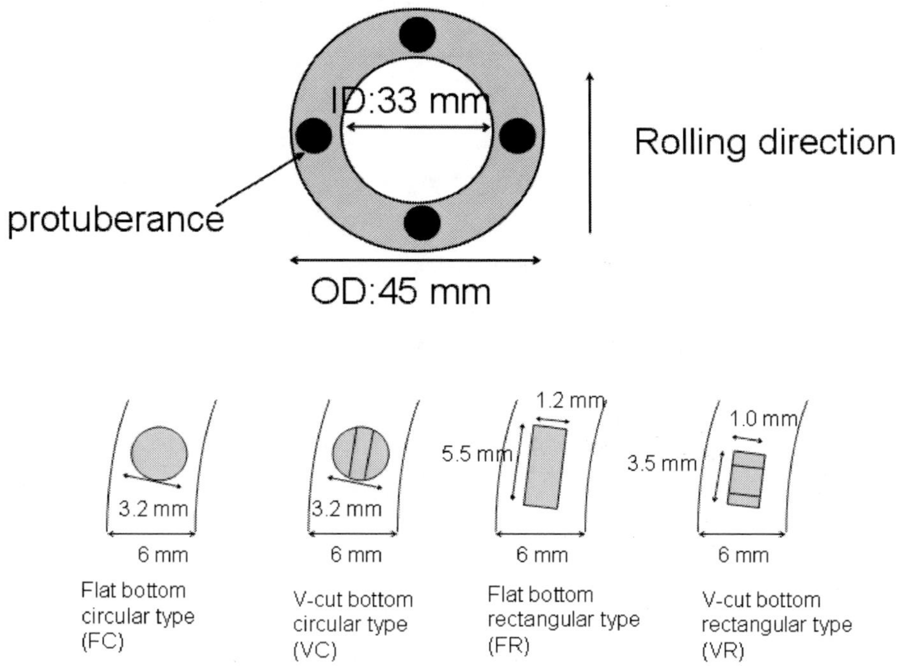

Figure 25. Dimension of ring core and dimensions of interlocking protuberances.

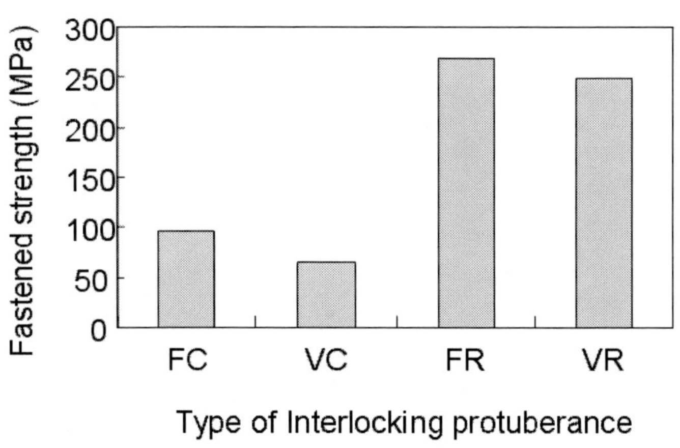

Figure 26. Comparison with protuberance types on fastened strength of the ring core (core material: 50SX400).

magnetic flux density near protuberance. Figure 28 shows the magnetic deterioration by interlocking among several protuberance shapes on the ring cores made of 50SX400. Core loss though laminations flux is estimated by the subtraction the calculated core loss from the measured core loss that annealed for stress relieving. Basically low silicon containing steels such as 50SX1300 is grown the grains and improved its magnetic properties by stress relief annealing at 1023 K. Although 50SX400, which contains 2.0 mass % silicon, is hardly growth grain and less improved its core loss by stress relief annealing at 1023 K without any stress, occasionally it occurred a suitable stress for grain growth at 1023 K annealing (FC). In that case, measured core loss is lower than the calculated one by improving its blank core loss after annealed.

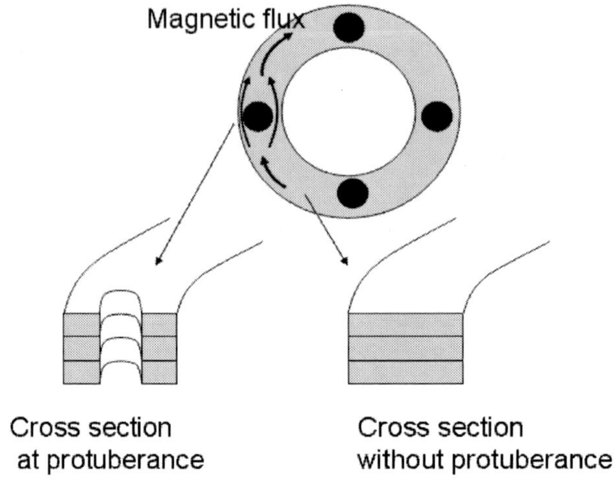

Figure 27. Magnetic flux model in the ring core.

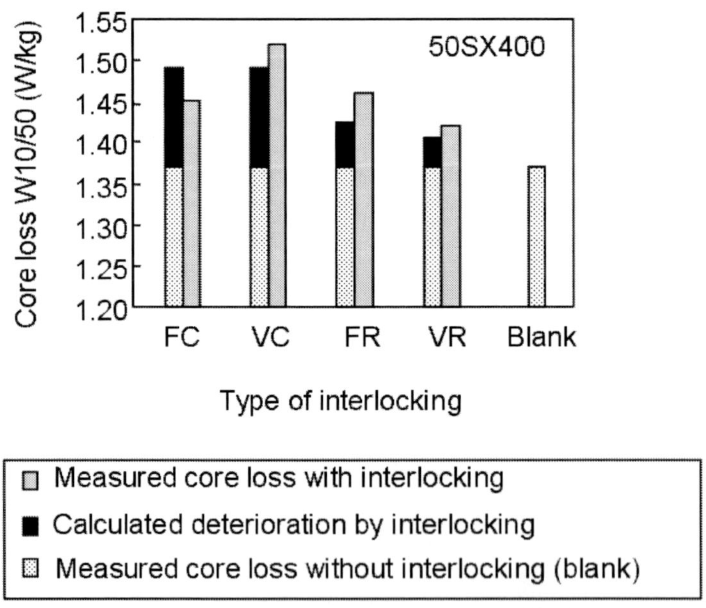

Figure 28. Comparison with calculated and measured core loss deteriorated by magnetic flux flow obstruction by interlocking protuberances of the ring core (core material: 50SX400).

Figure 29 shows the core loss comparison among 4 type protuberances on 50SX1300. This steel is grown the grain easily and is improved core loss by stress relief annealing at 1023 K. Figure 30 shows the near protuberance microstructures before and after annealing.

Measured core loss is less than calculated one on 3 type protuberances. This type of low silicon steel is unsuitable to investigate the deterioration effects.

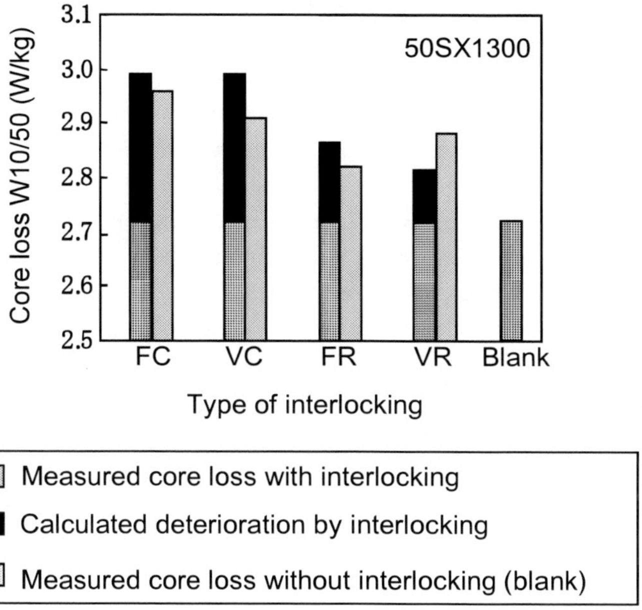

Figure 29. Comparison with calculated and measured core loss deteriorated by magnetic flux flow obstruction by interlocking protuberances of the ring core (core material: 50SX1300).

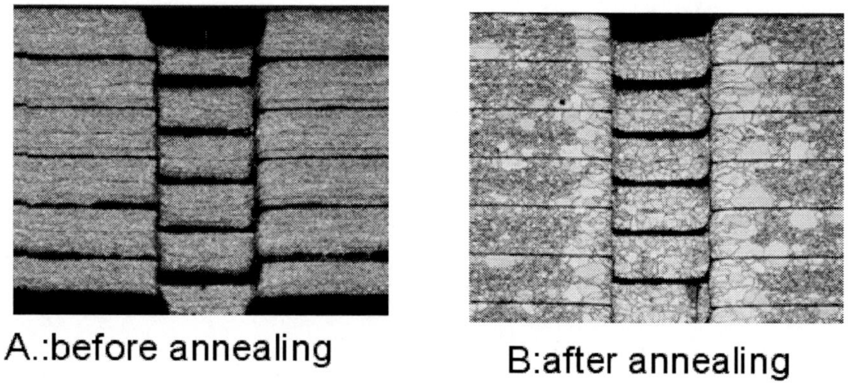

Figure 30. Cross-sectional microstructures of protuberances (A. before annealing; B. after stress relief annealing at 1023 K in nitrogen atmosphere for 2 hours, both core materials: 50SX1300).

V-cut interlocking protuberances are rammed into next two laminations, while flat bottoms are rammed into a half thickness of the next lamination. However, the core loss deterioration through lamination is hardly separated, due to unexpected grain growth near protuberances. This result indicates that the core loss increase by interlocking through laminations depends on not the overlap area ratio of the protuberance, but the cross-sectional obstacle by the interlocking protuberances and the stress.

2.2. Magnetic Properties Deterioration by Compressive Elastic Stress [27]

The compressive elastic stress deteriorates magnetic properties such as core loss or magnetic induction [47,48]. On manufacturing electrical motor, the core fits into the case by heat-shrunk. This shrink fitting introduces the compressive elastic stress into motor cores. For instance, industrial motors are fit by heat-shrunk into the die-casting frame. Or compressor motors for air conditioners are fit into the compressor shells. Those two cases are usually faced on the deterioration from the uncased motor to the cased motor as a final product.

To make this effect clear, single sheet magnetization test under several compression forces investigated. Figure 31 shows change of magnetic properties by compressive stress using 35SX230 as a core material. The magnetic induction deceases as increasing the compressive stress in the range of the lower magnetic force under 1500A/m. In the range over 1500 A/m, the affection by compressive force to the magnetic properties diminish.

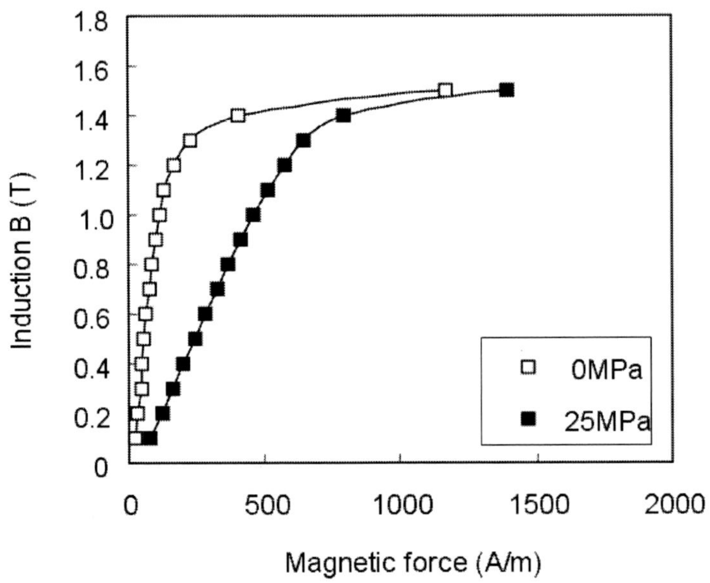

Figure 31. Effect of compressive elastic stress on the magnetic properties of 35SX230 at 400 Hz alternative current.

The influence on the core loss is shown in Figures 32 and 33. The core losses are separated to the hysteresis loss and the eddy current loss. The core loss deterioration by compressive stress is mainly on the hysteresis loss in the range of the magnetic induction lower than 1.2T and is mainly on the eddy current loss in the range of that over 1.2 T. This means compressive stress deteriorates not only the hysteresis but also the eddy current loss. This range of magnetic induction is most important range in operating the motors. The magnetic induction of the stator core yoke usually designed around 1.2-1.5 T, where is the most deteriorated range by the compressive stress. This eddy current loss increase derives from the domain structure change illustrated in Figure 34. Compressive stress generates the 90-degree domains, which increase the core loss [50].

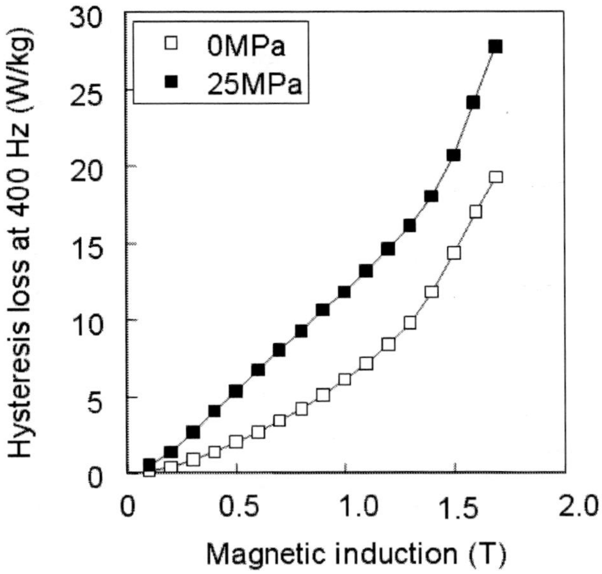

Figure 32. Compressive stress influence on the hysteresis loss of 35SX230 at 400 Hz alternative current.

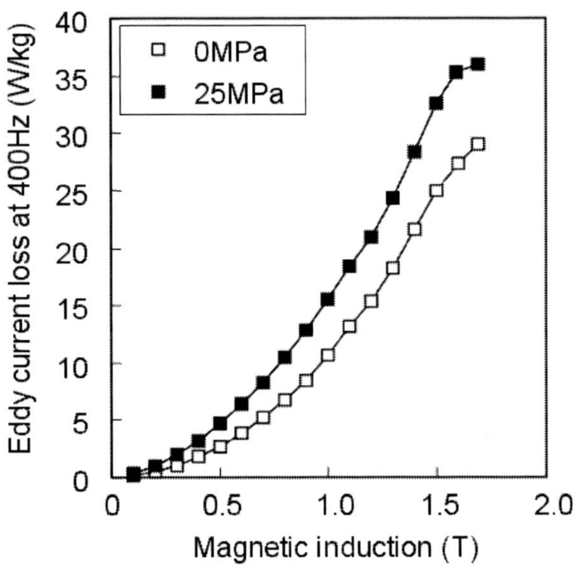

Figure 33. Compressive stress influence on the eddy current loss of 35SX230 at 400 Hz alternative current.

To reduce this deterioration the core shape especially stator core are designed by the calculations of stress analyses. Some stator cores have outer spins to intense the compression stress out of the main magnetic flux flows in the yoke. Other cores have widened the yoke where the stress is focused. Those straggle means how the compression stress makes worse to the motor efficiency.

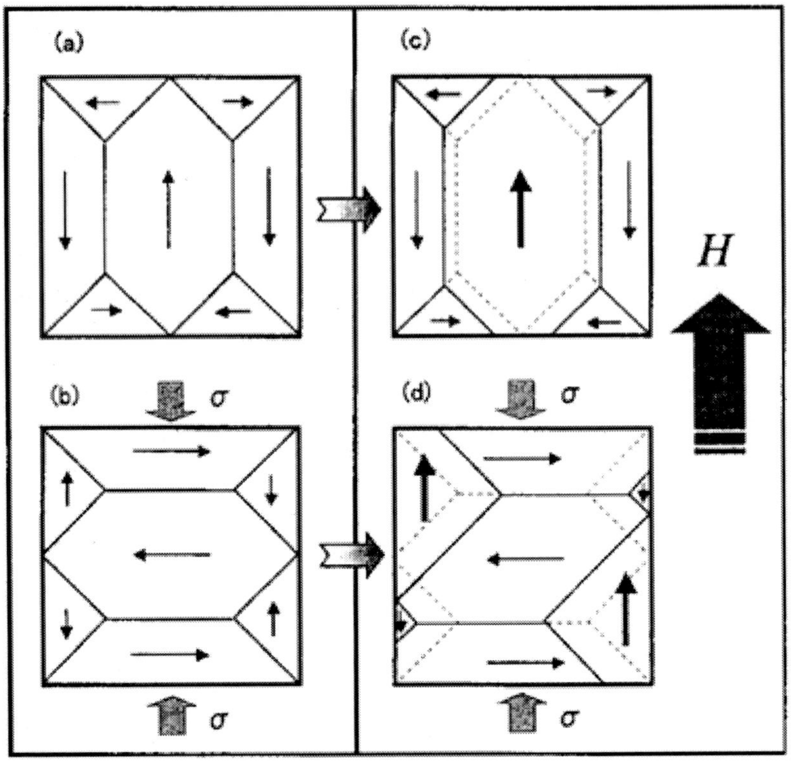

Figure 34. Domain structure under conditions of compressive stress and a magnetic field (a)without stress (b) with stress; (c) and (d) in a magnetic field.

2.3. Excellent Productivity Silicon Steel [29,30,32]

On manufacturing the motor core, productivity is very important factor for both production cost and product quality.

Motor cores are stamped from the rolled hoops or cut sheets continuously at high speed around 200 – 1500 strokes per minutes (s/min). Thus, stampability of the non-oriented electrical steels sheets is important factor for building the motors [35].

By controlling the chemistry, especially sulfur content, stampability of the steels is improved as shown in Figure 35. Regular steel 50SX1300 has 0.003 mass % carbon, 0.1 mass % silicon, 0.2 mass % manganese, and 0.08 mass % phosphorus with 0.005 mass % sulfur, while improved steel 50SXK1300 has 0.003 mass % carbon, 0.1 mass % silicon, 0.2 mass % manganese and 0.08 mass % phosphorus with 0.018 mass % sulfur. Figure 36 shows the SEM images of cut edge The improved steel has wider ductile fracture than the regular steel. This means knifes on the stamping dies touched less time and areas with the steels Therefore, the burr height / numbers of stamping as die-worn rate of the improved steel is less than that of the regular steel.

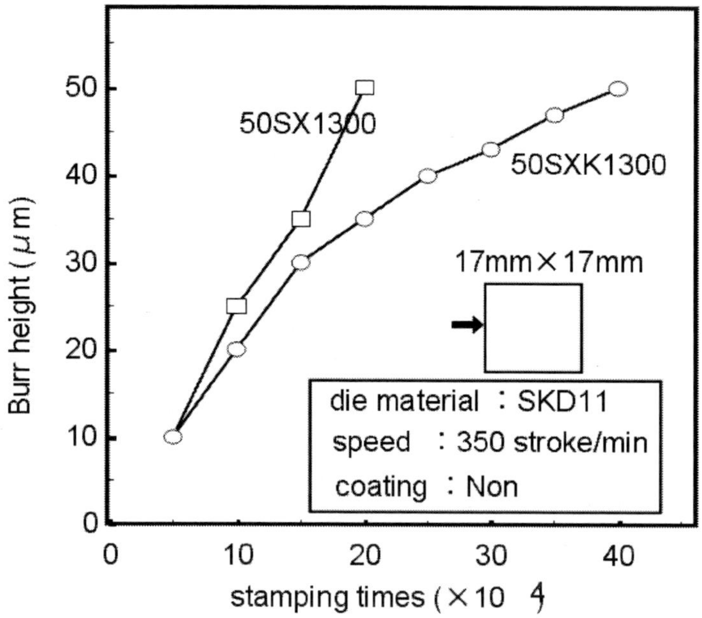

Figure 35. Stampability comparison between a regular steel 50SX1300 and the improved steel 50SXK1300.

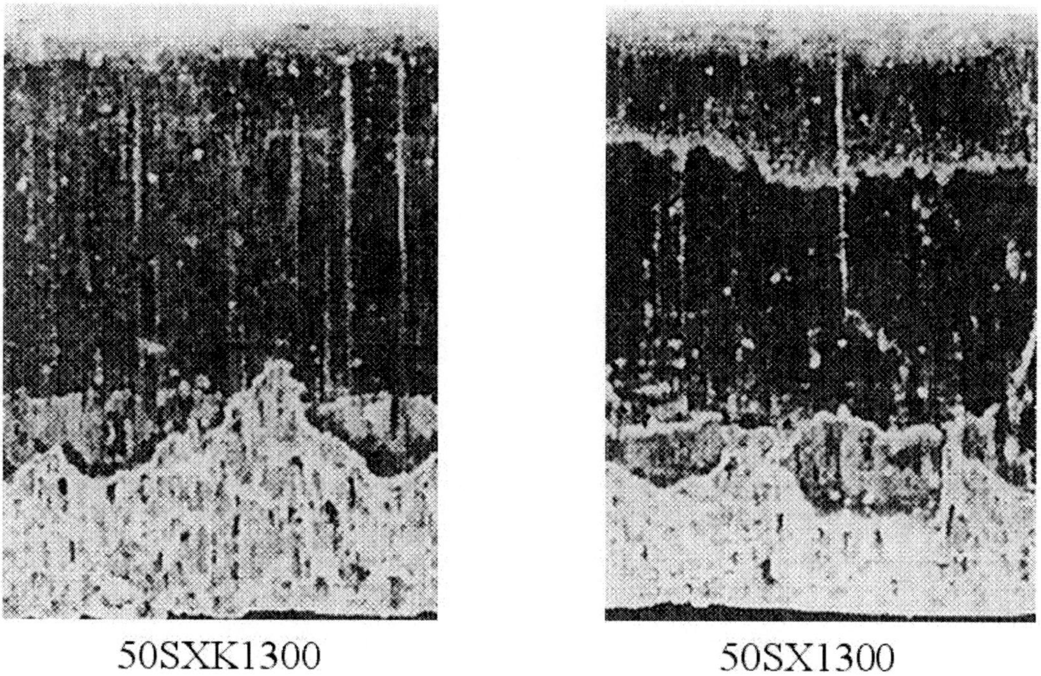

Figure 36. SEM images of the cut edge (A: regular steel 50SX1300, B: improved steel 50SXK1300).

This less sheared area of the improved steel leads to the less magnetic deterioration by the stamping stress. Figure 37 shows the magnetic deterioration by the shearing stress.

Introducing the shearing stress as a half cut or 3-pieces or 6-pieces cut to the improved steel magnetically less deteriorates than that to the regular one. Same as the results in Figure 36, Baudouin et al. [51] studied the relationship between the knife clearance and coercive force. The coercive force decreases as increasing the knife clearance, and the clearance increase makes the fracture increase. These deterioration occur mainly by the sheared zone stress, therefore, improved steel as 50SXK1300 has less deteriorations than regular steel 50SX1300.

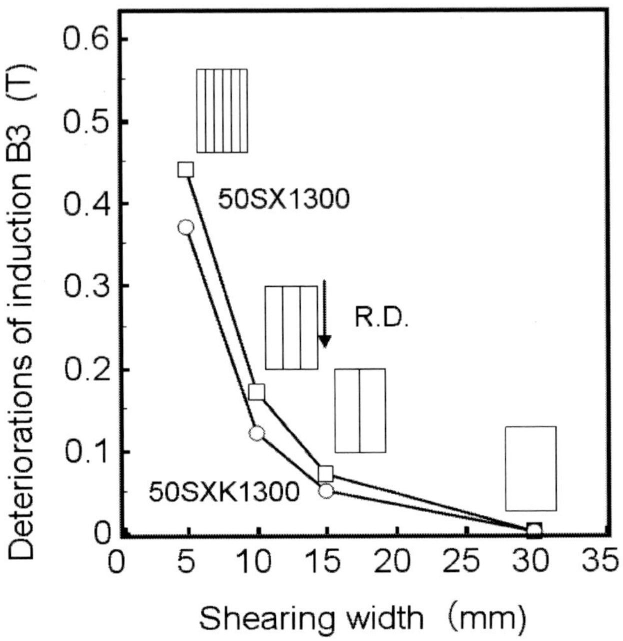

Figure 37. Deterioration of the magnetic induction at 300 A/m (B3) by the sheared stress. (A: regular steel 50SX1300, B: improved steel 50SXK1300).

Same as stampability, machinability is another important factor. Small motors with high-speed rotation require the balancing of the rotor. This balancing is usually done by curving of the rotor core surface. Figure 38 shows the bit worn rate comparison with regular steel (50SX1000) and improved one (50SXK1000). Same as high speed stamping die worn rate, improved steel bit worn rate is less than that of regular one.

The other characteristic of the improved steels is size deviations on stamping. For motor core stamping, inner diameter of the stator core and outer diameter of the rotor core is very important. The core size after stamping affects the air gap of the motor, which leads to the motor efficiency. Figures 39 and 40 show the stamped and interlocked ring core or disk diameter deviations. Both interlocked ring core simulated as a stator and stamped disk simulated as a rotor, reached the same result that the improved steel (50SXK700) is less deviated from the exact circle than the regular steel (50SX700). These phenomena derive from not only the mechanical properties such as elongation on the tensile test, but deviations of the area ratio between sheared area and ductile fracture (S/F). S/F ratio of the regular steel more deviated form the rolling direction than that of improved steel. Figure 41 shows the inclusions as manganese sulfides in the regular steel and improved steel. The manganese sulfides in the regular steel are liner shapes, which elongated to the rolling direction, while that of improved steels semispherical or rectangular shapes. These non-liner inclusions affect the mechanical properties and S/F ratio deviations from the rolling direction.

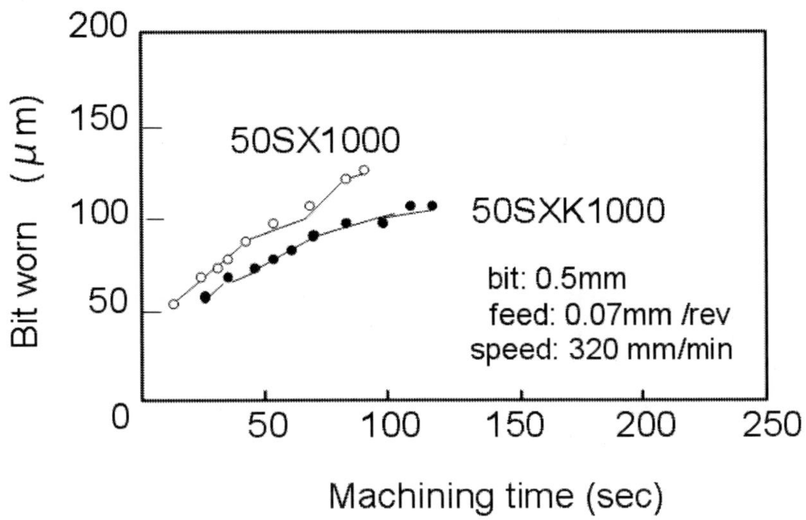

Figure 38. Bit worn rate comparison (A: regular steel 50SX1000, B: improved steel 50SXK1000).

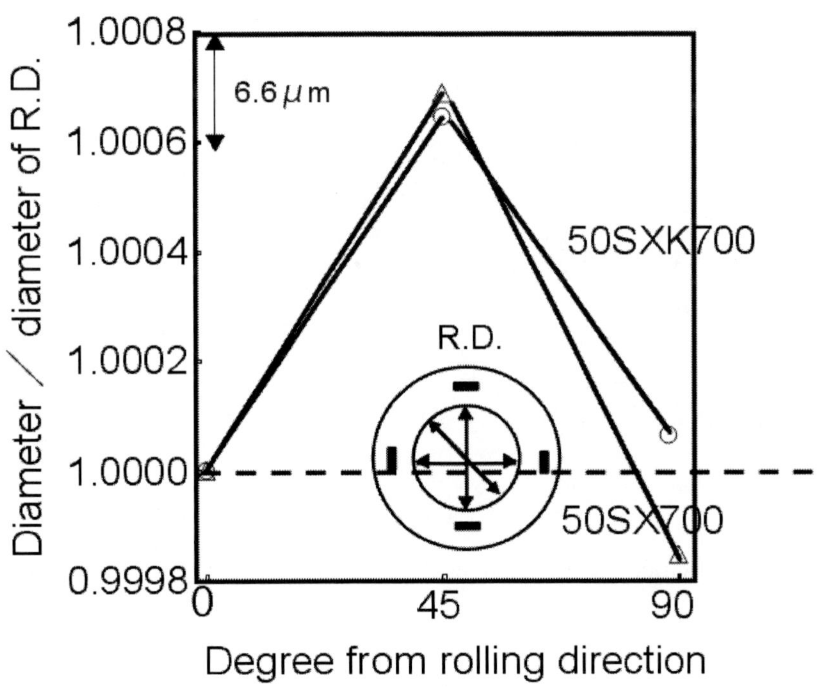

Figure 39. Interlocked ring core inner diameter deviations from the real circle. (A: regular steel 50SX700, B: improved steel 50SXK700).

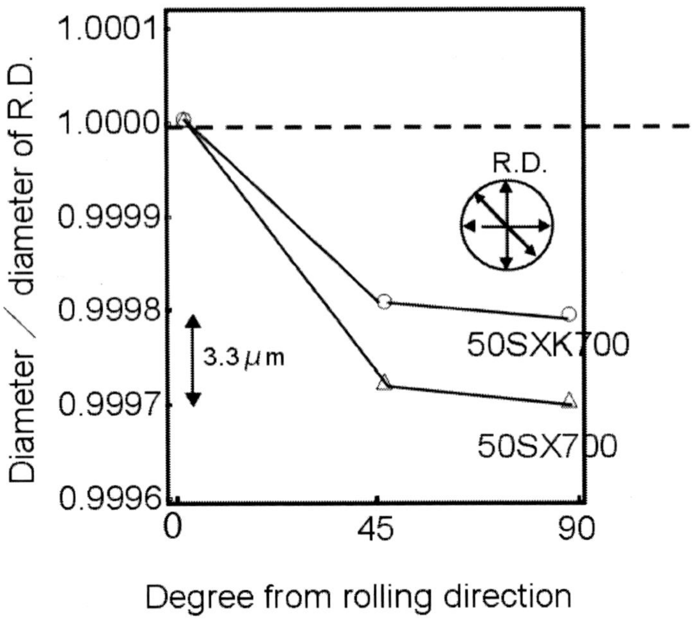

Figure 40. Stamped disk outer diameter deviations from the real circle. (A: regular steel 50SX700, B: improved steel 50SXK700).

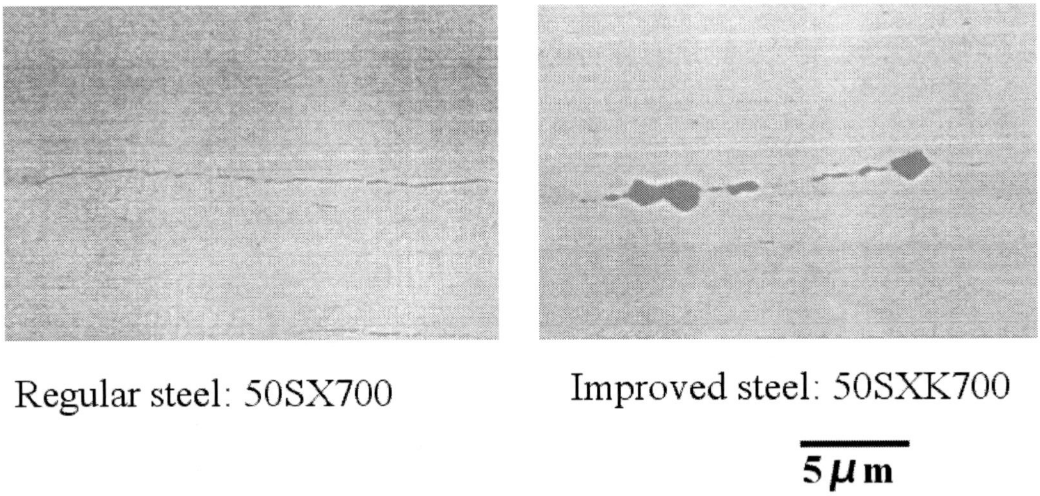

Figure 41. Manganese sulfides inclusions in regular steel(50SX1300) and improved steel(50SXK1300).

Conclusion

The effects of the additive elements as phosphorus, aluminum and manganese are improve the magnetic properties.

In the case of regular silicon steel (P=0.01mass %) the magnetic induction is decreased as increasing the cold rolled reduction for reducing core loss especially for eddy current loss, while the magnetic induction of phosphorus bearing silicon steel (P=0.1mass %) is less decreasing than that of regular silicon steel. The grains in the regular silicon steel are oriented the magnetically unfavorable texture such as {111}<112> in the case of high cold roll reduction (thin gauge) sheet, while that of phosphorus bearing steel are not.

The magnetic properties such as the core loss is not affected by the content of aluminum or nitrogen in the range of AlN <0.0024 mass %. In the case of steels with AlN > 0.0024 mass % and Al < 0.1mass %, the core loss was deteriorated by small grains which pinned down the grain boundaries by the 'harmful AlN' (< 0.5 μm in size), but not deteriorated the magnetic induction, due to the little effect on the grain size by those precipitates. In the case of steels with AlN > 0.0024 mass % and Al > 1.0 mass %, core loss was not deteriorated, but observed the 'harmless large AlN' (> 1.0μm) in steels. $MnSiN_2$ were formed in the steels with excess nitrogen and deteriorated the core loss.

The magnetic properties as the core loss are affected by the manganese and sulfur contents. The core loss of the steels increases with an increase of the sulfur in each steel with different manganese content. Among the same sulfur range steels, the core loss decreases with an increase of manganese. As increasing the sulfur content in each manganese level steels, the magnetic induction B50 decreased in any manganese level steels. To compare the steels processed in the condition of the SRT at 1273, 1423, and 1523 K, the core loss of the steels processed SRT at 1273 K is the lowest in each steels, and grains in each steels are same trend as decreasing in size as increasing the SRT. The deterioration of the core loss caused by the small grains pinned down by the 'fine MnS' (ca. 0.1 μm).

The grain size decrease with an increase in vanadium content < 0.016 mass %, and grow towards the lower vanadium content > 0.016 mass %. These phenomena caused by vanadium carbonitride along the grain boundary as pinning precipitates. In steel containing 0.001 mass % vanadium, no vanadium carbonitrides were observed, while large vanadium carbonitrides were observed in steel 0.124 mass % vanadium content steel. No pinning effects were observed in steels 0.001 mass % and 0.124 mass % containing vanadium. Magnetic properties were affected the vanadium content. The highest core loss was observed in 0.016 mass % vanadium containing steel, due to the smallest grain size after stress-relief annealing. The lowest induction was observed in steels containing vanadium 0.016 mass %, due to the smallest grain size after hot rolling, and magnetically unfavorable texture after stress-relief annealing.

The grain size decrease with an increase in titanium content < 0.11 mass %, and this phenomenon caused by titanium carbonitride along the grain boundary as pinning precipitates in steel containing < 0.016 mass % titanium, and numerous (Fe,Ti)P were observed in steels with titanium >0.016 mass % in grains and along the grain boundaries. Magnetic properties were affected the titanium content. The core loss increased as increasing titanium content, and induction B50 decreased as increasing the titanium content by the precipitates above.

The magnetic properties are not affected by the zirconium addition in the range of Zr <0.01 mass %. In steels containing Zr 0.01-0.13 mass %, the core loss are increasing with an increase of zirconium content by the numerous precipitates of Zr_3Fe, however, there is less effect on the magnetic induction, because of the no effect on the hot band grain size and texture by those precipitates.

Electrical steels sheets are usually coated with organic, inorganic or organic and inorganic mixture insulation on the surface. Stamping performance or die worn rate of the electrical steels with organic material as the surface insulation is better than that without organic insulations. Coating fragments pasted on the sheared area play a rubricating role at interlocking and affect the fastened strength. As increasing the coating thickness, stamping performance was improved due to the highly lubrication by thick organic layer. However this lubrication cause the interlocking performance to reduce the fastened strength. Magnetic flux is concentrated in the narrowed cross-sectional area beside protuberances. As increasing magnetic flux density, the core loss is increased at the narrowed point.

The core loss deterioration by compressive stress is mainly on the hysteresis loss in the range of the magnetic induction lower than 1.2T and is mainly on the eddy current loss in the range of that over 1.2 T. This means compressive stress deteriorates not only the hysteresis but also the eddy current loss. The anonymous eddy current loss increase derives from the domain structure change by compressive stress, which generates the 90-degree domains.

Stampability of the steels is improved by controlling the sulfur content. SEM images of sheared cross section show that the improved steel has wider ductile fracture than the regular steel. The improved steel magnetically less deteriorates by sheared stress and less bit worn rate than that to the regular one. Both interlocked ring core simulated as a stator and stamped disk simulated as a rotor, reached the same result that the improved steel is less deviated from the exact circle than the regular steel.

References

[1] Enokizono, M.; Tanabe, I.; Takahashi, K.; Fukunaga, H.; Yoshida Y.; Kubota, T. *J. Magnetism Magnetic Materials* 1999, 196-197, 335-337.

[2] Salsgiver,J.A. *Physica Scripta* 1989, 39, 775-779

[3] Yashiki ,H ; Kaneko ,H *J. Appl. Phys*. 1993, 73, 6606-6608.

[4] Beckley, P. *Electrical steels;* European Electrical Steels: New Port South Wales, 2000; pp1-307

[5] Hayakawa,Y.; Szpunar, J.A. *Acta Mater*. 1997, 45, 1285-1295.

[6] Oyama,K; *J. Inst. Electrical Engineers Jpn Trans 1A* 2003, 123, 63-66

[7] Yaguchi, H.; Sasaki, S; *SAE* 2005,01-0271, 11-15

[8] Kabasawa, A; Tahakashi, H; *SAE* 2005, 01-0276,49-54

[9] Hamada, M; *Nikkei Monozukuri* 2006, 8, 111-115.

[10] Lorenzo,J.B.; Ros-Yanes, T; De Wulf, M; Houbaert Y; *IEEE Trans Mag*. 2004, 40, 2739-2741.

[11] Chong, S.W.; Hilinski, E.J.; Rolett, A.D.; *Metallugical Mater. Trans. A* 2003, 34A, 1321-1327.

[12] Dzubinsky,M; Sidor, Y; Kovac,F; *Mater. Sci. Eng. A* 2004, 385, 449-454.

[13] Moseley, D.; Hu, Y.; Randle, V.; Irons, T; *Mater. Sci. Eng. A* 2005 , 392, 282-291.

[14] Park, J-T.; Szpunar J.A; Cha, S-Y.; *ISIJ International* 2003, 43, 1611-1614.

[15] de Campos, M.F.; Landgraf, F.J.G.; Tahanohashi, R.; Chagas, F.C.; Filleiros,I.G.S.; Fronzaglia, G.C.; Kahn, H.; *ISIJ International* 2004, 44, 591-597.

[16] Nakayama, T; in *Recent Research Development in Material Science* ; Pandalai S.G.;Ed., Research Signpost, Trivandrum, India, 2002; Vol.3-Part II, pp.811-824.

[17] Lindenmo, M; *J. Magnetism Magnetic Materials,* 2006, 304, 178-182.

[18] Kubota, T.; *Steel Research Int.* 2005, 76, 464-470.

[19] Nakayama, T., Honjou, N., Minaga, T. and Yashiki, H. *J. Mag. Mag. Mater.* 2001, 234, 55-61.
[20] Nakayama, T. and Honjou, N. *J. Mag. Mag. Mater.*:2000,213, 87-94.
[21] Tanaka, I.; Nitomi, Y.; Yashiki, H.; Kojima, H.; Nakayama,T.; *Materia Japan* 2006, 45, 225-227
[22] Tanaka, I.;Yashiki, Y.; *ISIJ International* 2007, 47, 1666-1671
[23] Nakayama, T. ; Takahashi, M.;*J. Mater. Sci.* 1995,30, 5979-5984
[24] Nakayama, T. ; Tanaka, T.; *J. Mater. Sci.* ,;1997, 32, 1055-1059.
[25] Nakayama, T. ; Honjou, N.; *J. Mater. Engineering Performance,* 2000, 9, 552-556
[26] Nakayama,T. ; Kojima, H.; *J. Mater. Engineering Performance*, 2007,16, 7-11.
[27] Fujimura, H.; Yashiki, H.; Kojima, H.; Nakayama, T.; *Paper Tech. Meeting Magnetics IEE Jpn.* 2003, MAG-03-190,9-14.
[28] Fujimura, H.; Yashiki, H.; Kojima, H.; Nakayama, T.; *Paper Tech. Meeting Magnetics IEE Jpn.* 2003, MAG-05-103,1-6.
[29] Nakayama, T; Honjou, N.; Yashiki, H.; *Paper Tech. Meeting Magnetics IEE Jpn.* 1998, MAG-98-43,23-26
[30] Yashiki, H.; Doi, M.; Ishikawa, H; Honjou, N.; Nakayama, T. *Materia Jpn.* 1999, 38,169-171.
[31] Nakayama, T. Honjou, N.; Nagai, A; Yashiki, H.; *Sumitomo Kinzoku*, 1996, 48, 39-44.
[32] Nakayama, T; Honjou, N; *Sumitomo Kinzoku*, 1998, 50, 58-63.
[33] Yensen, T.D.; *Metals Alloys* 1930, 1,493.
[34] Iguchi, N.; Sakanishi, S.; *Werkstatt Betribe* 1987, 120, 323-328
[35] EL-Refaire, A.M.; Manzke, R.; Jahns.M.; IEEE Trans. Industrial Application 2004, 40,717-725.
[36] Hou, C-K.; Hu,C-T. ; Lee, S.; *IEEE Trans Mag.* 1991, MAG-27, 4305-4309
[37] Yashiki, H.; Okamoto, A. ; *IEEE Trans. Mag.* 1987,MAG-23, 3086-3088.
[38] Sawamura, H.; Mori, T.; *Tetsu-to-hagane*, 1955, 41,1082.
[39] Darken, L.S.; Smith, R.P.; Filer, E.W.; *J Metals Trans AIME* 1951,3, 1174-1179..
[40] Yashiki, H.; Kaneko, T.**;** *Tetsu-toHagane*, 1994, 80,79-83**.**
[41] Matsumura, K.; Fukuda, B. ; *IEEE Trans. Mag.* 1984, MAG-20, 1553-1538.
[42] Shimoyama, Y.; Miyishi, K.; Tanino. M.; Wada, T.; *IEEE Trans. Mag.* 1983, MAG-19, 2013-2015.
[43] Stephenson, E.T.; Marder, A.R.; 1986, *IEEE Trans. Mag.* 1986, MAG-22, 101-106.
[44] Vanderschueren, J.; in *Physical metallurgy of IF steel*, ISIJ, Tokyo, 1994,pp.145-148
[45] Park, Y.B.; Kang, H.J.; Chang, S.K.; in *Physical metallurgy of IF steel*, ISIJ, Tokyo, 1994,pp.245-248
[46] Brun, C.; Patou, P.; Parniere, P.; in *'Metallurgy of continuous annealed sheet steel'*, Bramfitt, B.L. and Mangonon, P.L.; Eds., TMS-AIME, Warrendale, PA ,1982, pp173-197.
[47] Jeong, W.-C.; Chung, J.-H.; *Tetsu-to-Hagane*, 1989, 75 , 74-81
[48] Nakata, T; Nakano,M.; Kawahara, K.; *IEEE Trans. Mag.* 1992, MAG-7, 453-457.
[49] LoBue, M; Sasso, C; Bosso, V Fiorillo, F.; Bertotti ,S; , *J. Mag. Mag. Mater.* 2000, 215-216.
[50] Pry, R.H.; Bean, C.P; *J. Appl. Phys.* 1958, 29, 532-533.
[51] Baudouin, P; DeWulf, M.; Kestens, L.; Houbaert, Y. , *J. Mag. Mag. Mater.* 2003, 32-40.

In: Strength of Materials
Editors: G. Mendes and B. Lago, pp. 217-231
ISBN: 978-1-60741-500-8

Chapter 7

INFLUENCE OF LUTING CEMENT APPLICATION TECHNIQUE ON QUARTZ FIBER POST REGIONAL BOND STRENGTHS

Camillo D'Arcangelo[a,]*, Francesco De Angelis[a], Maurizio D'Amario[b], Simone Zazzeroni[a], Mirco Vadini[a] and Sergio Caputi[a]

[a] Department of Restorative Dentistry, School of Dentistry, University G. D'Annunzio - Chieti, Italy

[b] Department of Restorative Dentistry, Dental Clinic, University of L'Aquila, Italy

Abstract

The aim of this study was to investigate regional root canal push-out bond strengths for a fiber-reinforced post system varying the application method of the luting agent.
Recently extracted maxillary incisors (n=30) were sectioned transversally at the labial cemento-enamel junction, and the roots treated endodontically. Following post space preparations, fiber-reinforced posts (Endo Light-Post; RTD) were placed using adhesive system and resin cement provided by the manufacturer. Three equal groups (n=10) were assessed according to the technique used to place the luting agent into post space: using a lentulo spiral, applying the cement onto the post surface, injecting the material with a specific syringe. Each root was sliced into three discs (2 mm thick) representing the coronal, middle and apical part of the bonded fiber post. Push-out tests were performed for each specimen to measure regional bond strengths. Results were statistically analyzed using two-way ANOVA and Tukey tests ($\alpha = 0.05$). All fractured specimens were observed using a scanning electron microscope to identify the types of failure.

The results indicated that bond strength values were significantly affected by the application method of the resin cement ($p < 0.05$). The "syringe technique" and the "lentulo technique" showed higher bond strength values compared with the "post technique". No

* E-mail address: cdarcang@unich.it. Phone: +39(0)85.4549652. Fax: +39(0)85.4541279. Corresponding author: Prof. Camillo D'Arcangelo, Department of Restorative Dentistry, Dental School, University "G. D'Annunzio", Via dei Vestini 31, 66100, Chieti, Italy.

significant differences were recorded among the post space thirds. Microscopic analysis revealed a prevalence of post/cement and mixed failures.

The best performance in terms of push-out bond strengths for the post system tested was obtained when the luting agent was applied into the post space either with a specific syringe or using a lentulo spiral. There were not differences in bond strength among root thirds.

Introduction

The increasing popularity and widespread use of fiber-reinforced (FRC) posts is changing the restorative procedures for endodontically treated teeth. Fibre-reinforced composite posts are commonly used for the restoration of endodontically treated teeth with reduced crown structure [1]. Since fiber-reinforced posts (18-22 GPa) have a modulus of elasticity (E) similar to the dentin (18 Gpa) [2], they produce a stress field similar to that of natural teeth, thereby reducing the risk of root fractures [3]. FRC posts contain a high percentage of continuous reinforcing fibers embedded in a polymer matrix. Matrix polymers are commonly epoxy resins or other polymers with a high degree of conversion and a highly cross-linked structure [4],[5]. The benefits of adhesive techniques used for dental restoration are well documented, so the use of adhesive resin cements has been proposed for cementing endodontic fibre posts in non-vital teeth. FRC post and resin cement have similar moduli of elasticity to dentine enabling loading forces to be transferred consistently from the restoration to tooth structure [6]. The loss of bond at the fibre post/resin cement/root dentine interfaces still represents the main reason for which endodontically treated teeth reconstructed with fibre posts show clinical failures [7]-[9]. Retention of fibre-reinforced composite (FRC) posts within root canals is affected by several factors: type of post, its adaptation into the post space, type of adhesive and operative procedures [10]-[14]. The distribution of resin cement into the post space during the luting procedure and the anatomical and histological characteristics of the root dentine seemed to influence bond strength between resin luting agent and root canal regions [15]. An adequate polymerisation of luting agent is necessary to provide its mechanical properties, that clinically ensure post retention. Many current resin luting agents polymerise through a dual-curing process that requires light exposure to initiate the reaction. It has been reported that the mechanical properties of dual-cure type resin agents appear improved after photo-activation compared with chemical-activation alone [16]. Dual-cure resin cements are different in their handling characteristics, compositions and properties (such as polymerisation ability, flexural strength, hardness). The quality of adhesion to root dentine is also affected by the density and orientation of dentine tubules at different levels of the root canal walls [17] and the accessibility of the coronal, middle and apical third of the root during handling of the materials [18]. It has been demonstrated that the control of moisture after the application and removal of phosphoric acid as well as incomplete infiltration of the resin into the dentine significantly affect bond strengths [19].

A resin luting agent may create polymerization shrinkage stresses within the post space. Shrinkage stresses of luting materials in root canals are especially relevant due to the unfavourable factor of configuration (C-Factor) that restricts the flow of resin cement, which may affect the integrity of the adhesive interface at different root levels [8]. The contraction stress of resin cements in confined spaces depends upon the thickness of the cement layer [20], but this thickness changes with different root canal morphologies. Moreover, the post can be closer to the dentine on one side, which also influences cement behaviour [21] and explains the importance of inserting posts with identical pressure to achieve a standard cement thickness. Although the chemical-physical properties of resin cements have been evaluated, little information is available on the role of the application methods of dual-cure resin cements to the post space and their effect on regional bond strength of fiber post.

Aim of this study was to evaluate the bond strengths of quartz-fiber posts to coronal, middle and apical thirds of post space dentine varying the application method of the luting agent.

Materials and Methods

Specimen Preparation

Thirty freshly extracted human maxillary incisors were selected. External debris was removed (Suprasson P-max; Satelec/Acteon Equipment, Merignac, France). Selected specimens were stored in 0.5% chloramine T aqueous solution at 4°C. Crown surfaces of each tooth were sectioned at the labial cemento-enamel junction (CEJ) using a cylindrical diamond rotary cutting instrument (Intensiv 314, Ø ISO 014, L.8.0 mm; Intensiv, Grancia, Switzerland) mounted on a high-speed hand-piece (Bora L; Bien-Air, Bienne, Switzerland) with water-spray cooling.

Root canals were mechanically enlarged to ISO size 25, 0.06 taper (MTwo; VDW GmbH, Munich, Germany). Irrigants used were 5% sodium hypochlorite (Ogna, Muggiò, Milan, Italy) and 17% EDTA (Pulpdent, Watertown, MA). Enlarged canals were rinsed with distilled water, dried with paper points (Roeko, Langenau, Germany) and sealed with gutta-percha (Lexicon Gutta Percha Points; Dentsply Tulsa Dental, Tulsa, Okla) using the System-B HeatSource (Analytic Technology, Redwood City, CA) and endodontic sealer (Pulp Canal Sealer EWT; Kerr, Romulus, MI). Backfilling was performed with Obtura II (Spartan, Fenton, MO).

Bonding of Fiber Posts

After 24 hours, gutta-percha was removed with warm endodontic pluggers (Sybron Dental Specialties, Romulus, MI). Post spaces were prepared to a depth of 10 mm measured from the sectioned surfaces using Torpan drills ISO 100 Yellow (batch no. 042190611) provided by the manufacturer (RTD, St. Egrève, France). Post space preparations were rinsed with 5% NaOCl. A final irrigation was accomplished with distilled water, and post spaces were dried with paper points. Before cementation procedures, each post was marked at a distance of 10 mm from the apical end corresponding to the length of the post space preparation and sectioned horizontally with a water-cooled diamond rotary cutting instrument (R879.014; Diaswiss, Geneva, Switzerland). In this way, the complete seating of the posts has been verified. The root canal walls were etched for 60 s with 36% phosphoric acid, Conditioner 36, (batch no. 0507002142; Dentsply DeTrey, Konstanz, Germany), introduced into the spaces with a needle, rinsed using a water syringe and then gently dried with paper points. Bonding procedures were performed following the instructions provided by the manufacturers. XP Bond (batch no. 065001399; Dentsply DeTrey) and SelfCure Activator (batch no. 0510061; Dentsply DeTrey) were mixed for 2 s and applied to the root canal for 30 s with a microbrush (Microbrush X; Microbrush Corp, Grafton, WI). After 20 s, the excessive adhesive solution was removed with a paper point and then gently air-dried for 5 s.

The specimens were randomly divided into 3 groups (n=10) according to the technique used to place the luting agent (FluoroCore 2, batch no. 0610021; Dentsply DeTrey) into root canal: using a #30 lentulo spiral instrument (Dentsply Maillefer, Ballaigues, Switzerland) for 4 seconds before setting the post; applying the cement onto the post surface; injecting the material with a tube with needle and the appropriate plug (KerrHawe SA, Bioggio, Switzerland) using a specific Composite-Gun (KerrHawe SA). The posts used were Endo Light-Posts, size 2, 2% taper, with a maximum cross-section diameter of 1.36 mm and a minimum diameter at radicular end of 1.0 mm (batch no. 049520702; RTD). Endo Light-Posts are made of unidirectional pre-tensed quartz fibres (60% volume) embedded in an epoxy resin matrix.

The posts were seated to full depth in the prepared spaces using finger pressure. Excess of luting agent was immediately removed with a small brush. A constant axial load of 5 kg was applied for 60 seconds to stabilize the fiber posts in the post spaces. After the initial chemical polymerization, the resin luting agents were light polymerized (L.E. Demetron I, Sybron/Kerr, Orange, CA, with a 1200 mW/cm^2 output) for 40 seconds. Thirty minutes after the cementation procedures, all root specimens were stored in distilled water for 24 hours. Then, specimens underwent 10,000 thermal cycles between 5°C and 55°C, with a 30-second dwell time and a 5 second transfer between temperature baths. Specimens were also subjected to 300,000 cycles of mechanical loading parallel to the long axis of the post with a masticatory simulator (Willytec, Munich, Germany) at 30 N force and 1.6 Hz. The mechanical loading pattern was equivalent to 1 year of clinical function [22],[23]. Specimens were then preserved in a saline solution at room temperature for 1 week.

Push-out Testing

Specimens were then fixed to phenolic ring forms filled with a autopolymerizing acrylic resin (Technovit 4000; Heraeus Kulzer, Wehrheim, Germany). Specimens were attached to the arm of a low-speed saw (Micromet M; Remet S.p.A., Casalecchio di Reno, Italy) and sectioned perpendicular to the long axis under water cooling. Three slices per each root (Figures 1, 2), containing cross sections of coronal, middle and apical part of the bonded fiber post, were obtained by sectioning the root under distilled water coolant. The sections were 2.0 ± 0.1 mm thick. Each slice was marked on its apical side with an indelible marker. The thickness of each specimen was measured and recorded by a digital caliper with an accuracy of 0.001 mm (Figure 3). The sections were stored individually in black film canisters with sterile water. Push-out-test was performed by applying a compressive load to the apical aspect of each slice via a cylindrical plunger mounted on a Universal Testing Machine (Lloyd LR 30K; Lloyd Instruments Ltd., Fareham, UK) managed by PC software (Nexygen-Ondio Version 4.0; Lloyd Instruments Ltd.). With regard to the tapered design of the post, three different sizes of punch pins were used for the push-out testing. The diameter of the punch pin was 1.2 mm for the coronal slices, 1.0 mm for the middle slices and 0.8 for the apical slices [24]. Punch pin was positioned to contact only the post, without stressing the surrounding root canal walls [25] (Figure 4). Care was also taken to ensure that the contact between the punch tip and the post section occurred over the most extended area, to avoid

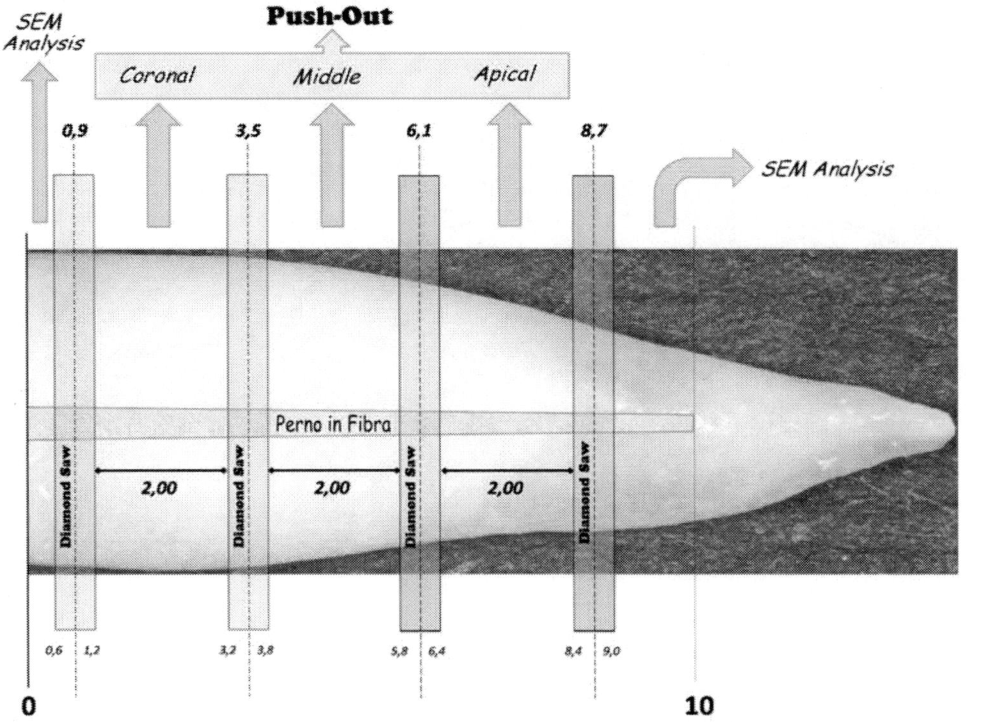

Figure 1. Root section performed by the Micromet M machine. The numbers express (millimeters) the distance from the arm of the machine and, thus, from the cervical end of the root. The diamond saw is 0.6 mm thick. The sections directed to the push-out-test are 2 mm thick.

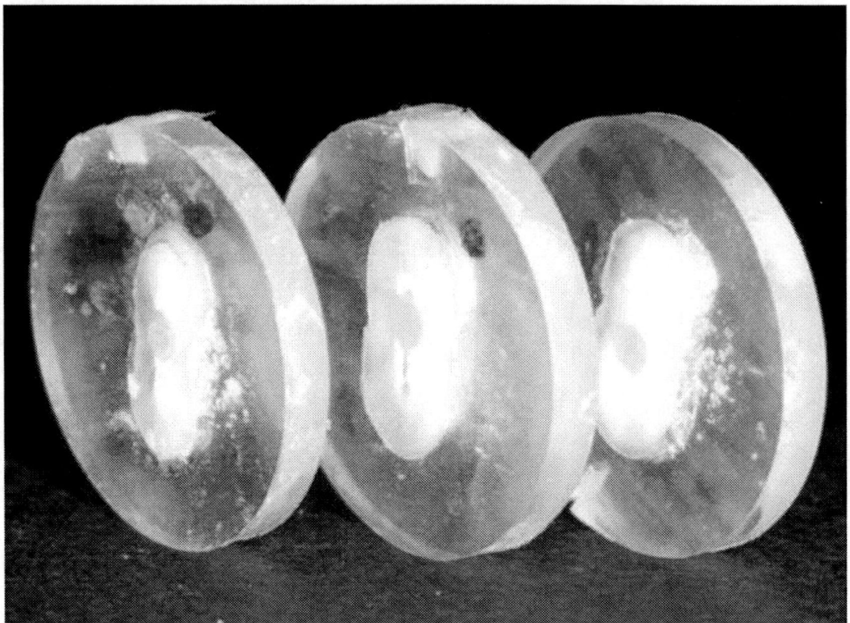

Figure 2. Coronal, middle and apical slices obtained from the same root and addressed to the push-out test.

Figure 3. The thickness of each sample was measured using a digital calliper with an accuracy of 0.001 mm.

Figure 4. Test specimen mounted on the Universal Testing Machine for push-out bond strength test.

notching of the punch tip into the post surface. The load was applied to the apical aspect of the root slice and in an apical-coronal direction, so as to push the post toward the larger part of the root slice, thus avoiding any limitation to the post movement. Loading was performed at a crosshead speed of 0.5 mm/min until the post segment was dislodged from the root slide

[14]. Maximum failure load values was recorded (N) and converted into MPa, considering the bonding area (mm^2) of the post segments. Post diameters were measured on each surface of the post/dentine sections using the digital caliper and the total bonding area for each post segment was calculated using the formula: $\pi(R+r)[(h^2+(R-r)^2]^{0.5}$, where $\pi = 3.14$, R represents the coronal post radius (mm), r the apical post radius (mm), and h the thickness of the slice (mm).

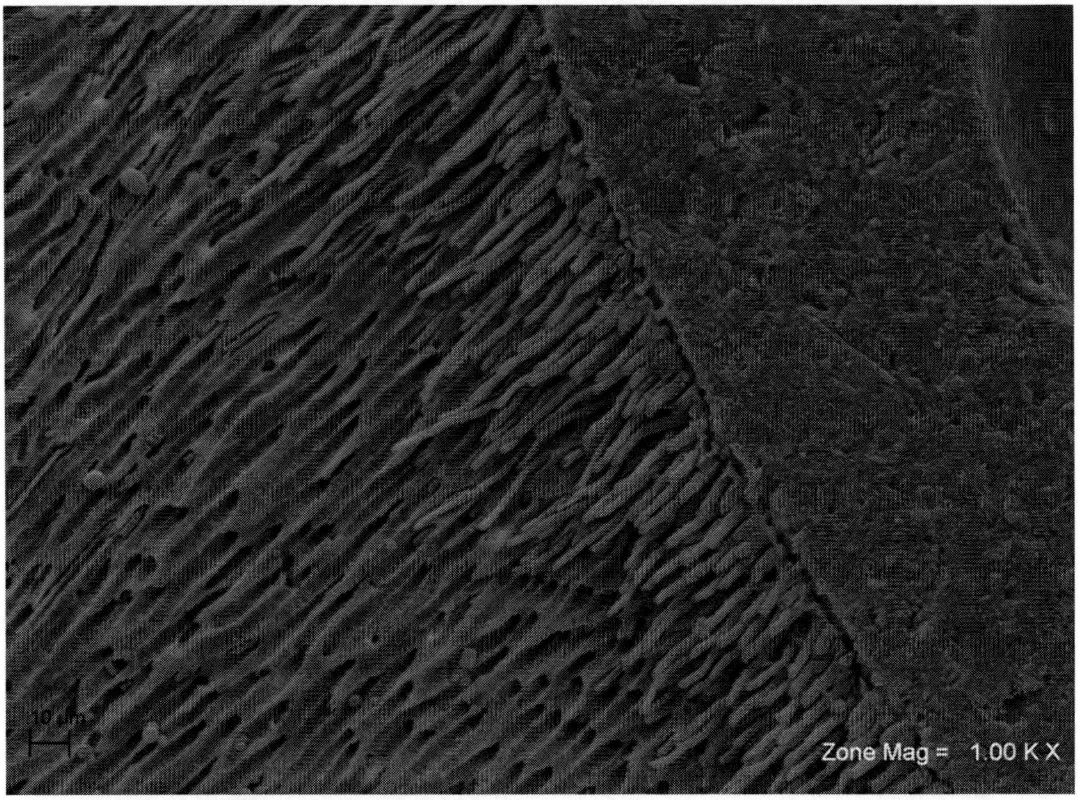

Figure 5. SEM micrograph showing the cervical root portion of a sample from the *POST* group (original magnification 100x). The arrow indicates a bubble embedded in the resin cement.

All fractured specimens were carefully removed and observed under stereomicroscope (Zeiss MC 80 DX; Zeiss, Jena, Germany) at 20X and 50X magnification from the coronal as well as from the apical direction to determine, for each root third, the mode of failure, which were classified into 5 types [26]: (1) adhesive between post and resin cement (no cement visible around the post); (2) mixed, with resin cement covering 0-50% of the post diameter; (3) mixed, with resin cement covering 50-100% of post surface; (4) adhesive between resin cement and root canal (post enveloped by resin cement); (5) cohesive in dentine. Furthermore, representative specimens from each group were analysed (Figures 5, 6, 7) using a scanning electron microscope (LEO 435 vp; LEO Electron Microscopy Ltd, Cambridge, UK).

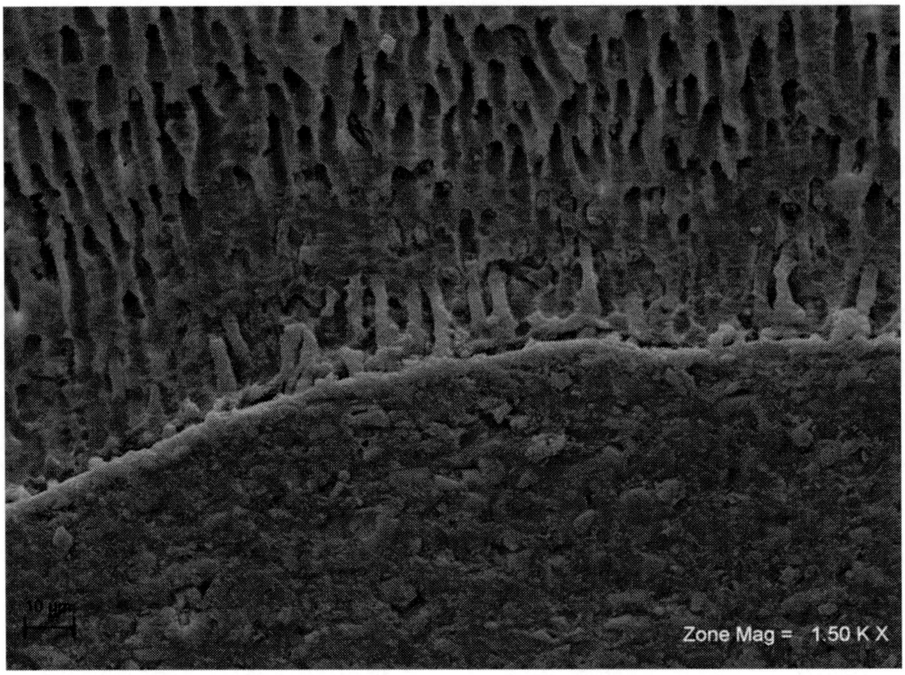

Figure 6. SEM view of interface between dentin (left side) and resin cement (right side) in cervical root segment of a specimen from the *LENTULO* group (original magnification 1000x).

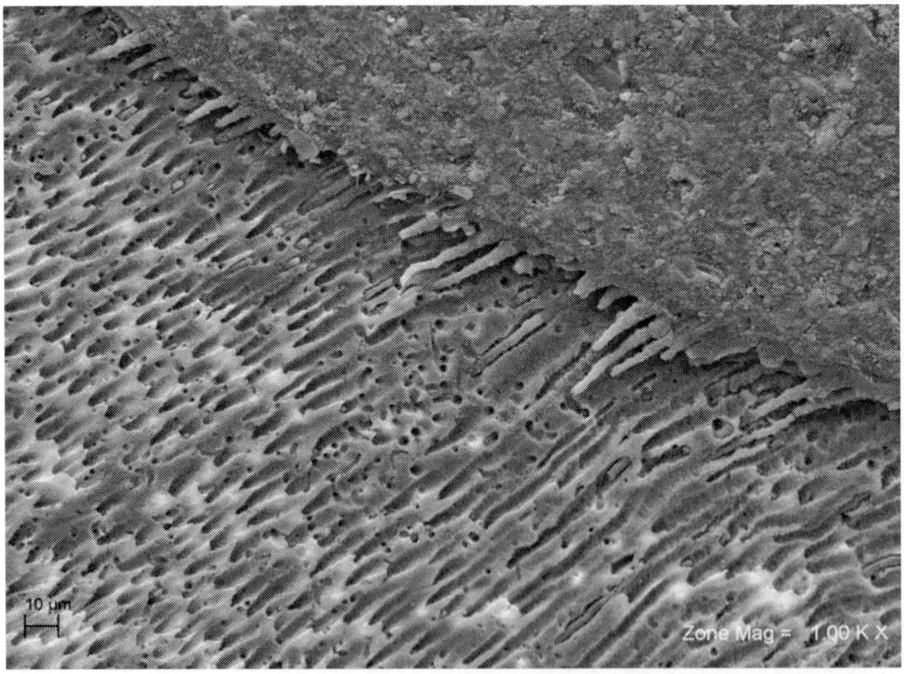

Figure 7. SEM view of interface between dentin (upper side) and resin cement (lower side) in cervical root segment of a specimen from the *POST* group (original magnification 1500x).

Results

Push-out test results are shown in Table 1. Statistical analysis displayed that the application method of the resin cement significantly affected the bond strength values ($p < 0.05$). No significant differences were recorded among the section levels (root thirds) ($p > 0.05$). The interaction between these two factors was not significant ($p > 0.05$). For the resin cement application method factor, the "syringe technique" showed the highest bond strength (13.51 ± 3.11 MPa) compared with the others methods. The "lentulo technique" exhibited high retentive strength (11.49 ± 2.33 MPa), while the "post technique" revealed significantly lower bond strength value (7.88 ± 2.08 MPa). For the section level factor, the coronal third showed the highest retentive strength (11.87 ± 3.21 MPa) but no statistically significant differences were found with the middle (10.79 ± 3.72 MPa) and the apical third (10.24 ± 3.27 MPa). Regarding failure types of tested specimens, the majority failed adhesively with a prevalence of post/cement interface and mixed failures (Figure 8). There were no cohesive failures in dentine (Table 2). The fracture pattern observed was very similar among the specimens of the tested groups.

Table 1. Mean push-out bond strengths (MPa) and Standard deviation for experimental groups according to the root thirds

	RTD ENDO LIGHT POST			*TOTAL*
	LENTULO	***POST***	***SYRINGE***	
CORONAL	12.62 (1.63)	8.70 (0.39)	14.28 (3.50)	11.87_1 (3.21)
MIDDLE	11.33 (3.09)	8.04 (2.95)	12.99 (3.50)	10.79_1 (3.72)
APICAL	10.53 (1.68)	6.91 (1.81)	13.28 (2.39)	10.24_1 (3.27)
TOTAL	11.49^b (2.33)	7.88^c (2.08)	13.51^a (3.11)	

Same numbers in pedex indicate not significant differences among the levels of the factor section-level. Different lower case letters represent significant differences with regard to the factor luting technique.

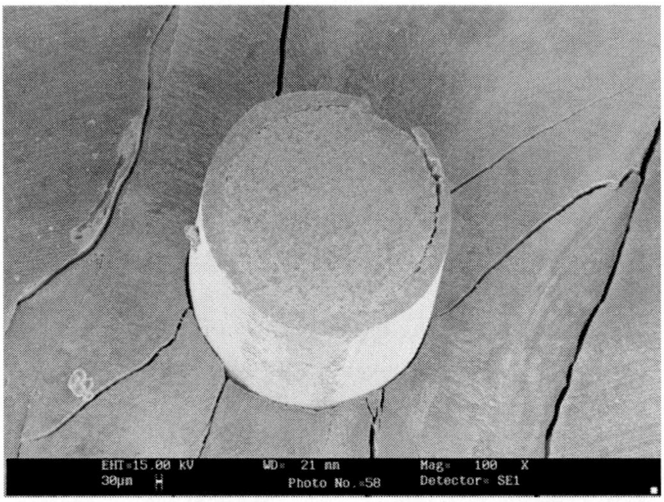

Figure 8. SEM view of interface between dentin (lower side) and resin cement (upper side) in cervical root segment of a specimen from the *SYRINGE* group (original magnification 1000x).

Table 2. Failure mode for experimental groups

Groups	1 Adesive : post-cement	2 Mixed : 0-50%	3 Mixed : 50-100%	4 Adhesive: cement- dentin	5 Cohesive
LENTULO					
Coronal	4	2	3	1	0
Middle	3	3	3	1	0
Apical	4	3	2	1	0
POST					
Coronal	4	2	3	1	0
Middle	3	4	3	0	0
Apical	2	3	4	1	0
SYRINGE					
Coronal	3	4	2	1	0
Middle	4	2	4	0	0
Apical	4	2	3	1	0

Discussion

This investigation was performed to evaluate if the application methods of the luting agent can influence bond strengths of a quartz post system to root canal dentin, using a push-out model. Push-out tests result in a shear stress at the interface between dentine and cement as well as between post and cement [27] and is comparable with the stress under clinical conditions. The push-out design is characterised by polymerisation stresses that would happen in the clinical situation. It has been suggested that, due to the small size of specimens, microtensile test permits a uniform stress distribution along the bonded interface [28]. Nevertheless, as observed previously [25], push-out test is a more reliable method for determining bond strengths between fiber posts and post space dentine because of the high number of premature failures occurring during specimen preparation and large data distribution spread associated with microtensile testing.

The effect of different resin-based luting agents on post retention has been investigated extensively, and various conclusions have been drawn [29],[30]. Regarding the adhesive/resin cement/fiber post system tested, the retentive strength was significantly affected by the luting technique. Irrespective of post luting technique, the interfacial strength was not significantly affected by the region of the root canal. Various in vitro researches revealed controversial results concerning bond strength values of different luting agents to FRC posts and root canal dentin [24],[25],[26]. Shear bond strengths depend on the degree and stability of interfacial micromechanical interlocking and chemical adhesion between root canal dentin, dentin bonding agent/resin-based luting cement and fibre post. Recent studies highlighted that retention of bonded fiber posts was contributed predominantly by friction [31],[32].

Bond strengths in the present study were not significantly affected by root canal thirds. This result confirms two previous studies [25],[33] that observed no influence of root canal region on fibre post retention. Gaston et al. [34] recording no significant differences in microtensile bond strength values between coronal and middle thirds of the post space, concluded that retentive strength may be related more to the area of solid dentine than to the density of the dentinal tubules. In contrast, Perdigao et al. [35] found that the coronal third resulted in statistically higher bond strengths than the apical third, while the middle and the apical thirds had no statistically significant different bond strengths. For the authors, the lower bond strength values found in the apical zone could be expected due to more difficult access to this third and the possible limitations of cement flow, but also due to a high and unfavourable C-factor. In a previous study [36], the most likely explanation for the higher resistance to post dislodgement in the coronal region of the root canal was identified in the decreasing effectiveness of light curing at greater distances from the light source. Moreover, the coronal portion of the canal seems to be the most accessible part of the canal space, making it easier to etch and more thoroughly apply the adhesive agents. Rinsing with water during the etching procedure, the difficulties of moisture control in the apical third of the post space probably result in the retention of remnant water within the dentine tubules, causing an incomplete infiltration of the resin agent. A reduction in strength in middle and apical thirds was also be related to the more difficult distribution of resin cement with voids formation [8],[37] or to traces of gutta-percha and endodontic sealer that may remain in these thirds after post space preparation. Bouillaguet et al. [8] reported that when endodontic posts are cemented inside root canals, the C-factor may exceed 200 (ratio of the bonded to the unbonded area). This is because there is a large area of resin cement bonded to the dental substrate and endodontic post, and there is little free area to allow for polymerization contraction. These findings seem to suggest that lack of direct viewing and luting agent application techniques may affect the bond strength in the apical region of the post space which will be inevitably lower.

Despite the promising results of dual-cure resin luting agents, existing literature regarding the role of the application methods of resin cements to the root canal and their effect on retentive strength of fiber post is scarce. It was reported that the application of luting agent with a lentulo spiral instrument permits a favourable distribution of resin cement throughout the post space and a formation of uniform, continuous cement layer [38]. Moreover this technique may guarantee the reduction of voids and bubbles within the luting agent (Figure 9). In this situation, if a dual-cure resin was the selected cementing material, the major recommendation is to avoid partial polymerisation before the adequate post seating. Even if it was reported that the presence of some porosity in the luting agent is not *per se* disadvantageous [11],some authors suggested that voids and air bubbles can impede an appropriate cementation of the post, thus causing its debonding [40]. The injection technique used for application of the resin cements is also reported as an effective technique for reducing voids and bubbles within the luting agent [6]. Fonseca et al. [41] in vitro evaluated the retention within root canals of posts cemented with dual-cure resin varying the application method of the primer/adhesive solution and luting agent. They reported that when the luting agent is placed into the root canal using only a lentulo spiral or when it is placed both using a lentulo and applying the cement on the post surface, post retention is increased. In a recent study [42], it was shown that bond strengths of three post systems in the apical post space third seemed to be not affected by luting agent application techniques. However, further researches were suggested to find out which luting protocol could be more suitable for each individual clinical situation. The present study revealed that bond strength to root canal dentine is significantly influenced by the application technique of luting agent for adhesive/resin cement/fiber post system tested. Statistical analysis showed that "syringe technique" exhibited the highest retentive strength to root canal dentin compared with the others luting methods. The "post technique" resulted in statistically lower bond strength value. The complete post system (post, adhesive, resin cement) supplied by the manufacturer was tested. Since it was demonstrated that the resin cement thickness significantly influences the pullout strengths of fibre-reinforced posts [13], the post spaces were prepared using the appropriate drill from the respective post manufacturer. Further researches would be necessary to clarify the influence of each component in the retentive values of the respective system group.

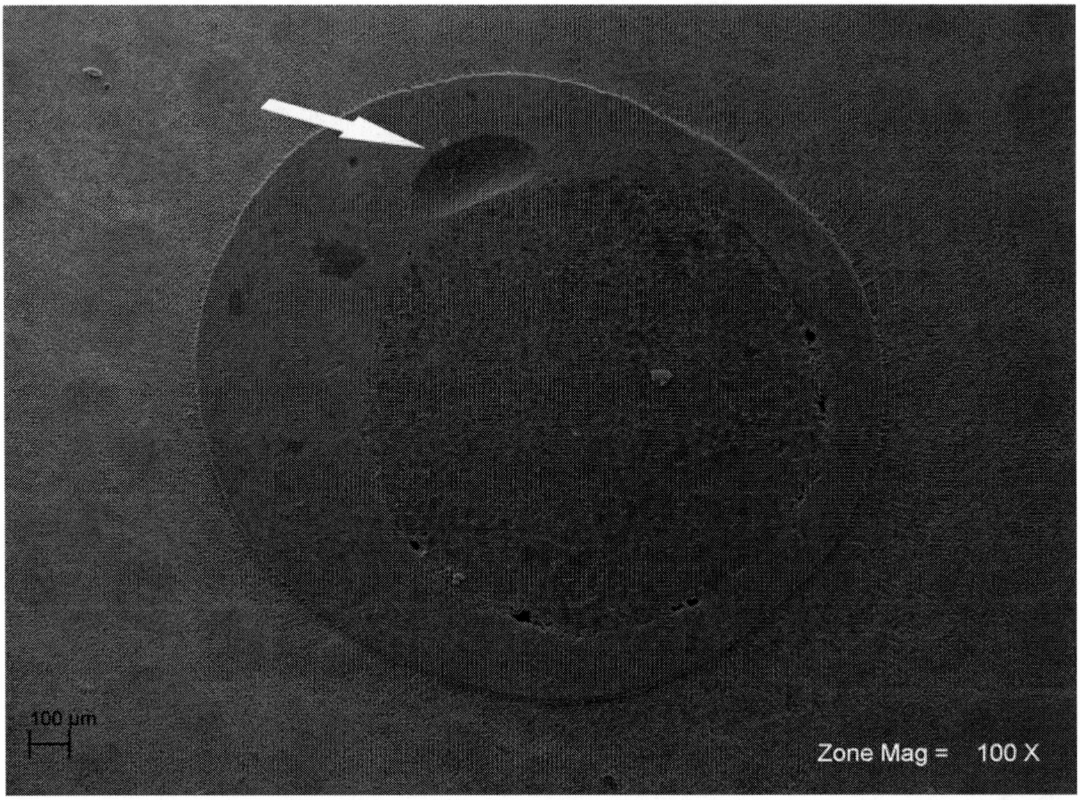

Figure 9. Representative SEM micrograph of a mixed (type 3) failure between post and cement from the apical slice of a *LENTULO* group specimen (Magnification 100x).

Analyses of failure mode demonstrated that most failures occurred at the cement-post interface or in a mixed mode. Polymerization shrinkage stresses that were generated because of the highly unfavourable cavity configuration factor (C-factor) of the post space probably accounted for the relatively higher percentage of mixed failure [8]. This finding suggested that the nature of the dentine surface of the canal wall or the tubule density might not be the basis for the difference in bond strengths between the coronal and middle/apical regions, which finding lends support to those investigations that aimed at improving the retention through various surface pretreatment procedures for the post [23],[43],[44].

Conclusion

Based on these findings, and within the limitations of an in vitro study, it may be concluded that, for the quarz fiber post system tested, the best performance was obtained when the luting agent was taken into the post space with a specific syringe. Moreover, push-out bond strengths were not statistically influenced by the root regions.

References

[1] D'Arcangelo, C; Prosperi, GD; Passariello, P; Caputi, S; and Malagnino, VA. Capacity of coronal dentin to increase fiberglass post retention: a pull-out test. *Am J Dent.* 2005, 18, 307-310.

[2] Asmussen, E; Peutzfeldt, A; and Heitmann, T. Stiffness, elastic limit, and strength of newer types of endodontic post. *J Dent.* 1999, 27, 275-278.

[3] Sirimai, S; Riis, DN; and Morgano, SM. An in vitro study of the fracture resistance and the incidence of vertical root fracture of pulpless teeth restored with six post-and-core systems. *J Prosthet Dent.* 1999, 81, 262-269.

[4] Terry, DA; Triolo, PT; and Swift, EJ. Fabrication of direct fiber-reinforced posts: a structural design concept. *J Esthet Restor Dent.* 2001, 13, 228-240.

[5] Grandini, S; Goracci, C; Monticelli, F; Tay, FR; and Ferrari, M. Fatigue resistance and structural characteristics of fiber posts: three-point bending test and SEM evaluation. *Dent Mater*. 2005, 21, 75-82.

[6] Pest, LB; Cavalli, G; Bertani, P; and Gagliani, M. Adhesive post-endodontic restorations with fiber posts: push-out tests and SEM observations. *Dent Mater.* 2002, 18, 596-602.

[7] Ferrari, M; Vichi, A; Mannocci, F; and Mason, PN. Retrospective study of the clinical performance of fiber posts. *Am J Dent*. 2000, 13, 9B-13B.

[8] Bouillaguet, S; Troesch, S; Wataha, JC; Krejci, I; and Meyer, JM; and Pashley DH. Microtensile bond strength between adhesive cements and root canal dentin. *Dent Mater.* 2003, 19, 199-205.

[9] Mannocci, F; Bertelli, E; Watson, TF; and Ford, TP. Resin-dentin interfaces of endodontically-treated restored teeth. *Am J Dent.* 2003, 16, 28-32.

[10] Peters, J; Zyman, G; Kogan, E; Kuttler, S; and García-Godoy, F. Retention of three endodontic post systems. *Am J Dent.* 2007, 20, 198-200.

[11] Vichi, A; Grandini, S; Davidson, CL; and Ferrari, M. An SEM evaluation of several adhesive systems used for bonding fiber posts under clinical conditions. *Dent Mater.* 2002, 18, 495-502.

[12] Vichi, A; Grandini, S; and Ferrari, M. Comparison between two clinical procedures for bonding fiber posts into a root canal: a microscopic investigation. *J Endod.* 2002, 28, 355-360.

[13] D'Arcangelo, C; Cinelli, M; De Angelis, F; and D'Amario, M. The effect of resin cement film thickness on the pullout strength of a fiber-reinforced post system. J Prosthet Dent, 2007, 98, 193-198.

[14] Vano, M; Cury, AH; Goracci, C; Chieffi, N; Gabriele, M; Tay, FR; and Ferrari M. The effect of immediate versus delayed cementation on the retention of different types of fiber post in canals obturated using a eugenol sealer. *J Endod.* 2006, 32, 882-885.

[15] Mannocci, F; Pilecki, P; Bertelli, E; W and atson, T. Density of dentinal tubules affects the tensile strength of root dentin. *Dent Mater.* 2004, 20, 293-296.

[16] Hofmann, N; Papsthart, G; Hugo, B; and Klaiber, B. Comparison of photoactivation versus chemical or dual-curing of resin-based luting cements regarding flexural strength, modulus and surface hardness. *J Oral Rehabil.* 2001, 28, 1022-1028.

[17] Ferrari, M; Mannocci, F; Vichi, A; Cagidiaco, MC; and Mjör IA. Bonding to root canal: structural characteristics of the substrate. *Am J Dent.* 2000, 13, 255-260.

[18] Ferrari, M; Vichi, A; Grandini, S. Efficacy of different adhesive techniques on bonding to root canal walls: an SEM investigation. Dent Mater, 2001, 17:422-429.

[19] Tay, FR; Gwinnett, AJ; and Wei, SH. The overwet phenomenon: an optical, micromorphological study of surface moisture in the acid-conditioned, resin-dentine interface. *Am J Dent.* 1996, 9, 43-48.

[20] Alster, D; Feilzer, AJ; De Gee, AJ; and Davison, CL. Polymerization contraction stress in thin resin composite layers as a function of layer thickness. *Dent Mater.* 1997, 13, 146-150.

[21] Ceballos, L; Garrido, MA; Fuentes, V; and Rodríguez, J. Mechanical characterization of resin cements used for luting fiber posts by nanoindentation. *Dent Mater.* 2007, 23, 100-105.

[22] Stegaroiu, R; Yamada, H; Kusakari, H; and Miyakawa, O. Retention and failure mode after cyclic loading in two post and core systems. *J Prosthet Dent.* 1996, 75, 506-511.

[23] D'Arcangelo, C; D'Amario, M; Prosperi, GD; Cinelli, M; Giannoni, M; and Caputi S. Effect of surface treatments on tensile bond strength and on morphology of quartz-fiber posts. *J Endod.* 2007, 33, 264-267.

[24] Bitter, K; Meyer-Lueckel, H; Priehn, K; Kanjuparambil, JP; Neumann, K; Kielbassa, AM. Effects of luting agent and thermocycling on bond strengths to root canal dentine. *Int Endod J*, 2006, 39, 809-818.

[25] Goracci, C; Tavares, AU; Monticelli, F; Raffaelli, O; Cardoso, PC; Tay, F; et al. The adhesion between fiber posts and root canal walls: comparison between microtensile and push-out bond strengths measurements. *Eur J Oral Sci.* 2004, 112, 353-361.

[26] Perdigao, J; Gomes, G; and Lee, IK. The effect of silane on the bond strengths of fiber posts. *Dent Mater.* 2006, 22, 752-758.

[27] Van Meerbeek, B; De Munck, J; Yoshida, Y; et al. Buonocore memorial lecture. Adhesion to enamel and dentin: current status and future challenges. *Oper Dent,* 2003, 28, 215-235.

[28] Cardoso, PE; Sadek, FT; Goracci, C; and Ferrari, M. Adhesion testing with the microtensile method: effects of dental substrate and adhesive system on bond strength measurement. *J Adhes Dent.* 2002, 4, 291-297.

[29] Cohen, BI; Pagnillo, MK; Newman, I; Musikant, BL; and Deutsch, AS. Retention of three endodontic post cemented with five dental cements. *J Prosthet Dent.* 1998, 79, 520-525.

[30] O'Keefe, KL; Miller, BH; and Powers, JM. In vitro tensile bond strength of adhesive cements to new post materials. *Int J Prosthodont.* 2000, 13, 47-51.

[31] Goracci, C; Fabianelli, A; Sadek, FT; Papacchini, F; Tay, FR; and Ferrari M. The contribution of friction to the dislocation resistance of bonded fiber posts. *J Endod.* 2005, 31, 608-612.

[32] Pirani, C; Chersoni, S; Foschi, F; Piana, G; Loushine, RJ; Tay, FR; and Prati, C. Does hybridization of intraradicular dentin really improve fiber post retention in endodontically treated teeth? *J Endod.* 2005, 31, 891-894.

[33] D'Arcangelo, C; D'Amario, M; Vadini, M; Zazzeroni, S; De Angelis, F; and Caputi, S. An evaluation of luting agent application technique effect on fibre post retention. *J Dent.* 2008, 36, 235-240.

[34] Gaston, BA; West, LA; Liewehr, FR; Fernandes, C; Pashley, DH. Evaluation of regional bond strength of resin cement to endodontic surfaces. *J Endod.* 2001, 27, 321-324.

[35] Perdigao, J; Geraldeli, S; and Lee, IK. Push-out bond strengths of tooth-colored posts bonded with different adhesive systems. *Am J Dent.* 2004, 17, 422-426.

[36] D'Arcangelo, C; Zazzeroni, S; D'Amario, M; Vadini, M; De Angelis, F; Trubiani, O; and Caputi, S. Bond strengths of three types of fibre-reinforced post systems in various regions of root canals. *Int Endod J.* 2008, 41, 322-328.

[37] Mallmann, A; Jacques, LB; Valandro, LF; Mathias, P; and Muench, A. Microtensile bond strength of light- and self-cured adhesive systems to intraradicular dentin using a translucent fiber post. *Oper Dent.* 2005, 30, 500-506.

[38] Akgungor, G; and Akkayan, B. Influence of dentin bonding agents and polymerization modes on the bond strength between translucent fiber posts and three dentin regions within a post space. *J Prosthet Dent.* 2006, 95, 368-378.

[39] Alster, D; Feilzer, AJ; De Gee, AJ; Mol, A; and Davidson, CL. The dependence of shrinkage stress reduction on porosity concentration in thin resin layers. J Dent Res, 1992, 71, 1619-1622.

[40] Ferrari, M; Vichi, A; Grandini, S; Goracci, C. Efficacy of a self-curing adhesive/resin cement system on luting glass-fiber posts into root canals: An SEM investigation. *Int J Prosthod,*.2001, 14, 543-549.

[41] Fonseca, TS; Alfredo, E; Vansan, LP; Silva, RG; Sousa, CS; Saquy, PC; and Sousa-Neto, MD. Retention of radicular posts varying the application technique of the adhesive system and luting agent. *Braz Oral Res*. 2006, 20, 347-352.

[42] D'Arcangelo, C; D'Amario, M; De Angelis, F; Zazzeroni, S; Vadini, M; and Caputi, S. Effect of application technique of luting agent on the retention of three types of fiber-reinforced post systems. *J Endod.* 2007, 33, 1378-1382.

[43] D'Arcangelo, C; D'Amario, M; Vadini, M; De Angelis, F; and Caputi, S. Influence of surface treatments on the flexural properties of fiber posts. *J Endod.* 2007, 33, 864-867.

[44] Monticelli, F; Osorio, R; Albaladejo, A; Aquilera, FS; Tay, FR; Ferrari, M; and Toledano, M. Effect of adhesive systems and surface treatment of methacrylate resin-based fiber posts on post-resin-dentin bonds. *Am J Dent.* 2007, 20, 231-234.

In: Strength of Materials
Editors: G. Mendes and B. Lago, pp. 233-245
ISBN: 978-1-60741-500-8

Chapter 8

MICROSTRUCTURAL INFLUENCE ON FLEXURE STRENGTH OF A CEROMER REINFORCED BY TWO TYPES OF FIBERS (POLYETHYLENE AND GLASS)

Silvana Marques Miranda Spyrides*[1],* *and Fernando Luiz Bastian*[2],*
*Department of Prosthesis and Dental Materials, School of Dentistry,
Federal University of Rio de Janeiro, Rio de Janeiro, RJ, Brazil
**Program of Metallurgical and Materials Engineering,
COPPE, Federal University of Rio de Janeiro, Caixa Postal 68505,
Rio de Janeiro, RJ, Brazil, CEP 21941-972COPPE/UFRJ

Abstract

In this study, the microstructures of a ceromer (Artglass®) reinforced by either glass fibers (GlasSpan®) or polyethylene fibers (Connect®) were characterized and compared and the influence of the fiber reinforcement on the flexural strength of the resulting products evaluated. With this objective, seven bars of each material were produced. One bar of each material was separated for microstructural analysis. The microstructural samples were subjected to metallographic polishing and finishing, and then analyzed using optical microscopy at different magnifications. The images obtained were treated using an image processing computer program (Image Pro Plus) in order to quantify the microstructure by calculating the mean diameter and mean volume fraction of fibers. The flexure tests were made by three-point bending, using six samples of each material. After statistical analysis, the results showed that the mean diameter of the glass fibers (4μm) was smaller than the polyethylene ones (23.6 μm). The mean volume fraction of glass fibers (0.42) was larger than that of the polyethylene fibers (0.28) and the mean center-to-center distance between fibers was smaller in the glass fibers material (33 μm) than in the polyethylene fibers material (61 μm). The flexural strength of both glass and polyethylene fiber-reinforced materials was

[1] E-mail address: Av. N. S. de Copacabana, 195/310, Copacabana CEP 22020-000 Rio de Janeiro – RJ – Brasil.
[2] E-mail address: silmms@globo.com.

statistically equal, despite the fiber volume fraction being statistically larger in the fiber glass material.

Keywords: Fiber Reinforced Composites, Ceromer, Microstructure, Mechanical Properties, Metal Free Fixed Prosthesis

1. Introduction

Fiber-reinforced composites were developed as an alternative to conventional fixed partial prosthodontic, metaloplastic or metaloceramics, to fit small edentulous spaces. This system, comprised of a fiber-reinforced ceromer allows a conservative, less invasive and more esthetic preparation of the supporting teeth [1,2].

These prostheses have a substructure formed by long fibers impregnated with resin matrix which provide strength and rigidity, and a particulate composite cover that improves wear resistance and esthetics [3].

Several types of fibers have been used and studied for dental purpose, and it seems that glass and polyethylene fibers are the most widely used and studied due to their transparency, and ability to keep the original color of the covering composite [4,5,6,7].

Polyethylene fibers have shown high mechanical properties if the load is applied in tension; however, if the load is in compression it shows lower resistance. On the other hand, the glass reinforced material have shown the same properties regardless of the direction of loading so their bending properties are higher than those of polyethylene fibers [6].

Fibers act as reinforcement, increasing the flexural strength, the fracture resistance and the tensile strength of the polymer matrix, but they do not increase the compressive strength [8,9]. Failures of fiber-reinforced ceromers are due to their poor adhesion or cohesion at the fiber/matrix interface.

Mechanical properties of the fiber-reinforced composites are influenced by some factors listed by BEHR [10] as follows: the orientation and the quantity of fibers, the impregnation of fibers within the polymeric matrix, the adhesion of fibers to the polymeric matrix and the properties of fibers versus properties of the polymeric matrix. Another factor that may affect the properties is the position of the fiber reinforcement [11].

The properties of the reinforced fibers differ according to the type of fibers and their architecture [12]. The properties of the covering composites vary according to their composition, not only in the organic matrix but also in the dispersed inorganic phase (type, size and distribution of particles). According to FREILICH et al. [12], ELLAKWA et al., [13] this variation in the composite composition and the type and architecture of fibers affects the properties of fiber-reinforced composites.

ELLAKWA et al., [13] verified that different composite compositions did not produce significant differences in flexure strength of polyethylene fiber-reinforced composites. In a previous work, SPYRIDES and BASTIAN [14] observed that the variation between glass and polyethylene fibers did not produce significant changes in the flexure strength of composites reinforced with these two types of fibers.

The impregnation of the fibers is required to provide close union of the fiber to the polymeric matrix and thus for the strength of the composite [15]. Poor impregnation leads to failure in transferring the load from the matrix to the fiber. On the other hand, spaces between them may occur, which may lead to a considerable decrease of tensile strength and the elasticity modulus of the fiber composites when compared with the theoretical strength and modulus of such composites [16]. These regions increase the water absorption and, thus, decrease the mechanical strength of the composite [17]. Porosity between matrix and fibers diminishes the capacity of load support of fiber-reinforced composites [5,13,15].

According to FREILICH [12], due the necessity of having all fibers fully wetted by the resins, the fiber volumes are generally limited to less than 50%. Pre-impregnated unidirectional dental fiber composites incorporating approximately 45% of glass fibers have flexure strength in the range of 600 and 1,000 MPa. These values are about ten times higher than those of composites without fiber reinforcement, and this represents the primary mechanical benefit of using fiber reinforcement in a prosthetic framework for dental purpose.

VALLITTU and NARVA [5], VALLITTU [15] demonstrated that there is a relationship between the amount of fibers within the polymeric matrix and the increase of flexure strength on the fiber reinforcements.

Larger amounts of fibers result in better properties. However, an excess can lead to an insufficient wetting of the resin, which represents less resistance and poor characteristics of manipulation. To reach an optimum strength, it's necessary to have a high volume fraction of fibers, thorough wetting and a uniform distribution of fibers within the matrix resin [6].

Due to the mechanical demands that fiber-reinforced fixed partial dentures are submitted during the masticators function, it is known that their clinical behavior depends directly on their microstructure and components. As a consequence, the purpose of this study was to evaluate, quantify and compare the microstructure and its influence on flexure strength of a ceromer reinforced by either glass or polyethylene fibers.

2. Materials and Methods

A microstructural and flexure strength chacterization of fiber reinforced ceromer was performed. Manufacturers and compositional information of the tested materials are summarized in table 1.

Table 1. Investigated materials

Material	Composition	Manufacturer
Artglass Dentine*	Second-generation laboratory composite resin. Multifunctional metacrilic ester matrix (30% by weight) filled with glass particles of silicium, barium and aluminum oxides and silanized (70% by weight with 1 μm medium particle width) and photoiniciators.	Heraeus Kulzer Germany
Connect*	Braided polyethylene fiber (UHMW-PE) 2mm width, coated by plasma gas.	Kerr Corporation USA
GlasSpan*	Silicium, aluminum and boro oxide braided silanized glass fiber with 2mm width.	GlasSpan Inc. USA
Connect Resin*	Fluid resin composed of non cured metacrilic ester monomer, photoiniciators, inorganic fillers and additives for stabilization.	Kerr Corporation USA

*Brand name.

Fourteen rectangular specimens (30 mm length, 4 mm height and 4 mm width) of Artglass, seven reinforced by polyethylene fibers and seven by glass fiber were produced by inserting the material into a silicone mold (Stern Tek, Sterngold) in three increments. One impregnated fiber layer was placed between the first and second layer of material. Each layer was photocured in the *UniXS* unit for 90 seconds and after the last cure the bar was removed from the silicone mold and photocured for 180 seconds.

2.1. Microstructural Characterization

2.1.1. Samples Preparation

Two bars previously prepared, one with glass fiber and the other with polyethylene reinforcement were randomly selected and cut in small blocks each measuring 5 X 4 X 4 X mm. These blocks were embedded in epoxy resin (Ciba Geisy 311 with 10% of 24 catalizer) with the transversal, longitudinal and longitudinal base surface exposed as shows Figure 1.

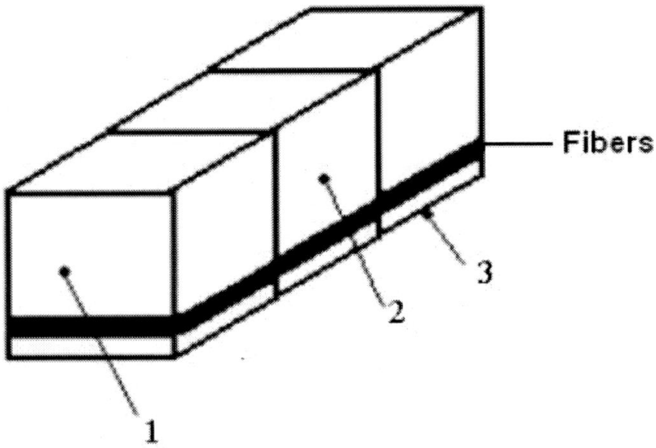

Figure 1. Schematic representation of produced material bar showing the position of the fibers layer and the cuts: transversal (1), longitudinal (2) and longitudinal base (3).

After the resin cure, the samples was finished in an industrial polishing machine (DP - NAP - Struers model Knut Rotor) with a series of emery papers (600, 800 and 1,200 grit - 3M Company), under current tap water refrigeration.

After that, fine metallographic polishing was carried through by pan cloth with 1μm and 0,5μm diamond paste (AP paste F - Struers Metallographic Equipment) in industrial polishing machine (Prazis APL – 4). Final cleanness after the polishing process was performed by an ultrasonic device.

2.1.2. Acquisition, Treatment of the Images and Quantitative Characterization

After polishing and finishing, samples of Artglass/Connect and Artglass/GlasSpan were observed at the fiber region using optical microscopy (Olympus BX60M) in three sections with magnifications of 100X, 200X and 500X. Digital photographs in JPEG format were acquired with a photographic camera (CCD COMU/Snappy), with the intention of obtaining the mean volume fraction, diameter and centre-to-centre spacing of the fiber reinforcement region of those materials.

The acquired pictures were then treated with Adobe Photoshop 5.0 program (Image/Adjustments/Equalize) in order to compare shape, size and distribution of fibers, and also detect if there was any void or porosity next to fibers or in the fiber/matrix interface.

Six acquired pictures (500X) of the glass fibers region and another six (200X) of the polyethylene fiber region were subjected to a new treatment with Corel Photo Paint program (Magic wand, brush, Eraser, Fill tool and Zoom) with the objective of leaving the fibers with white and the matrix with black color. After that these pictures were processed with an Image Pro Plus program with the purpose of obtaining statistically the average diameter and volume fraction of each type of fiber.

Using the obtained values of fibers mean diameter (μm) and the fibers mean volume fraction (%) the mean centre to centre distance between fibers (λ) were calculate with the following equation [18]:

$$\lambda = L \times \frac{1\text{-}Vv}{Vv}$$

where λ is the mean centre to centre distance between fibers, L the fiber mean diameter (μm) and Vv the fibers mean volume fraction (%).

2.2. Flexure Strength

Flexure testing was performed according to ANSI/ADA No. 27 – 1993 specification 7.8.1.1. Six bars of each material were used. The bars were subjected to manual grinding with 800 grade emery paper (3M Co.). The tests were performed in an Instron testing machine, model 4204, at a crosshead speed of 0.5 mm/min. The specimens were placed on a three-point bending test device with 20mm between the supports as illustrated in figure 2.

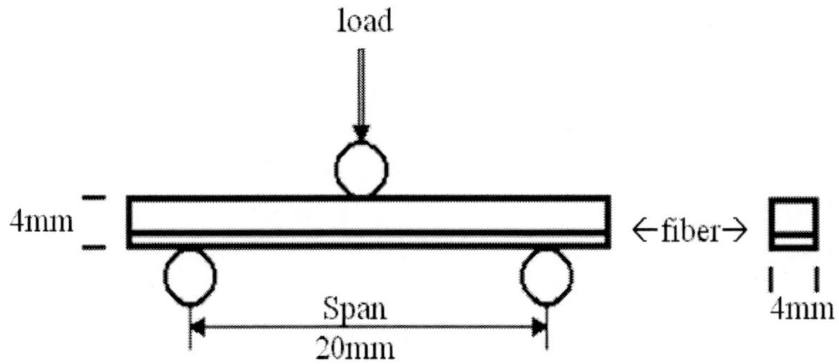

Figure 2. Diagrammatic representation of three point flexure testing device.

From the results of the flexure tests it was possible to calculate the flexural strength, deflexion and elasticity modulus of the test specimens using the following equation:

$$\sigma = \frac{3Pl}{2bh^2}$$

where: σ is the flexure strength, l the span between the supports, b the width, h the height of the sample and P the fracture load.

Means, standard deviations and coefficients of variation were computed (Excel 2000 Microsoft Office). All data were statistically analyzed with t student test for averages comparison between the groups. The level of significance was set at $\alpha = 0.05$.

3. Results and Discussion

3.1. Microstructural Characterization

The observations of the acquired images in transversal, longitudinal and longitudinal base cuts (figures 3 and 4) shows that the shape, size and distribution of fibers were different between the two studied materials and that the fibers had a reasonable impregnation by the polymeric matrix in both glass and polyethylene reinforced composites.

In transversal sectional views, figures 3a, 4a 5a, 5b, 6a and 6b, the glass fibers showed circular transversal shape, a relative uniform distribution and appearance of being totally involved by polymeric matrix, meanwhile the polyethylene fibers showed irregular and elliptical shapes, lager size, less uniform distribution, with some areas without fibers alternating with areas of high concentration of fibers. This irregular distribution of polyethylene and glass fibers can be attributed to the braid architecture of both. In places where fibers cross each other, greater proximity between fibers occurs.

In present study, the shape and distribution of fibers in transverse section were similar to the observed in the study of GOLDBERG and BURSTONE [19], BEHR et al [10] and BAE et al [20]. It also appeared that wetting of glass fibers to the polymeric matrix was similar to that of study of BAE et al [20] which used GlasSpan fibers impregnated with Aelitefil.

In longitudinal section views, figures 3b and 4b, the glass fibers showed a range of ovalized shapes. This happened due to the position of cutting. In this same section, the polyethylene fibers presented varied shapes with tendency to elongated forms.

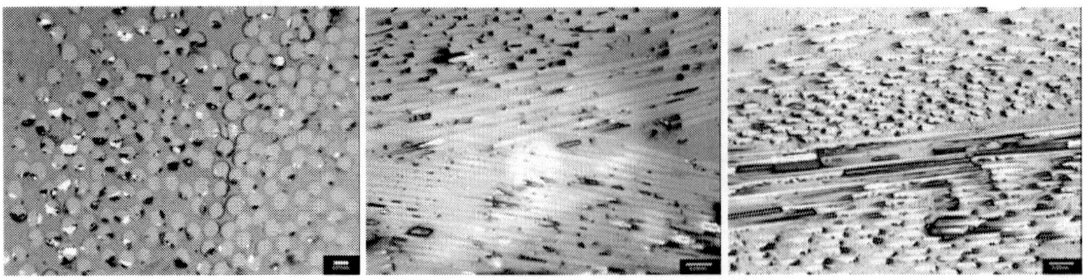

(a) (b) (c)

Figure 3. Optical microscopy views of Artglass/GlasSpan material (a) transversal section X 500, (b) longitudinal section X 200 and (c) longitudinal base section X 200.

In longitudinal base section, figure 3c and 4c, it was possible to observe the braid architecture of glass fibers. The polyethylene fibers didn't show the braid architecture probably due to the larger diameter of its fibers in relation to the glass fibers.

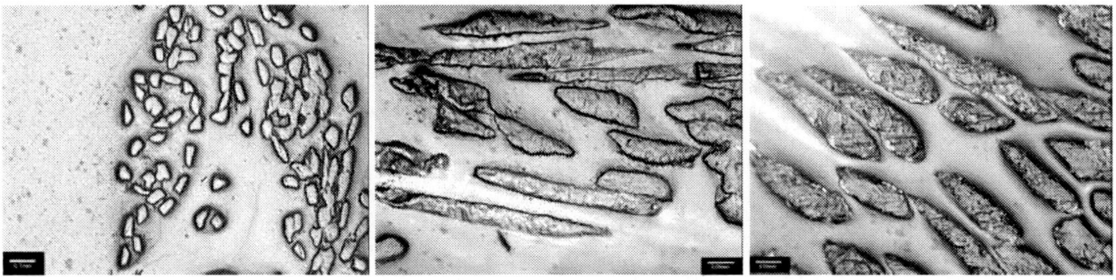

(a) (b) (c)

Figure 4. Optical microscopy views of Artglass/Connect material (a) transversal section X 100, (b) longitudinal section X 200 and (c) longitudinal base section X 200.

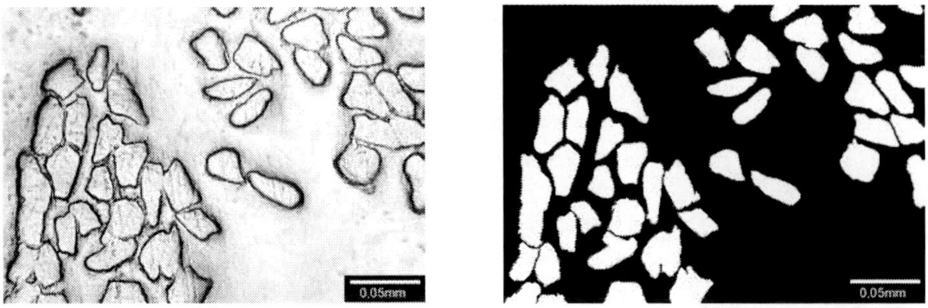

(a) (b)

Figure 5. Optical microscopy views of Artglass/Connect material (fibers region) with magnification X 200 in transversal section, (a) treated image using Photoshop 5.0 program and (b) treated image using the Corel Photo Paint program.

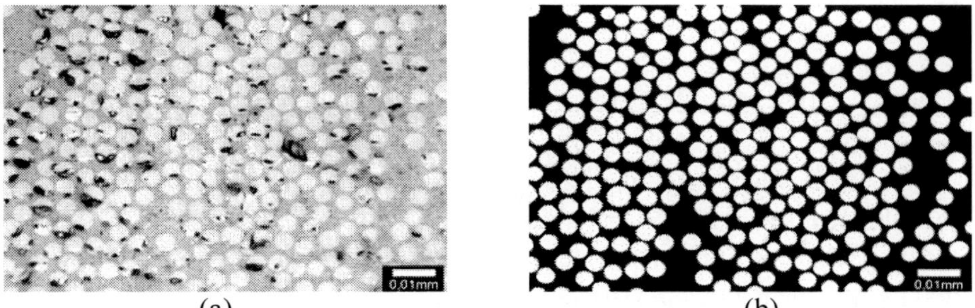

(a) (b)

Figure 6. Optical microscopy views of Artglass/GlasSpan material (fibers region) with magnification X 500 in transversal section (a) treated image using 5.0 Photoshop program and (b) treated image using the Corel Photo Paint program.

The values of fiber diameter, fiber number (per square mm) and fiber volume fraction are shown in table 2.

The Student t-test exhibited significant differences ($p < 0.05$) for all means values.

Table 2. Values of fiber diameter (μm), fiber number (per mm^2) and fiber volume fraction (%). Means values, standard deviations (sd), p-values and coefficients of variation (CV)

	Material	Mean	p-value	s.d.	C.V.
Diameter (μm)	*GlasSpan*	4.0		0.5	12.5
			0.000004*		
	Connect	23.6		6.0	25.4
Number in 1 mm^2	*GlasSpan*	31385.2		2051.2	6.53
			0.00000028*		
	Connect	640.8		141.9	22.1
Volume fraction (%)	*GlasSpan*	42		3.44	8.2
			0.0002*		
	Connect	28		4.84	17.3

* significant difference ($\alpha = 0.05$).

According to CHAWLA [21], smaller diameter of the fibers leads to lower probability of defects. In present study (Table 2) the mean diameter of the glass fibers (4μm) was smaller than the polyethylene ones (23.6 μm).On the other hand, the mean centre to centre distance between fibers was smaller in the glass fibers material (33 μm) than in the polyethylene fiber one (61 μm). Also, the mean volume fraction of glass fibers 0.42 (42%) was larger than that of the polyethylene fibers 0.28 (28%)

GOLDBERG and BURSTONE [19], VALLITU [15], BEHR et al. [10] performed microstructural evaluation by the incinerating or dissolution methods. The weight was calculated before and after the incinerating or dissolution process in order to get the total weight of fibers and thus the total fibers weight fraction. These methods seem to be imprecise, due to the fact that by incinerating or dissolving the materials there is a possibility of finding particles and resin rests together with the fibers themselves. Besides, these two methods do not allow observing inhomogeneous regions in the composite materials. Another problem is that these two methods consider that all fibers have an uniform distribution in the composite matrix which is not always true.

In our investigation, the fiber glass content (42%) was very close to the reported by GOLDBERG and BURSTONE [19] (43% and 45%) which used the dissolution method. However, the present results (42%) were completely different from the results of BERH et al. [10] (28% and 12%) which used the incineration method.

The results of BAE et al. [20] of mean volume fraction for both glass and polyethylene fibers was far smaller than the results found in present study. This probably happened due to the method used in the BAE et al [20] study. In that work, the authors weighed the material as a whole, the fiber reinforcement plus the composite. As a result, the mean volume fraction of fibers was very small. However, this method is not used in engineering because the mean volume fraction must represent the structural region of the reinforcement and not the whole material itself.

One of the main reasons for different fiber content in the literature is the method of measurement. However, different methods of sample preparation, other types of fibers or fiber wetting also can cause differences in fiber content [10].

A problem with the obtained values of volume fraction of fibers, as already pointed by BEHR et al [10], is that the fibers are not cut completely axially or transversally during the preparation of the samples. As a result, "the sectional view of the fiber is not circular but instead elliptical, which results in incorrect volume content of fibers". The same authors (BEHR et al) also pointed that "the distribution of fibers in the matrix is not always homogeneous. A sectional view with a representative low or high fiber content could falsify the fiber volume calculation". The first point could not be avoided in present work, so there is some error in the obtained values of volume fraction, mainly in the polymeric fiber material. The second point was avoided through measurements on different regions of the samples and subsequent averaging of the values.

3.2. Mechanical Behavior Characterization - Flexure Strength

The values of flexure strength obtained are listed in table 3.

The t student test exhibited no statistical differences for flexure strength (p = 0.20) between groups.

Artglass/Connect presented the largest coefficient of variability in all tests indicating a higher dispersion.

During the flexure tests Artglass/GlasSpan presented an elastic behavior with only one maximum load peak before breaking, while Artglass/Connect presented an elastoplastic behavior and the occurrence of two load peaks before breaking as can be seen in Figure 7. The load used to calculate the flexure strength of this material was the first load peak from which the material was already in process of cracking by delamination process.

During the tests, the bars showed different behavior: The glass fiber bars (Figure 8a) developed cracks in the tensile part of the central region of the bar with propagation into de compression side; while the polyethylene fiber bars (Figure 8b) showed not only these cracks in the central portion but also fracture in other parts of the bar, mainly next to one of the supports.

Table 3. Values of flexure strength (σ) of Artglass/GlasSpan and Artglass/Connect. Mean values, standard deviations (sd), p-values and coefficients of variation (CV)

	Material	Mean	p-value	s.d.	C.V.
Flexure Strength (σ) (MPa)	Artglass/GlasSpan	131.02	0.20	9.95	7.59
	Artglass/Connect	116.81		23.51	20.12

* significant difference ($\alpha = 0.05$).

After the tests, it was observed that Artglass/GlasSpan did not show delamination between the particulate composite and fibers, which means that this material probably reached its maximum and possible flexure strength. On the other hand, Artglass/Connect showed cracks at the interface with delamination that could had resulted from porosities, impurities, air captured or even lack of adhesion at the interface. This can indicate that this material had reached its limit of shear bond strength at this interface, with consequent cracking, as shown in Figure 9

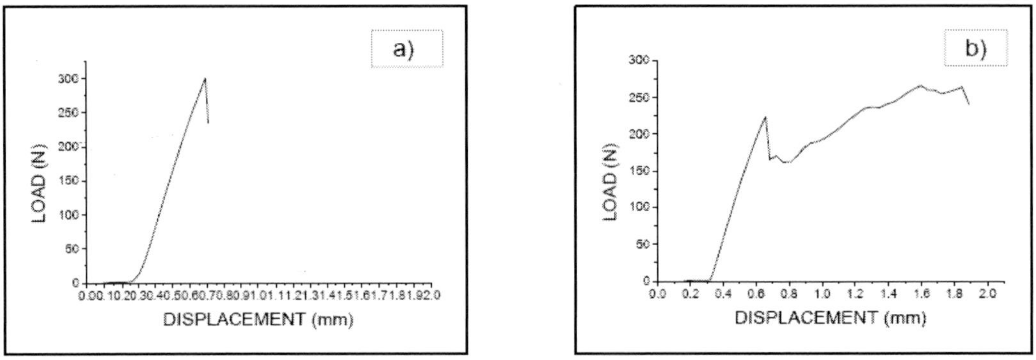

Figure 7. Typical load vs. displacement records in the flexure test a) Artglass/GlasSpan and b) Artglass/Connect.

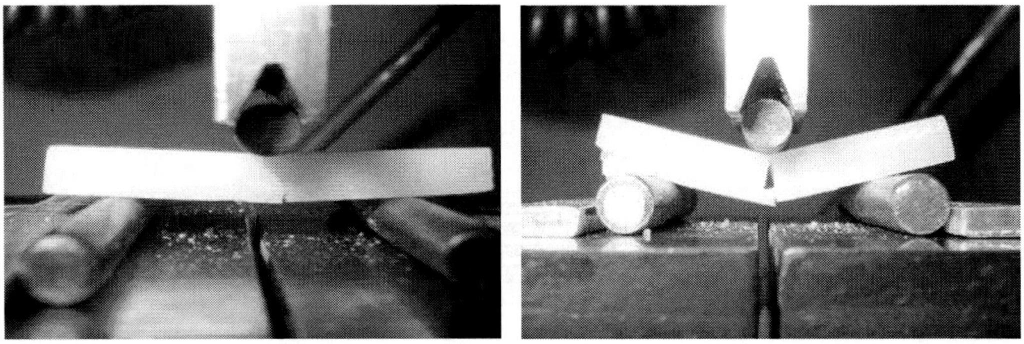

(a) (b)

Figure 8. Flexure testing (a) fiber glass reinforced bar (b) polyethylene fiber reinforced bar.

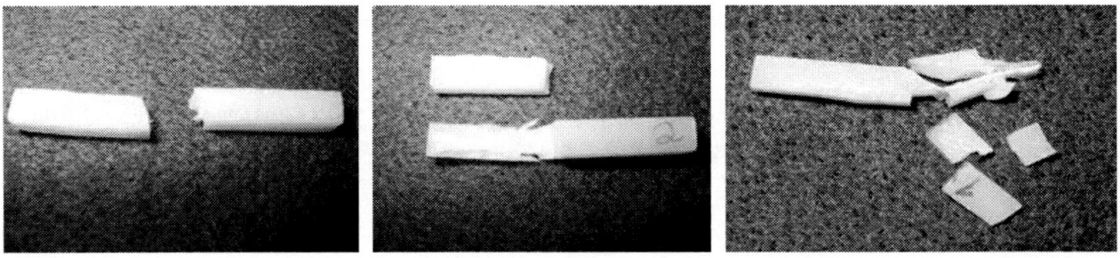

(a) (b) (c)

Figure 9. Cracked bars after flexure testing (a) glass fiber reinforced, (b) and (c) polyethylene fiber reinforced.

The flexure strength of the two materials was considered statistically equal despite of Artglass/Connect had presented delamination and supported a slightly larger load. In terms of material development, it can be presumed that this material still presents potential for improvement through further studies on its fiber/matrix interface.

The mean values of flexure strength of ceromer reinforced with polyethylene fiber (116.81 MPa) found in present work was smaller than the mean value of ELLAKWA et. al.

[22] work for the same material (261.1 MPa). This probably happened due to their fibers position which was also in the tensile side of the bars but closer to the base.

In BEHR et. al. [10] work the flexure strength values found for composites reinforced with previously impregnated glass fibers Vectris, 618.0 MPa, and manually impregnated Fibrekor, 585.0 MPa, are very high not only in relation to the values found in present work but also to the values found by ELLAKWA et al.[22] and BAE et al [20]. This great difference may be due to the fact that in BEHR et al.[10] study the authors tested bars with fibers impregnated without coverage of particulate composite. This fact can be evidenced comparing the values found by BAE et al. [20] and CHONG and CHAI [23] for the fiber glass Vectris with coverage of composite Targis, respectively 296.0 MPa and 84,0 MPa, and for fiber glass Fibrekor with coverage of composite Sculpture, 203.0 MPa and 165.0 MPa.

VALLITU [5,12] demonstrated that there is a relationship between fiber content and the flexure strength. The same fact had been reported by GOLDBERG and FREILICH [6] which found that higher fiber content caused higher strength. Despite this, FREILICH [12] limited the fibers volume fraction to 50% in order of having all fiber wetted by the resin.

GOLDBERG and BURSTONE [19] also found that the highest fibers content lead to an increase of flexure strength contradicts the results of BEHR et al. [10] and BAE et al. [20] which did not obtain any increase in flexure strength when the amount of fibers had increased.

In the present study the mean volume fraction of glass fibers 0.42 (42%) was larger than that of the polyethylene fibers 0.28 (28%). Moreover, when the flexural strength results of these two types of fiber reinforced materials are compared, we can see that despite the larger mean volume fraction of the glass fiber material and the higher tensile strength of the glass fibers according to FREILICH et al. [12], the flexural strength of both glass and polyethylene fiber reinforced materials was statistically equal.

6. Conclusion

The flexural strength of both glass and polyethylene fiber-reinforced materials was statistically equal, despite the fiber volume fraction being statistically larger in the fiber glass material. The elastic behavior and the type of cracking, without any delamination observed in the glass fiber reinforced ceromer, indicates that the fibers had a good distribution and adhesion within the ceromer, probably reaching the maximum possible strength. On the other hand, the elastoplastic behavior and cracking by delamination of the polyethylene reinforced ceromer indicates that the development of this material needs improvement, by increasing the fiber adhesion within the ceromer, better distribution of fibers and a higher mean volume fraction of fibers.

Acknowledgements

The authors thank Mr. Renan Hufnagel Bella for the session of his dental prostheses laboratory. Also, to CNPq for continous support and IMA-Macromolecule Institute for the use of equipments.

References

[1] Shannon, A. (1997). Fiber-Reinforced Composite Bridge: Inlay-to-Inlay Technique. *Dentistry Today*, 48-53.

[2] Winters, K.L. (1999). Using a fiber-reinforced ceromer for fixed restorations. *Dent. Today*, 70-73.

[3] Freilich, M.A; Duncan, J.P; Meiers, J.C; Goldberg, A.J. (1998). Preimpregnated, fiber-reinforced prostheses. Part I Basic rational and complete-coverage and intracoronal ficed partial denture designs. *Quint. Int. v.*29, n.11, 689-696.

[4] ramos, v; runyan, d.a; christensen, L.C. (1996). The effect of plasma-treated polyethylene fiber on the fracture strength of polymethyl methacrylate. *J. Prosthet. Dent.,* v..76, n.1, 94-96.

[5] Vallitu, P.K and Narva, K. (1997). Impact strength of a modified continuous glass fiber-polymethyl methacrylate. *Int. J. Prosthod.,* v.10, n.2, 142-148.

[6] Goldberg, A.J., Freilich M.A., (1998, march).Fiber-reinforced Composites. *Dental Advisor*, v.15, n.2, 2-5.

[7] Honda, M.I; Vieira, D; Guedes, A.P.P. (2001, nov.) Restaurações com reforço de fibras. Dental Special [on-line serial]. 1-8, www.Dentalspecial.com.br

[8] Stipho, H.D. (1998). Effect of glass fiber reinforcement on some mechanical properties of autopolimerizing polymethyl methacrylate. *J. Prosthet. Dent.,* v.79, n.5, 580-584.

[9] Belvedere, P.C. (1998). Single-sitting, fiber-reinforced fixed bridges for the missing lateral or central incisors in adolescent patients. *Dent. Clin.of Nor. Am.*, v.42, n.4, 665-682.

[10] Behr, M; Rosentritt, M; Lang,R; Handel,G. (2000). Flexural properties of fiber reinforced composite using a vacuum/pressure or a manual adaptation manufturing process. *Journal of Dentistry,* v.28, n.7, 509-514.

[11] Ellakwa, A.E; Shorthall, A.C; Shehata, M.K; Marquis, P.M. (2001). The influence of fibre placement and position on the efficiency of reinforcement of fibre reinforced composite bridgework. *J. Oral Rehabil.,* v.28, n.8, 785-91.

[12] Freilich, M.A., Meiers, J.C., Duncan, J.P.,Goldberg,A.J.(2000), *Fiber-Reinforced Composites,* Hong Kong, ed. Quintessence Publishing Co, Inc.

[13] Ellakwa, A.E., Shorthall, A.C., Shehata, M.K., Marquis, P.M. (2001), Influence of veneering composite composition on the efficacy of fiber-reinforced restorations (FRR). *Oper. Dent,* v. 26, n.5, 467-75.

[14] Spyrides, S.M.M. And Bastian F.L. (2004). In vitro comparative study of the mechanical behavior of a composite matrix reinforced by two types of fibers (polyethylene and glass). *Materials Science and Engineering C*, v. 24, 671-677.

[15] Vallitu, P.K. (1999).Flexural properties of acrylic resin polymers reinforced with unidirectional and woven glass fibers. *J. Prosthet. Dent.,* v.81, n.3, 318-326.

[16] Vallitu, P.K. (1998). Some aspects of the tensile strength of unidirectional glass fiber polymethyl methacrylate composite used in dentures. *J. Oral Rehabil.,* v.25, 100-5.

[17] Vallitu, P.K; Ruyter, I.E; Ekstrand, K. (1998). Effect of water storage on the flexural propertires of E-glass, and silica fibers acrylic resin composite. *Int. J. Prosthodont*, v.11, 340-350.

[18] Underwood, E.E. (1970). *Quantitative Stereology*. Massachusetts, Addison-Wesley Publishing Company.

[19] Goldberg, A.J. and Burstone, C.J. (1992).The use of continuous fiber reinforcement in dentistry. *D. Material, v.8*, 197-202.

[20] Bae, J. M; Kim, K. N; Hattori, M; Hasegawa, K; Yoshinari, M; Kawada,E; Oda, Y. (2001). The flexural properties of fiber-reinforced composite with light-polimerized polymer matrix. *Int. J. Prosthodont,* v.14, n.1, 33-39.

[21] CHAWLA, K. K. (1987). *Composite Materials Science and Engineering*. Germany. ed. Springer-Verlag New York Inc.

[22] Ellakwa, A.E; Shorthall, A.C; Shehata, M.K; Marquis, P.M. (2001). Influence of veneering composite composition on the efficacy of fiber-reinforced restorations (FRR). *Oper. Dent.,* v. 26, n.5, 467-75.

[23] Chong, K.H. and CHAI, J. (2003). Strength and mode failure of unidirectional and bidirectional glass fiber-reinforced composite material. *The International Journal of Prosthodontics*, v.16, n.2, 161-166.

In: Strength of Materials
Editors: G. Mendes and B. Lago, pp. 247-259

ISBN: 978-1-60741-500-8

Chapter 9

INFLUENCE ON STRENGTH PROPERTIES OF ANISOTROPY PLANES IN SLATES SAMPLES IN THE NW OF SPAIN

M.A. Rodríguez-Sastre**, *M. Gutiérrez-Claverol, M. Torres-Alonso and L. Calleja
Geology Department, Oviedo University. c/Jesús Arias de Velasco, s/n, Oviedo 33013 Asturias, Spain

Abstract

The purpose of this paper is to describe the influence of anisotropy on the geomechanical strength properties of two Spanish slates with different chemical and physical characteristics. From laboratory testing results of slates under point load and uniaxial compression and the use of indirect methods, as it is the measurement of P velocities, principal parameters were calculated for this rock material. As it is well known under uniaxial compressive strength slates are strong and also very strong rock when loading is parallel (90°) o perpendicular (0°) to the main anisotropic planes. In contrast it is a weak rock with minimum strength values for angles between 45 to 60° of inclination of anisotropy planes. The correlation equations were calculated between different parameters. Despite weak correlation between different geotechnical properties were found and when all lithologies are considered together correlation of geomechanical properties is weak. However when each lithology is considered separately the geomechanical properties can be coherently defined. Linear and polynomial equations were found for the point load and uniaxial strength correlations with the inclination of the anisotropy. Different strength fields were calculated when uniaxial strength and point load test plot and its comparison include the inclination of the anisotropy planes on slates. Uniaxial compressive strength and P wave velocity appears to be strongly influenced by uniaxial strength and good polynomial correlations resulted. Plots of slates with other sedimentary type of rocks from Cantabrian Zone, CZ, revealed the hardness and highest strength of slates when loading is perpendicular to the main anisotropy planes.

Keywords: slates, anisotropic rocks, rock strength, uniaxial compressive strength, point load test.

---

* Corresponding author: Geology Department, Oviedo University, c/Jesus Arias de Velasco, s/n, Oviedo 33005 Asturias, Spain

1. Introduction

Slates under investigation in this study are one of the problematic geological materials due to the presence of anisotropic planes, which represent weak planes, were the failure of the rocks could suddenly occur.

In previous studies, researchers had been focused on core samples from boreholes and had investigated most of the important aspects of rock strength related with the anisotropy with the loading axis perpendicular to any planes of weakness and mainly considerations on rock strength were provided in relation with different conditions of saturation, cored diameters, different sample shapes, and different geotechnical tests (Rodríguez-Bouzo, 1993; González-Buelga, 1995; Hawkins, 1998; Bell, 2000; Tsiambaos & Sabatakakis, 2004; Vásárhelyi & Ván, 2005). These studies have also described the influence of strata orientation when measuring the strength under unconfined compressive conditions.

Another extensive work was based on the influence of these cleavage planes on strength properties and it was supported in factual information on metamorphic foliated rocks and demonstrated that cores cut perpendicular to the cleavage planes possessed the highest strength whilst those cores cut at 30 or 60° exhibit the lowest (Griggs, 1951; Donath, 1961; Brown et al, 1977; Goshtasbi, et al. 2006). All studies were related to engineering works developed in a metamorphic area where the influences of the anisotropy under triaxial conditions were calculated.

In other way, the failure of anisotropic rocks had been analyzed between others by Sheorey (1997), Mustschler & Natau (1991) and Natau et al (1995), and an exhaustive description of anisotropy and its influence on uniaxial compressive strength, modulus of Young and confining pressure was described by Ramamurthy (1993). Although Ramamurthy had established the classification of the main types of anisotropy for different types of rocks, during his investigation all properties were related with the inclination of the anisotropy planes under confined conditions and a rock anisotropy index was proposed, in which the maximum and minimum values of one geotechnical property were resolved.

Further investigations on artificial anisotropy rock samples were proposed by Singh et al. (2002) who had differentiated several types of failures depending on the cleavage anisotropy and the influence of different space between two anisotropy planes with normal disposition between both planes and foliation.

Truchas slates are a planar anisotropic rock due to the millimetre spaced cleavage planes developed during Variscan Orogeny on these Ordovician rocks within a geological structure known as Truchas syncline in the NW of Spain. Casaio and Rozadais formations are an abundant geological sequence in this syncline dominated by silty slates, fine grain slates and quartzite rocks that are characterized by wide variations in their engineering properties. These rocks are problematic material because of the presence of cleavage planes. There is a little data on the engineering performance of Casio and Rozadais slates, and therefore the current study is focused on the determination of the strength related to the inclination of the anisotropy planes on this rocks. This formation is a main source of roof slate. These slates were extensively investigated by Barros-Lorenzo (1989) and Taboada (1993) and related to the organization of mining works Taboada et al., (1997, 1998 and 2006) or to a new design and planning techniques of mining as was proposed by Bastante et al. (2004). The investigation only covered the geology, some geotechnical data and mining aspects of the rock.

The geomechanical properties discussed are essential to guide future geotechnical developments as noted by Pine & Harrison (2003). These data have a great value for the investigations of anisotropic rocks, for example when selecting different engineering works in Truchas area that can be designed on these slate materials or related to new methods of mining extraction.

The geomechanical properties of these slates for the entire range of inclination of anisotropy related to the load application are the main concern of this paper. We compare with the strength values of other data from factual reports in the NW of Spain (Labrada-Rubio et al., 1982).

Values of the uniaxial compression tests (UCS) on these rocks are presented because this is undoubtedly the main geotechnical property used for determining the strength of intact rock ISRM (1981). These test results are directly applicable to different engineering works as slope excavation and tunnelling and also it is included as a main input parameter for rock mass characterization and classification. More often the point load testing (PLT) is used to determine rock strength indexes in geotechnical practice, following the proceed of the ISRM (1985), and the results of this test are to its correlation with the UCS values on these rock and to establish the influence of the inclination of the anisotropic rocks on it.

The main objective of this study is to determine main strength properties to investigate the influence of cleavage anisotropy and the control of the lithological compositions on the geomechanical properties of the slates in the two main formations of the Truchas syncline.

2. Slates Under Study

The slates evaluated in this study are located in the Centro Iberian Zone (CIZ) in the NW of Spain, and belong to the Truchas syncline area that has been described as a D1 synclinorium, with more than 20 km wide, and consists of many individual recumbent folds (Martínez-Catalán et al., 2004). They belong to two different types: Casaio (C) and Rozadais (R) slates. From chemical analysis on slates the main differences between Casasio and Rozadais slates have been established as it is shown in Table1. Those slates showing a planar anisotropy are Ordovician metapelite rocks metamorphosed under low grade conditions with a fine or silty grain with a slaty cleavage microstructure and a lepidoblastic texture, having some of them intercalated millimetre sandstone layers (Rodríguez-Sastre, 2003), main petrographical data for the fine grain slates are given in Table 2. Chlorite and opaque minerals are slightly more abundant in Rozadais than in Casaio slates.

Table 1. General chemical analyses characteristics of the slates studied

Mayor Elements	Chemical composition (%)	
	Casaio	*Rozadais*
SiO_2	76-57	56-52
Al_2O_3	20.8-8.6	22-20
Fe_2O_3	7.7-3.2	9.8-8.9

Table 2. General petrographical characteristics of the fine grain slates studied

<table>
<tr><th>Sample</th><th>Minerals (% Estimated)</th><th>Texture</th><th>Microe-structure</th><th>Minor minerals</th><th>Opaque minerals</th><th>Matrix</th></tr>
<tr><td>C2</td><td>Quartz (45%)
phyllosilicate (40%)
Chlorite (15%)
Others (<1%)</td><td rowspan="2">Lepido-blastic</td><td rowspan="2">Slaty cleavage</td><td rowspan="2">Tourmaline
Rutile
Zircon</td><td rowspan="2">Pyrhotite
Chalcopyrite
Magnetite
Illmenite
Pyrite</td><td rowspan="2">Chlorite+ muscovite+o paque minerals</td></tr>
<tr><td>R2</td><td>Quartz (25%)
phyllosilicate (40%)
Chlorite (20%)
Calcite (10%)
Others (5 %)</td></tr>
</table>

Three main types of slate have been recognised in Casaio Formations, fine grain slates (C2) comprise a fine grained slate with variable quartz content and abundance of sulphurous minerals concentrated in zones, varying in size from millimetre to centimetre scale and concentrated on the cleavage planes, silty slates (C3 and C4) are a grey slate with 1-2 cm nodules of quartz and sulphurous or sulphide minerals related with the cleavage formation and with the presence of thin millimetre laminations, and sandstone laminated slates (C1a and b) include well foliated slates with millimetre interlayer sandstones of 3-5 cm thick, which are related with the initial stratification planes at the original sedimentary basin deposit. Sulphur minerals often are found on the slaty cleavage planes. Two types have been identified and related to the millimetre interlayer sandstone and slaty cleavage planes. C1a type presents a sub horizontal orientation of the interlayer sandstone (S_o) that shows an angle of 20° dip in relation with the cleavage planes, whereas C1b shows angles between interlayer sandstones (S_o) and cleavage planes (S_p) of 50°. From field observations the C1a type appears as a more homogenous rock than the C1b type, which has weaker cleavage planes. In Rozadais Formation two main types were identified, fine grained (R2) is a bluish grey slates with widespread sulphurous minerals of a millimetre scale, and siltly slates (R1 and R5) samples which consist of silty grey slate with bluish colours and sulphur minerals widespread.

3. Test Procedure

Rock samples were obtained from unweathered quarry outcrops. Also all these samples were kept submerged in water a minimum of 24 hours to ensure saturation or wet conditions. Thus tests were carried out under water saturation conditions, which are the natural conditions for rocks under the water level surface in the outcrop. Therefore these results will have shown the lowest values of strength for these slates, because as demonstrated by Hawkins (1998), the strength values decrease with the water content in samples. Thus laboratory core drill for NX cylindrical samples and saw machines were used to cut the samples with a length-to-diameter ratio of 2.0 to 2.5 and end faces were ground in order to provide specimens with size, shape and ends geometries according to ISRM (1981) specifications. The test specimens were obtained from the same block of rock in order to decrease the effect of natural variations in the intrinsic characteristics of the slates and so as to compare the values for the various properties in the different slate types differentiated, to allow a coherent interpretation to be made.

The rock samples examined in this study were prepared and tested in uniaxial compression and point loading with the loading axis from parallel to perpendicular through intermediate positions of the plane of weakness. The fracture created by the point load test was always fresh and through the rock material at parallel and perpendicular positions of loading whereas at the intermediate positions fracture had a little interference of the planes of weakness. Axial and diametral tests were conducted on rock core samples and during the axial test, for cores with inclination of the anisotropy from 0 to 60°, the core is loaded parallel to the longitudinal axis of the core and this test it most comparable to a UCS test. In the diametral test, for cores with inclination of the anisotropy of 90°, the core is loaded perpendicular to the longitudinal axis of the core and in case of anisotropic rocks it is parallel to the anisotropy planes and this test is most comparable to an indirect tensile strength test and also in this study it is compared with the UCS test.

To finish and to characterize as a preview method the influence of the anisotropy planes on the properties of this rocks ultrasonic velocities were recorded using an ultrasonic NEW Sonic Viewer model 5217A from OYO (Japan) and correlated with the UCS values.

The better comparison of the results was as follows: PLT- inclination of anisotropy, UCS- inclination of anisotropy and UCS-PLT, UCS-V_p and V_p-inclination of anisotropy. These results have to be considered as an input parameter to rock classification rather than as a means of predicting mechanical properties of rocks because these parameters are not well correlated as it has been shown by Hawkins (1998) and Tsiambaos & Sabatakakis (2004).

4. Physico-Mechanical Properties

Physical properties for slates were calculated following the procedures for laboratory tests in the UNE (Aenor, 1999) and ISRM suggested methods (1981) from 5 unweathered specimens. Uniaxial compressive strength (UCS) and the Point load test (PLT) were determine on 26 samples. The propagation rate of seismic waves was calculated on the basis of determination performed on 65 specimens of each rock type identical to those used in the uniaxial compressive strength test.

The studied slates have very low natural moisture content (<0.004%) with very low void index ratio and very low porosity (<0.01%), which it is characteristic in this type of rocks. The bulk density calculated varies between 2.47 and 2.89 kN/m^3. Rozadais slates had recorded the lowest values in porosity and in the saturation of moisture content and also in the natural moisture content when it is related to Casaio slates.

The values of UCS and $Is_{(50)}$ for the 26 slate rocks which we tested are given in Table 3.

The Casaio slates show a uniaxial strength varying with the lithological type of slate and classified as strong to very strong. Thus these highest values of strength were observed on the slates with millimetre interlayer sandstones up to 80 MPa, whilst intermediate uniaxial values between 60 and 80 MPa were found in the fine grain slates and the lowest values in silty slates were recorder under 60 MPa. Those values are consistent with the values measured in

Schist, slate and phyllite rocks from the Toros Mountains tectonic group in Turkey (Őzsan & Karpuz, 1996).

The point load strength index values ranged from 14.23 for the axial test perpendicular to the cleavage planes on slates to 0.34 MPa for the diametral test. To compare these results with the values from other sedimentary rocks in the Cantabrian Zone (CZ) from Labrada-Rubio et al. (1982), and all these data were plotted together from the results recorded in P-wave velocity results (V_p) range from 2.691 (perpendicular to anisotropy planes) to 6.991 m/s (parallel to anisotropy planes). These data are consistent with the values obtained by several authors for slates in the NW of Spain (Pernia et al., 1986). On the other hand these ranges are wider than those recorded of P wave velocity in granites from the NE of Portugal (Sousa et al., 2005). In other cases measurements made along and across the cleavage planes in metamorphic rocks like quartzitic phyllite, carbonaceous phyllite and micaceous phyllite varies in same range or even lower values than found in our study (Bell, 2000; Johnson & De Graff, 1994).

Table 3. Data from laboratory testing of slates from ZCI and compared with WALZ values of sedimentary rocks

Zone	Formation	lithology	Numbers of samples	σ_c (MPa)	Is ($_{50}$) (MPa)
ZCI (Truchas Syncline slates) loading at 0, 10, 25 and 90	Casaio (C1a, C1b, C2, C3, C4)	Slates	20	143.2-18.42	14.23-0.34
	Rozadais (R1, R2, R5)	Slates	6	161.1-36.6	6.07-0.78
CZ loading perpendicular at any anisotropy plane	Herreria, Lancara, Ermita, Griotte, Montaña, Villamanin, Oville, San Pedro, Santa Lucia, La vid, Nocedo Portilla, Pola de Gordon, Levinco-Llanon-Tendeyon, Volcanic rocks	Limestones, sandstones, quartzites and volcanics	83	152.8-1.8	15.1-0.4
	Huergas and Formigoso	Shales	10	22.9-2.6	2.53-0.64

5. The Influence of Anisotropy on Strength and Other Properties of Slates

As shown in Figure 1, expected uniaxial compressive strength decreases from high values for anisotropy inclination of zero to a minimum for an inclination of 60°. The UCS then increases with further increases in inclination to 90°. These findings concur with those of Griggs, 1951; Donath, 1961; Brown et al., 1977; Ramamurthy, 1993; Goshtasbi et al., 2006 and others. In terms of the ISRM (1981) classification the values varies betweenstrong and very strong for inclination of 0 and 90° to weak at 60°.

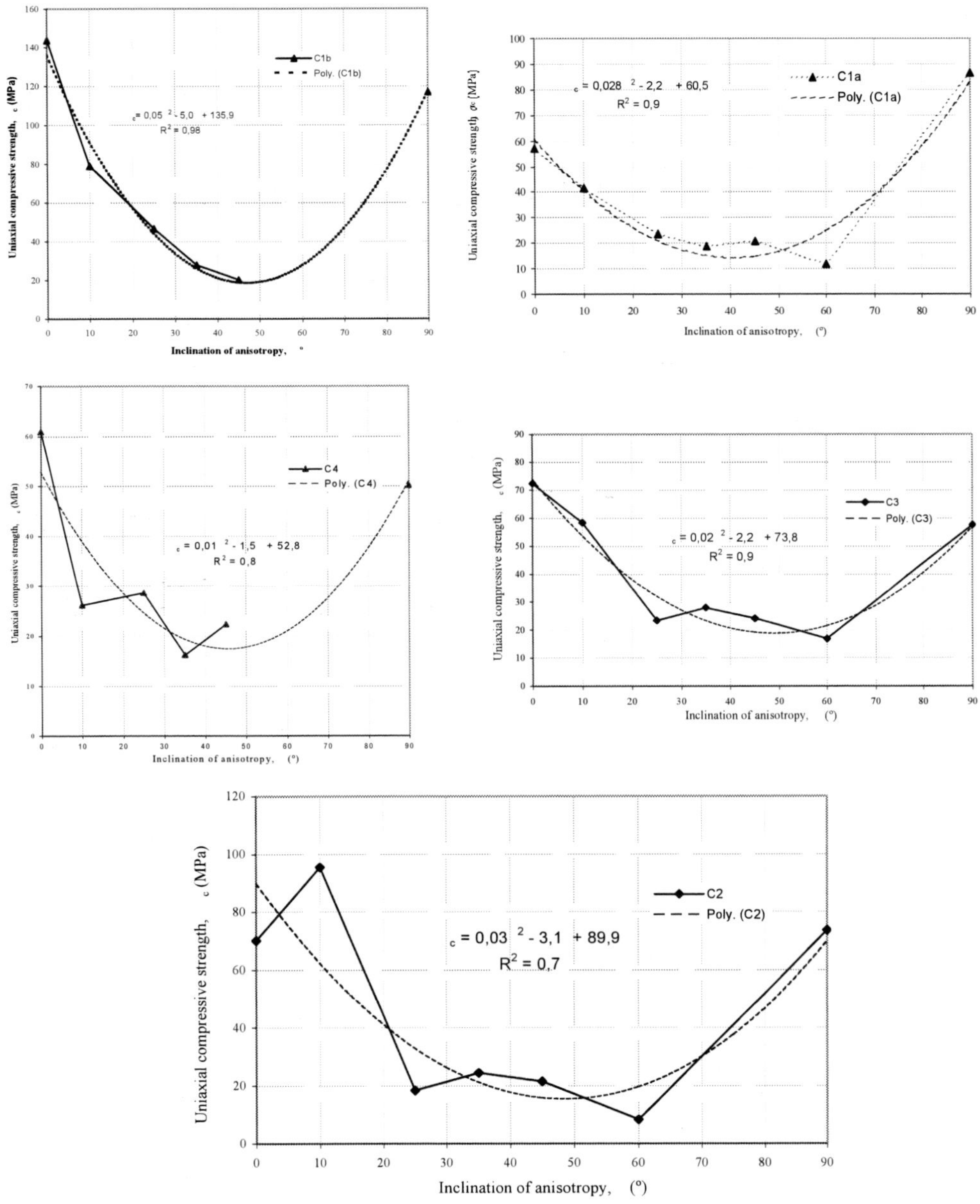

Figure 1. Polynomial function relationship between uniaxial compressive strength and inclination of anisotropy for Casio slate Formation.

As shown in Figure 2, two trends were found relating P wave velocities to the anisotropy inclination, namely approximately linear and polynomial. In both formations it was found that maximum P wave velocities occur along the cleavage planes, parallel to the maximum grain length growth direction. However when waves travel along the cleavage planes but perpendicular to the grain length growth direction the velocities are reduced by 6% in Casaio Slate and 2% in Rozadais Slate. The lowest values were obtained perpendicular to the cleavage planes where velocity drops around 55-36% in Casaio Slates and 34% in Rozadais Slate.

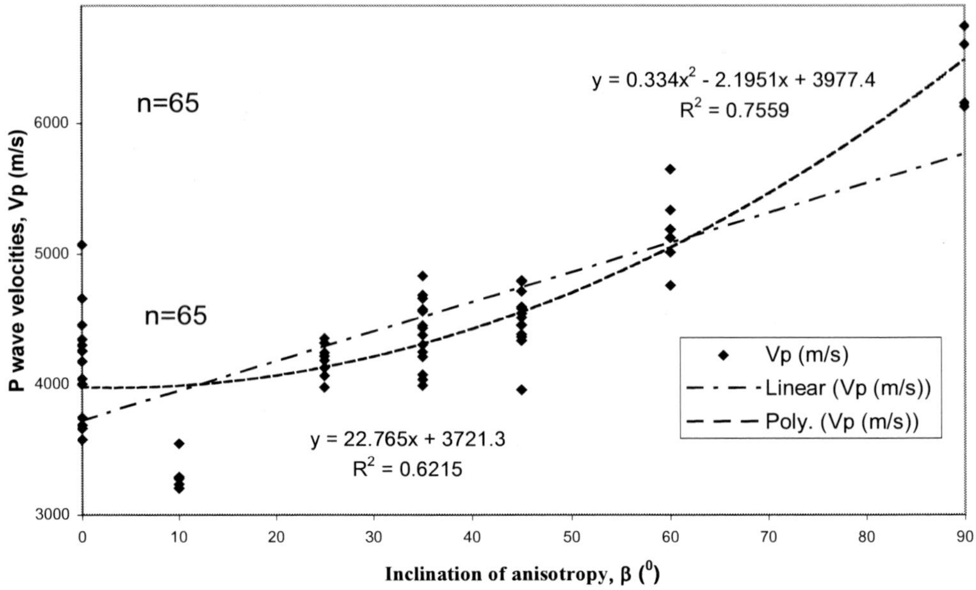

Figure 2. Linear and polynomial function relationship between P wave velocities and the inclination of anisotropy.

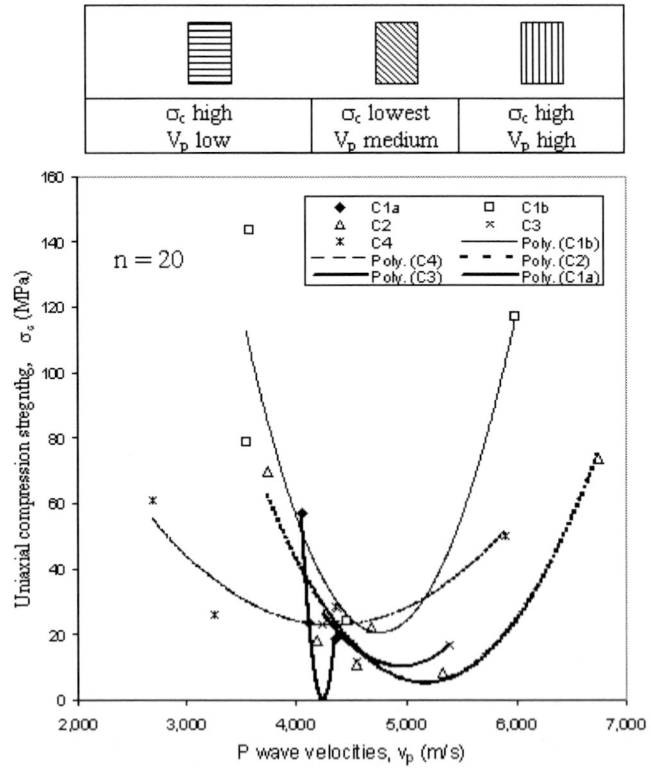

Figure 3. Polynomial function relationship between uniaxial compressive strength and P wave velocities for Casaio slate Formation.

The relationship between uniaxial compressive strength and compressional P-wave velocity appears in Fig. 3. Although the P-wave velocity increases linearly with the angle of inclination of the anisotropy the UCS shows a good polynomial correlation as described above (Table 4), which means that polynomial functions are found for the different slate types. The P wave velocity increases and strength decreases as the inclination of the anisotropy increases from 0° to 60°. As the inclination increases to a maximum at 90° but the strength decreases.

Good linear correlation was found between P-wave velocity and cleavage inclination where the P-wave velocity varies between 2,000 and 4,000 m/s for an inclination of 0°, 4,000 and 5,000 m/s for inclinations between 25 and 45° and over 5,500 m/s or an inclination of 90°.

Good correlation was also found between point load strength and inclination of anisotropy for all the Casaio slate types (Fig. 4). As seen in Fig. 4, the lowest correlation was found for the C1b type, which is the most anisotropic specimen, whereas good correlations were obtained for the other types.

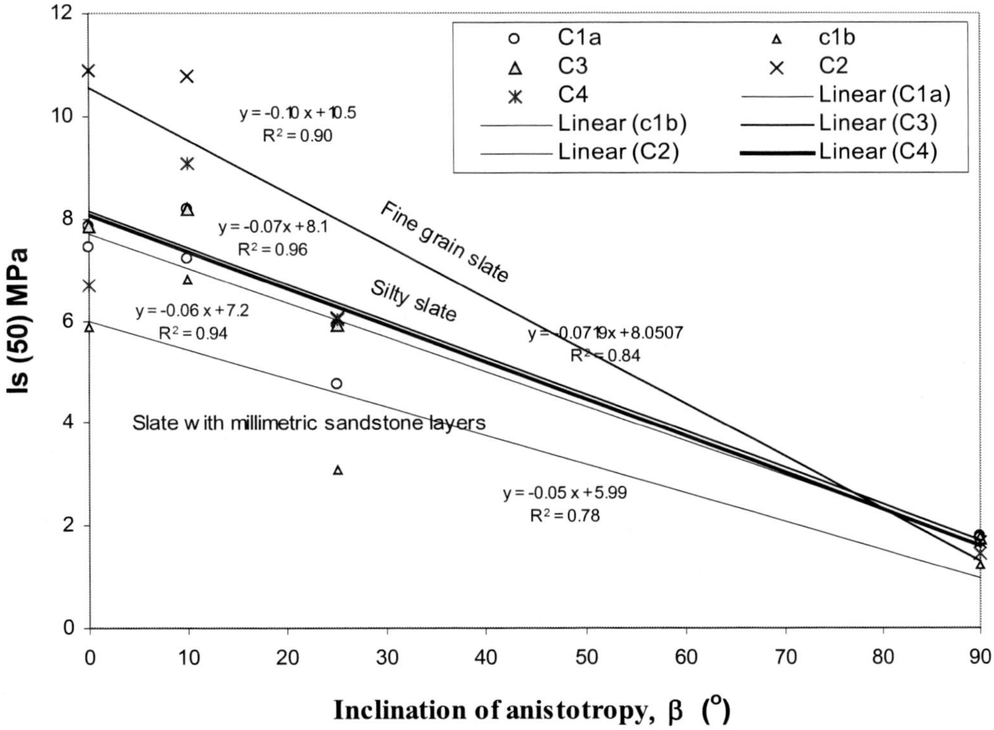

Figure 4. Linear function relationship between point load index and inclination of anisotropy in Casaio slates.

In Figure 5, Point Load Strength is plotted against UCS for different inclinations. Thus it would appear that the relationship between these to parameters depends on the inclination of the cleavage.

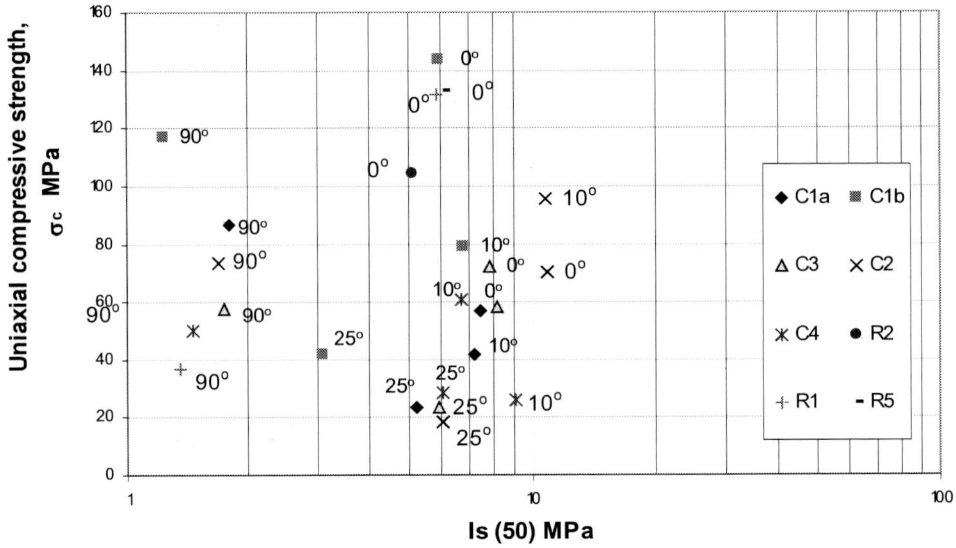

Figure 5. Fields correlating point load index with uniaxial compressive strength and inclination of anisotropy in slates from Truchas syncline.

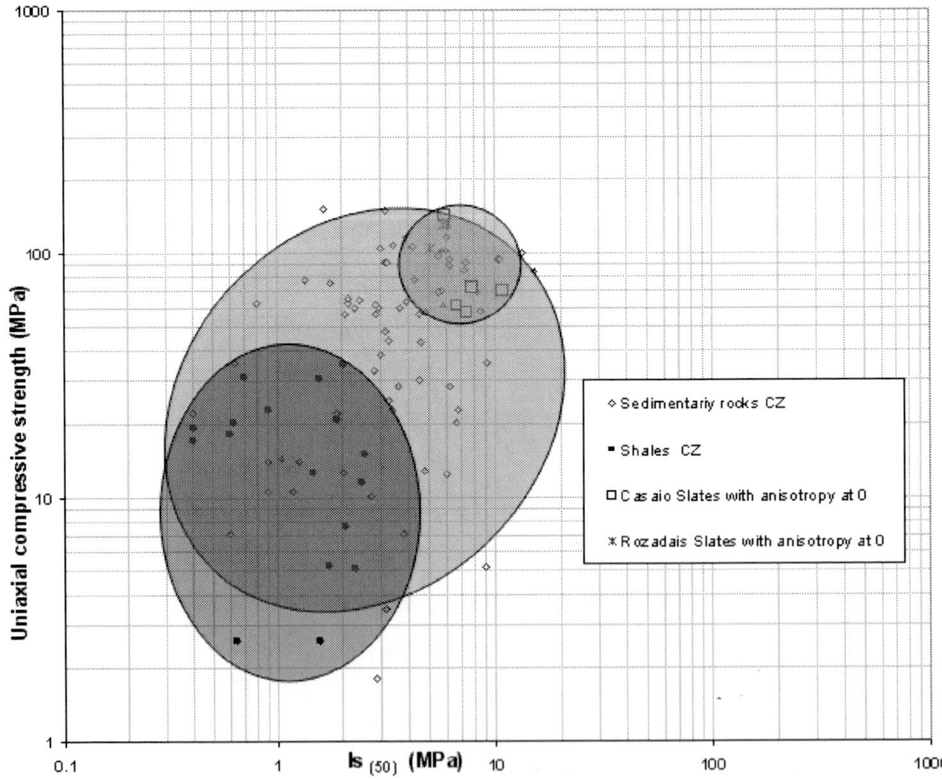

Figure 6. Comparison of plots values of different rock class and slates in correlation between uniaxial compressive strength and point load index.

Figure 6 shows the relationship between UCS and PLS for various rocks, including limestones and slates that were loaded perpendicularly to the cleavage. These data demonstrate that the metamorphic rocks have higher strength. The difference in behaviour between Casaio and Rozadais formations appears to be due to variation of porosity and mineralogy.

6. Conclusion

It has been confirmed that anisotropy orientation strongly influences the strength properties of the slates studied. Polynomial correlations for relationships between uniaxial compressive strength and inclination of anisotropy, and also with P- wave velocity have been established. The strength is lowest for an inclination of 60°, and highest at 0 and 90°.

As would be expected for rocks of this type, PLS varies over a very range although good linear correlations were found with inclination of the foliation. In other words the PLS decreases with increasing angle of cleavage inclination between 60 and 90°. However, it was also influenced by lithology. Thus, Point load tests on foliated rocks would be able to discriminate between differences on lithological compositions of slate sequences and tectonic setting could have an influence on this test results. This could help to interpret sequences and tectonic setting at outcrop scale.

Good linear correlations are found between P-wave velocity and the inclination of anisotropy planes. However relatively high values of Uniaxial Compressive strength may occur for either high or low velocities values and are at a minimum for intermediate velocities.

Although poor correlation was obtained between Point Load Strength and UCS data, different rock types group in different areas of this graph.

As expected, the geomechanical properties of the slates under study leads us to conclude that the anisotropic fabric gives rise to relatively weak planes and which influence the strength and elastic properties of the material.

Acknowledgements

Many thanks go to H. Stoll who is acknowledged for their critical reviews and correction of the manuscript. J. Cripps kindly read through the text and provided useful comments.

References

[1] AENOR (1999): *Geotecnia. Ensayos de campo y de laboratorio*. Ed. AENOR, 385 pp.

[2] Barros-Lorenzo, J. C. (1989): Nuevos datos geológicos y cartográficos sobre el flanco Sur del Sinclinorio de Truchas (Ourense-León, NW de España). Cuadernos do Laboratorio Xeolóxico de Laxe, 14: 93-116.

[3] Bastante, F. G., Taboada, J. & Ordoñez, J. (2004): Design and planning for slate mining using optimization algorithms. *Engineering Geology*, **73**: 93-103.

[4] Bell, F. G. (2000): Engineering properties of Soils and Rocks. Ed Blackwell, London.

[5] Brown, E. T., Richards, L. R. & Barr, M. V. (1977): Shear strength characteristics of the Delabole Slates. In: *Proc Conf Rock Engineering Newcastle upon Tyne*: 35-51.
[6] Davis, J. C. (1986): Statistics and data analysis in Geology. John Wiley and Sons, New York.
[7] González Buelga, M. T. (1995): Caracterización geomecánica y valoración geotécnica de materiales rocosos del Paleozoico inferior en la cuenca alta del Río Bernesga (León, Noroeste de España). Thesis. Geology Department, University of Oviedo, 445 pp.
[8] Goshtasbi, K., Ahmadi, M. & Seyedi, J. (2006): Anisotropic strength behaviour of slates in the Sirjan-Sanandaj zone. *South African Institute Mining and Metallurgy*, **106** (1): 71-75.
[9] Hawkins, A. B. (1998): Aspects of rock strength. *Bull. Engineering Geology Env.* **57**: 17-30.
[10] ISRM SUGGESTED METHODS (1981): Description of Rock Masses. *Int. Jour Rock Mech. and Min. Sci. y Geomech. Abst. Pergamon Press*, **18** (1): 85-110.
[11] ISRM SUGGESTED METHODS (1985): Determining Point Load Strength. *Int. Jour Rock Mech. and Min. Sci. y Geomech. Abst. Pergamon Press*, **22** (2): 51-60.
[12] Johnson, R. B. & Degraff, J. V. (1994): *Engineering Geology*. Ed. Macmillan, 190 pp.
[13] Labrada-Rubio, J. A., Martínez- Álvarez, J. A., Torres M. & Gutiérrez-Claverol, M. (1982): Estudio geológico-geotécnico el paso por ferrocarril de la divisoria Astur-Leonesa (La Robla-Ujo- Pola de Lena). Factual Report. E.T.S.I.M.O, University of Oviedo.
[14] Martínez-Catalán, J. R., Fernández-Suárez, J., Jenner, G. A., Belousova, E. & Díez-Montes, A. (2004): Provenance constraints from detrital zircon U-Pb ages in the NW Iberian Massif: implications for Paleozoic plate configuration and Variscan evolution. *Journal of the Geological Society*, **161**: 463-476.
[15] Natau, O., Bühler, M., Keller, S. & Mustchler, Th. (1995): Large scale triaxial tests in combination with a FEM analysis for the determination of the properties of a transversal isotropic rock mass. Proc. 8th Int. Congr. ISRM: 635-643, Tokio.
[16] Őzsan, A. & Karpuz, C. (1996): Geotechnical rock-mass evaluation of the Anamur dam site, Turkey. *Engineering Geology*, **42**: 65-70.
[17] Pernia, J. M., Fresno, F., Mena, J. M., Lucena, A. & Lechosa, R. (1986): Registro de datos en sondeos de reconocimiento. Ed. IGME, 144 pp.
[18] Pine, R. J. & Harrison, J. P. (2003): Rock mass properties for engineering design. *Quarterly Journal of Engineering Geology and Hydrogeology*, **36**: 5-16.
[19] Ramamurthy, T. (1993): *Strength and modulus responses of anisotropic rocks*. In: Comprehensive rock engineering, principles practice and projects. vol.3.
[20] Rodríguez Bouzo, L. F. (1993): *Caracterización geomecánica de materiales carbonatados del Carbonífero en la zona de Pajares-Villamanín*. Thesis, Explotación y Prospección de Minas Department, University of Oviedo, 416 pp.
[21] Rodríguez-Sastre, M. A. (2003): *Caracterización geomecánica de materiales pizarrosos del Sinclinal de Truchas (León-Orense).* Thesis, Geology Department, University of Oviedo, 388 pp.
[22] Sheorey, P.R. (1997): *Empirical Rock Failure Criteria*. Ed. A. A. Balkema, Rotterdam. 176 pp.
[23] Singh, M., Rao, K. S. & Ramamurthy, T. (2002): Strength and Deformational Behaviour of a Jointed Rock Mass. *Rock Mechanics and Rock Engineering*, **35** (1): 45-64.

[24] Sousa, L. M. O., Suarez del Río, L. M., Calleja, L., Ruíz de Argandoña, V. G. & Rodríguez Rey, A. (2005): Influence of microsfractures and pososity on the physico-mechanical properties and weathering of ornamental granites. *Engineering Geology*, **77**: 153-168.

[25] Taboada, J. (1993): *Diseño de minas subterráneas de pizarra*. Thesis, Explotación y Prospección de Minas Department, University of Oviedo, 173 pp.

[26] Taboada, J., Vaamonde, A., Saavedra, A. & Alejano, L. (1997): Application of geostatistical techniques to exploitation planning in slate quarries. *Engineering Geology*, **47**: 269-277.

[27] Taboada, J., Vaamonde, A., Saavedra, A. & Argüelles, A. (1998): Quality index for ornamental slate deposits. *Engineering Geology*, **50**: 203-210.

[28] Taboada, J., Matias, J. M., Araujo, M. & Ordoñez, C. (2006): Assessing the viability of underground slate mining by combining an expert system with a GIS. *Engineering Geology*, **87**: 75-84.

[29] Tsiambaos, G. & Sabastakakis, N. (2004): Considerations on strength of intact sedimentary rocks. *Engineering Geology*, **72**: 261-273.

INDEX

A

B

C

D

E

F

G

H

I

J

K

L

M

N

O

P

Q

R

S

T

U

V

W

X

Y

Z